THE HANDBOOK OF PROJECT-BASED MANAGEMENT

Other books by Rodney Turner published by McGraw-Hill

Turner, J.R., Grude, K.V. and Thurloway, L., 1996, (eds), *The Project Manager as Change Agent*, McGraw-Hill, London, 264p, ISBN: 0-07-707741-5.

Turner, J.R., (ed), 1995, *The Commercial Project Manager*, McGraw-Hill, London, 408 p, ISBN: 0-07-707946-9.

THE HANDBOOK OF PROJECT-BASED MANAGEMENT

Leading Strategic Change in Organizations

J. Rodney Turner

Third Edition

New York Chicago San Francisco Lisbon London Madrid
Mexico City Milan New Delhi San Juan Seoul
Singapore Sydney Toronto

The McGraw-Hill Companies

Library of Congress Cataloging-in-Publication Data

Turner, J. Rodney (John Rodney), date.
The handbook of project-based management : leading strategic change in organizations / J. Rodney Turner.—3rd ed.
p. cm.
Includes index.
ISBN 978-0-07-154974-5 (alk. paper)
1. Project management. I. Title.
HD69.P75T85 2009
658.4′04—dc22 2008026994

McGraw-Hill books are available at special quantity discounts to use as premiums and sales promotions, or for use in corporate training programs. To contact a special sales representative, please visit the Contact Us page at www.mhprofessional.com.

The Handbook of Project-Based Management, Third Edition

1 2 3 4 5 6 7 8 9 0 DOC/DOC 0 1 4 3 2 1 0 9 8

ISBN 978-0-07-154974-5
MHID 0-07-154974-9

This book is printed on acid-free paper.

Sponsoring Editor
Larry S. Hager

Acquisitions Coordinator
Alexis Richard

Editorial Supervisor
David E. Fogarty

Project Manager
Aparna Shukla

Copy Editor
Ragini Pandey

Proofreader
Upendra Prasad

Indexer
Broccoli Information Management

Production Supervisor
Richard C. Ruzycka

Composition
International Typesetting and Composition

Art Director, Cover
Jeff Weeks

To Edward, now 18

ABOUT THE AUTHOR

RODNEY TURNER is Professor of Project Management at the Kemmy Business School of the University of Limerick and at the Lille School of Management. He is also an Adjunct Professor at the University of Technology, Sydney and Educatis University, Zurich, and was a Visiting Professor at Henley Management College and George Washington University.

Rodney Turner is the author or editor of fourteen books. He is editor of *The International Journal of Project Management*, and has written articles for journals, conferences, and magazines. He lectures on and teaches project management worldwide.

From 1991 to 2004, Rodney was a member of Council of the UK's Association for Project Management, with two years as Treasurer and two as Chairman. He is now a Vice President. From 1999 to 2002, he was President and then Chairman of the International Project Management Association, the global federation of national associations in project management, of which APM is the largest member. He has also helped to establish the Benelux Region of the European Construction Institute as foundation Operations Director. Rodney is director of several SMEs and a member of the Institute of Directors. He is also a Fellow of the Institution of Mechanical Engineers and of the Association for Project Management.

CONTENTS

Part 2: Managing Performance

Part 4: Governance of Project-Based Management

PREFACE

One of my aims in writing successive editions of this book has been to maintain the book's length. That means that as I include new ideas, I have to drop some material. I don't want a book that gets fatter and fatter to the point where I have to start dividing it into two or more separate books. Project management is a dynamic and developing topic, and that means that there are new ideas that need to be included in the book. But also some ideas that were included in the first and second edition are now past their sell-by date and so can be dropped. I have aimed to produce a book that covers the key topics of project management as people see it at the moment, and to leave out some of the concepts that have not proved so effective.

The book is one part shorter than the previous edition, at four parts rather than five. The first three parts cover the same ground as the first three parts of the previous two editions.

Part 1 describes the context of projects. In particular it considers how the strategy of the parent organization and the desire to achieve performance improvement through strategic change drive the creation of projects. It then looks at project success strategy and describes the criteria by which we judge success, the factors by which we increase the chance of success, and how we combine the two into a strategy for our projects. The third chapter in the part considers the people involved in the project. It takes a different perspective from the previous two editions where the equivalent chapter looked at the position of projects in the parent organization. In this edition that chapter focuses much more on how to lead the stakeholders to gain their support for the project.

Part 2 covers the same ground as the previous two editions, describing the functions of project management, how to manage the scope, project organization, quality, cost, time, and the risk that pervades them all.

Part 3 also substantially covers the same ground as the previous editions, describing three stages of the project life cycle: start, execution, and close-out. However, I have included a new chapter at the start of the part, describing the project life cycle, and different versions for different types of project. This chapter covers much of the ground of what was previously the fifth part, on applications, but in a more focused way.

Although these three parts cover very much the same ground, I have incorporated new thinking, and so in places the material is different from the previous editions.

It is in Part 4 where I have taken a radically different approach. In the previous two editions, Part 4 described administrative support given to the project by the parent organization. Now, in accordance with the modern style, I take a governance perspective. As a result, it covers some of the same ground, because the administrative support described in the previous editions is governance support, but it also introduces many new ideas. I start by defining what we mean by governance and describe the governance of the individual project, and the governance roles that imply. In the next two chapters, I describe the governance of the context, particularly program and portfolio management and the development of organizational project management capability. I then describe the project governance role of the executive board, and the interest they should take in projects.

I have retained the chapter on international projects as the last main chapter, and as in the previous two editions close with an epilogue.

I have updated the references throughout the book. I think the main purpose of references is to point to further reading for readers who want to find out more about the topics. I think that only books that are readily available are useful for the purpose, so I tend not to cite academic research journals or magazine articles for that purpose, and definitely not obscure conferences. The other main purpose for references is to acknowledge source materials, and for that purpose I may cite an academic research journal article.

Rodney Turner
East Horsley, Surrey

ACKNOWLEDGMENTS

My main thanks in writing this third edition continues to go to the now nearly 30,000 people who have bought the previous editions, and thereby give me encouragement to continue to spread the good word of project management. I would also like to thank Wade Ren and Vladimir Voropayev who led the translation of the book into Chinese and Russian, respectively.

I wish to thank the people with whom I have worked and whose ideas have contributed to the material of this book. The first and second editions drew on distance learning material in project management at Henley Management College. Elements of the third edition still draw on the ideas of Mahen Tampoe, Susan Foreman, Svine-Arne Jessen, Peter Morris, Nick Aked, Roger Sharp, Richard Morreale, David Topping, and Anne French. There are also people with whom I have written research papers over the years, particularly Bob Cochrane, Anne Keegan, Martina Huemann, and Ralf Müller. I also wish to acknowledge the ongoing inspiration I receive from my work with Kristofer Grude, with whom I wrote my first book over 20 years ago.

I have received significant help in the process of writing the book from Judy Morton. Judy has proofread all the material, and helped me prepare the figures.

And finally, I must thank my family for putting up with all the travel that spreading the good word of project management seems to entail.

Rodney Turner
East Horsley, Surrey

THE HANDBOOK OF PROJECT-BASED MANAGEMENT

CHAPTER 1
LEADING CHANGE THROUGH PROJECTS

Change, and the need to manage change through projects, touches all our lives, in working and social environments. Twenty years ago most managers were not directly involved in the management of projects. Bureaucracies were viewed as providing an efficient, stable, and certain environment in which to conduct business. Change was mistrusted. Managing change was limited to specialist, technical functions. That has now changed. Change is endemic, brought on by an explosion in the development of technology and communications. Rather than being the preferred style of management, bureaucracies are viewed as restricting an organization's ability to respond to change, and thereby maintain a competitive edge.

The last 50 years has characterized this changing emphasis. The 1960s were a decade of mass production. Manufacturing companies strove to increase output. Production methods were introduced to facilitate that. High production rates were achieved, but at the expense of quality. During the 1970s, to differentiate themselves companies strove for quality. By imposing uniformity and restricting their product range, managers could achieve quality while maintaining high production. In the 1980s, the emphasis shifted to variety. Customers wanted their purchases to be different from their neighbours'. No two motor cars coming off the production line were the same, and nonsmokers would rather have a coin tray in place of the ashtray. Companies introduced flexible manufacturing systems to provide variety, while maintaining quality and high production. In the 1990s, customers wanted novelty. No one buying a new product wanted last year's model. Product development times and market windows shrank, requiring new products to be introduced quickly and effectively. Now customers want functionality. They don't just want their cell phones to make telephone calls; they want to send text messages and e-mails, surf the Internet, take photographs, and store their music library. (My son Edward describes products with excessive functionality as being Gucci.) Organizations must adopt flexible structures to respond to the changing environment. To gain competitive advantage, they need to be in a constant state of flux to improve their business processes. Many clients expect every product to be made to a bespoke design, and so every product becomes a mini project.

The project-oriented organization is now common; project-based management is the new general management[1]; 30 percent of the global economy is project-based.[2] Project management is a skill required of all managers. This book provides general managers in project-oriented companies with a structured approach to the management of projects, so they can achieve performance improvement through the management of change.

In this chapter, I describe the structured approach and its three dimensions: the project, the process of managing the project, and the levels over which it is managed. I then explain the importance of the process approach and introduce a model for the strategic management of projects. Next, I cover two issues, one dealing with the nature of projects, and one the nature of project management. The first is a classification of projects based on how well defined the project's goals are and the methods of achieving those goals, which influences the choice of strategy for managing the project. The second is an analogy of project management as sailing a yacht, which challenges traditional concepts of management. I end the chapter by explaining the overall structure of the book.

1.1 PROJECTS AND THEIR MANAGEMENT

Projects come in many guises. There are traditional major projects from heavy engineering, or WETT, industries, (water, energy, transport, telecommunications). These are significant endeavours involving large dedicated teams, often requiring the collaboration of several sponsoring organizations. But the projects with which most of us are involved are smaller. Projects at work include engineering or construction projects to build new facilities; maintenance of existing facilities; implementation of new technologies or computer systems; research, development, and product launches; or management development or training programs. Projects from our social lives include moving to a new house; organizing the local church fête; or going on holiday. So what do we understand by projects and project management?

Project management is about converting vision into reality. We have a vision of some future state we would like to achieve. It may be a new computer system, a new production process, a new product, a new organization structure, or more competent managers. We foresee that the operation of that new state will help us improve performance of our business, by solving a problem or exploiting an opportunity, and so provide us with benefit that will repay the cost of achieving it. Project-based management is the structured process by which we successfully deliver that future state (I discuss in a later chapter what is understood by "successfully"). In this section I define what I mean by projects and their management, and describe the three key components of project-based management: the project, the management of the project, and the levels over which they are managed.

The Project

Previously, I used the following definition of a project:

> A project is an endeavour in which human, financial, and material resources are organized in a novel way to undertake a unique scope of work, of given specification, within constraints of cost and time, so as to achieve beneficial change defined by quantitative and qualitative objectives.

One of my former MBA students objected to this definition (see Example 1.1). Although I think he was missing the point, his objection had some validity in that this definition is rather prescriptive, and unnecessarily so. Now I have chosen to adopt a less prescriptive definition which focuses on the key features (Fig. 1.1):

> A project is a temporary organization to which resources are assigned to do work to deliver beneficial change.

Example 1.1 Maintenance "projects" in BT

I had an MBA student who took exception to my definition. He worked on projects, he said, that were repetitive, and neither unique nor novel. They were maintenance projects in

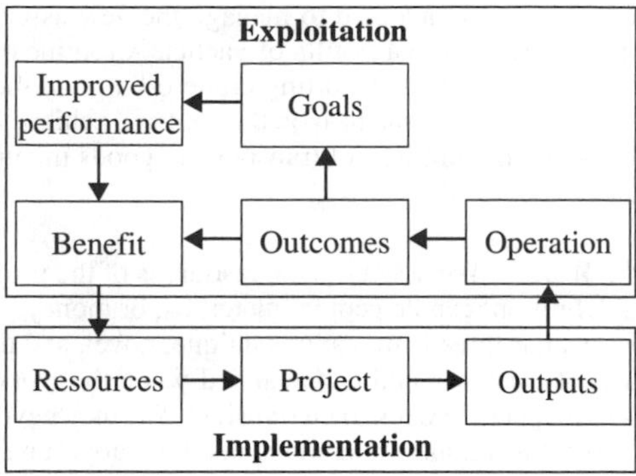

FIGURE 1.1 The definition of a project.

British Telecom. He said my definition was wrong; he did not have the humility to see that his application of the word "project" might be wrong. But of course that was not the point; his maintenance projects had some features of projects and some of routine operations, and therefore needed a hybrid management approach. He found some value from looking on his work as projects, but he did not see that the purpose of a definition is to aid understanding, not to be precise and prescriptive.

A Temporary Organization. A project is a temporary organization. We have a vision of a future state we wish to achieve, and we need resources to do work to deliver it. So we create a new organization within which those resources can work. That organization will have only a temporary existence, being disbanded when the new state is achieved. For me the concept of the project as a temporary organization in which we assemble resources to do the work to achieve our desired future state is key. Many people define a project as a temporary task, or temporary endeavour (for example, see the PMI® PMBoK®[3]). I do like to differentiate between a temporary task given to the routine organization and a temporary organization specifically created to deliver the project. I would not describe the maintenance tasks in Example 1.1 as projects; they are temporary tasks undertaken by the routine organization. (If someone finds value in labelling what they do as a project I would encourage them to do so, but I would also advise not to label things as projects when more routine management approaches may be more appropriate for their delivery.)

How temporary is temporary? All organizations are permanent on some time scales and temporary on others. The oldest organization I know about is the Church of Rome, which is 2000 years old. The longest project I am aware of from first work to eventual completion is the Rhine to Danube Canal. The first work was done by Charlemagne about 792 and it was completed 1200 years later in 1992. Fortune 500 companies have an average life of 50 years, and so are temporary on that timescale. Permanent and temporary are social constructs. The parent organization views itself as permanent, and creates a project that it expects to have shorter existence to achieve specific objectives.

Carroll[4] suggests that the success of an organization form depends on its ability to attract resources. Projects as an organizational form are very effective at attracting resources because they are an effective way of managing change. They can deliver change in a fast and flexible way, in ways that cannot be achieved in the routine organization. They can also be used to prototype new ways of working. Carroll also suggests that an organization's longevity is an indication of its efficiency. Projects are effective at delivering change, but an inefficient way of working, so as soon as the change is delivered the project should be disbanded and routine management adopted to manage the new asset delivered. I use the analogy of comparing a supertanker to a flotilla of yachts: A routine organization is like a supertanker—a very efficient way of transporting crude oil around the world, but it takes three miles to turn. A flotilla of yachts can turn on the spot and achieve things a supertanker cannot achieve, but it is an inefficient way of transporting goods in bulk. We return to this analogy later in the chapter.

The Resources and the Work. We assemble the resources of the project to do work. The resources (per my old definition) can be people, materials, or money, or all three.

The work of the project has three features: it is unique, novel, and transient (Table 1.1). The project has a transient existence and is disbanded when the work to deliver the new asset is finished. Thus we expect the work to be transient. We may never have built an asset like this before—the project is unique—and so we need to adopt novel work processes. It annoys me when project managers try to grab the moral high ground by saying projects are about delivering objectives within constraints of time, cost, and quality. All of business—all of life—is about trying to deliver objectives within constraints of time, cost, and quality. By trying to grab the moral high ground in this way project managers do themselves no favors because they fail to focus on what is special about their discipline, the uniqueness, novelty, transience, and implied risk. In business there are repeat objectives, which require us to do repetitive things, and there are new objectives which require us to do unique, novel, and transient things. With the latter, it is more difficult to achieve the constraints of time, cost, and quality, because there is less previous experience on which to base our plans, and therefore greater risk of failure.

TABLE 1.1 The Features of a Project

Goal	Features	Pressures	The Plan
Unitary	Unique	Uncertainty	Flexible
Beneficial	Novel	Integration	Goal oriented
Change	Transient	Urgency	Staged

What do we mean by unique and novel? The student in Example 1.1 thought his projects were very repetitive. There is a way of categorizing projects, into runners, repeaters, strangers, and aliens that recognizes that projects range from the familiar to the unknown:

Runners: These are very familiar. They almost count as batch processing. The projects in Example 1.1 (if they are projects) would fall in this category. Routine processes can be used.

Repeaters: These are fairly familiar. There is knowledge in the organization about how they should be managed, on which the project team can draw during the project start-up process.

Strangers: The organization has undertaken similar projects in the past but there are unfamiliar elements. I would classify the construction of the Channel tunnel as this type of project: it wasn't the first undersea tunnel ever built; it wasn't the first time a high speed railway line had been put in a tunnel; but it was the first time that such a tunnel had been built between England and France. There were many familiar elements to draw on but the overall project was completely novel.

Aliens: The organization has never done anything like this before. These projects are high risk. You may try to identify familiar elements, and if you cannot, seriously consider not doing the project. But many projects like this are mandatory, brought on by a change in legislation.

Projects create several pressures (Table 1.1) that require the project plan to have certain features. The transience creates urgency, a need to complete the work and obtain the benefit to repay the money spent. The novelty requires us to create new ways of working, and hence to integrate the working of people from across established organization structures. The uniqueness creates uncertainty; you cannot predict the future, and therefore you cannot be certain that the planned ways of working will deliver the objectives you want. This uncertainty creates the first dilemma of project management: how much planning to do. There are those who say there is no point doing any planning; you cannot predict the future, so you might as well " knife-and-fork" your way through the project. Well, there are two sayings about this approach:

If you fail to plan, then plan to fail; and

We never seem to have time to plan our projects, but we always have time to do them twice.

You must have a plan; you need a framework to coordinate people's activities and the use of materials and money. However, one thing you can guarantee about your plan is it is wrong, that is not the way the project will turn out. You must have it as the framework for coordination, but you must be ready and willing to change it as the project progresses.

There are those, on the other hand, who think they can eliminate all uncertainty by planning in minute detail; by developing a highly detailed plan they can cover every eventuality, they can predict the future. There are two problems with this approach. The first is it costs time and effort to plan. There is an empirical rule that says if a certain amount of effort, x, is required to produce a plan of a given accuracy, then to double the accuracy requires four times as much effort, $4x$, and to double it again requires four times as much effort again, $16x$. Further planning gives decreasing returns, and you reach a point where you are putting more effort into planning than is warranted by the value of the information you get out. You have to stop planning and start managing the risk. The second problem is you cannot eliminate the risk entirely, you cannot predict the future; if you make the plan too complicated, too sophisticated, it becomes inflexible and less able to respond to changes as they occur. Thus, we must have a plan, but we must accept that it will not be completely accurate and so will need to be flexible to change. We will see later it must be goal oriented to be flexible.

The Beneficial Change. The project is a temporary organization where resources are assembled to do work. But we do not do the work for its own sake; we do it to deliver some output, a new asset (which I often refer to as "the facility"). The asset may be a new building, manufacturing plant, computer system, organization structure or a new design, and is called the *output* in Fig. 1.1. It is something we want. However, we do not produce the asset for its own sake; we make it so we can operate it to satisfy some purpose or produce some benefit. As we operate the facility it will do something for us, which is called the *outcome* in Figure 1.1, and the use of that outcome will provide benefit. The aim is to solve a problem or exploit an opportunity to help us improve the performance of our business. The performance improvement is the desired outcome of the project, the asset (the change we have introduced) is simply the desired output from the project that will enable us to achieve the outcome, the *desired performance improvement.* The long-term use of the outcome may also help us achieve higher order objectives, referred to as the *impact* in Fig. 1.1, and may help us achieve our strategic goals (see Example 1.2). In the next chapter I describe how to identify the desired performance improvement and the asset or change that will help achieve that. In Part 1, I also describe how we judge the success of the project and develop a strategy for its delivery.

Example 1.2 A bridge across the Yangtze River

The Chinese Government wanted to achieve economic development on the north side of the Yangtze river, just across the river from Shanghai. On the south side, around Shanghai, people were relatively well off, but they were poorer on the north side of the river. So the government built a bridge. The project's output was the bridge. The desired outcome was faster traffic flows (compared to the old ways of crossing the river). The benefit was cheaper distribution of goods. The cheaper distribution of goods encouraged economic development and so the government achieved the desired impact and strategic goal.

In routine operations, the plant is operated to produce a product, which is sold to provide benefit. However, here projects and routine operations differ again. In the routine operation, the plant is operated today to produce a product tomorrow, which is sold for the next day. We have instant feedback about how well we are doing, and we can make small adjustments to the plant, small touches on the tiller, to bring the process back on course and achieve the profit we want. On a project we do the work today, to produce the asset next year, and achieve the benefit the year after. By the time we achieve the benefit, the project team is disbanded, and it is not possible to make minor adjustments to achieve the benefit we actually wanted. This reemphasises the risk. It means that on a project, rather than focusing on the work, you must focus on the desired results, continually reminding yourself of the purpose of what you are doing, to try to ensure that all the work done delivers essential project objectives which are necessary to achieve the purpose or expected benefit.

Figure 1.2 illustrates that there are two groups of people involved on the project, the *owner* and the *contractor*. The owner pays the contractor to do the work, and in the process buys the asset. They then operate it to achieve the benefit. They achieve their value from the difference between the benefit they receive from operating the asset and the price they pay the contractor. The benefit may be nonfinancial, so I have purposefully used the term *value* and not *profit*. The contractor does the work of the project. The contractor receives

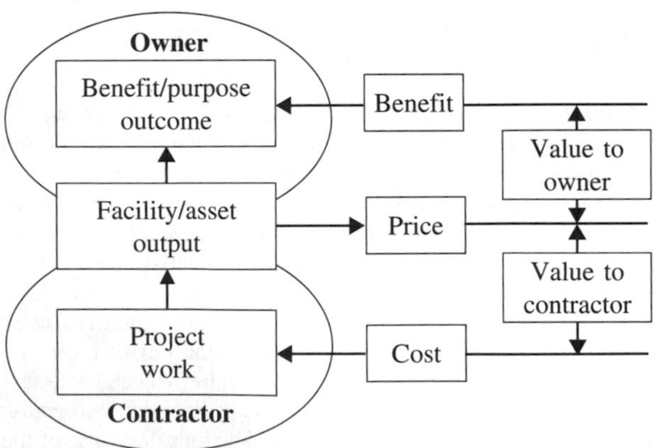

FIGURE 1.2 The owner-contractor model.

money from the owner to do the work, and profits from the difference between the price they receive and the costs they pay to do the work. Here we see for the first time that different people working on the project can have conflicting objectives, different views about what constitutes success. Owners increase their profits if they can get the price down, and contractors can increase theirs if they get the price up. If the owner and contractor are separate client and contractor organizations, we understand that conflict. Its resolution is part of contract negotiations between the two parties. However, if they are part of the same parent organization, the production and engineering departments of one company, for example, you may assume that they are all part of the same organization and share the same objectives. They don't!

The Functions of Project Management. The above definition of a project implies that there are several functions of project management, five of which are illustrated in Fig. 1.3. These five core functions can be explained as follows:

1. The project entails work, and that scope of work must be managed.
2. We assemble the resources into a temporary organization which must be managed.
3. In order to deliver the desired benefit, the asset must function in certain ways, and at required levels of performance. Therefore, the performance, or quality, of the asset must be managed. But to deliver a quality asset the work of the project must also meet certain quality standards. Quality needs to be managed.
4. In order for the project to be of value to both the client and contractor, it must cost less than the value of the benefit. Thus cost needs to be managed. This involves managing the consumption of all resources, including people and material, not just money.
5. Time needs to be managed for several reasons. In order for the work of the project to take place effectively and as efficiently as possible, the input of the various resources needs to be coordinated. Also there will be a time value associated with the benefit from the asset. The later it is delivered, the less its value, so the timing of the work needs to be managed to deliver the asset within a time frame that will give the desired benefit. On

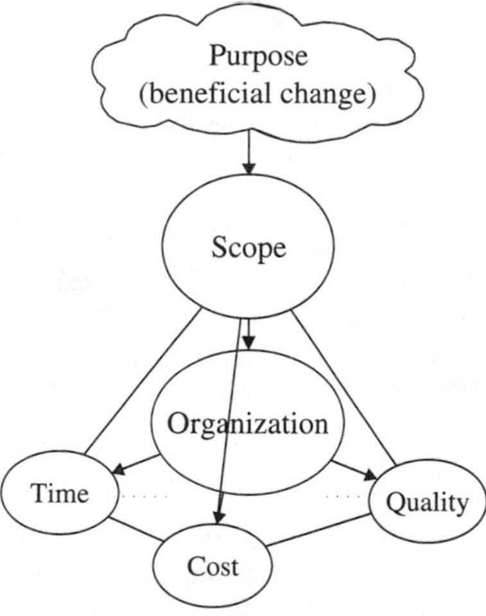

FIGURE 1.3 Five functions of project management.

some projects, the Olympic Games for instance, the project must be completed to the nearest minute. But on others the time value of the asset must be balanced against its performance levels (quality) and the cost of delivering it.

Two additional functions, not illustrated in Fig. 1.3, are as follows:

6. As previously stated, the uniqueness, novelty, and transience of the work of the project create risk. That risk needs to be managed.

7. Figure 1.2 illustrates at least two stakeholders to the project with different objectives. There are a wide variety of stakeholders to a project, all with differing objectives. The commitment of these stakeholders to the project needs managing.

The description of the management of the first six of these functions comprises Part 2 of this book. Stakeholder management and communication with them is described in Chap. 4, where I discuss the people involved in the project.

Those of you familiar with the PMI® PMBoK®[3] will know that it contains nine "body of knowledge" areas. These are the management of integration, scope, human resources, quality, cost, time, risk, communication, and procurement. The middle seven are equivalent to my seven functions. I don't overtly mention integration, but in fact it pervades everything I do in this book. I have used the term project organization rather than human resources, but the intention is the same. Communication between the project manager and client is included in discussions control in Chap. 13 and governance in Chap. 15.

Table 1.2 summarizes various tools and techniques used to manage the five core functions, and shows where in the book they are covered.

TABLE 1.2 Tools and Techniques of Project-Based Management

Method	Techniques	Tools	Chapter
Managing stakeholders	Stakeholder analysis	Stakeholder register	4
		Communication	
Managing scope	Product breakdown	Milestone plans	5
	Work breakdown	Activity schedules	5
	Configuration management		7
Managing organization	Organization breakdown	Responsibility charts	6
Managing quality	Quality assurance	Quality plans	7
	Quality control	Reviews and audits	7
	Configuration management	Procedures manuals	17
Managing cost	Cost control cube	Estimating techniques	8
	Earned value		8
Managing time	PERT/CPA	Networks/bar charts	9
Managing risk	Risk management		10
Feasibility	Startup workshop	Definition report	12
Design	Definition workshop	Project manual	12
Execution	Baselining	Work-to-lists	13
Control	Forward-looking control	Turnaround documents	13
		S-curves	
Close-out		Checklists	14

1. Scope is managed through product and work breakdown. The definition of a project, Fig. 1.1, initiates the product breakdown: impact-outcome-output. But the project is fractal; every bit of the project has the features of a project. So the hierarchy of objectives continues down with the output, or deliverables, from work areas, work packages and activities. This hierarchy is called the *product breakdown structure (PBS)*.

2. Organization is managed through an *organization breakdown*, by which we break down the skill sets of the people who will do the work. This is called the *organisation breakdown structure* (*OBS*). At any level of breakdown, the products to be delivered and the skill sets involved define a two-dimensional matrix, called a *responsibility chart*, which indicates who will do what work to deliver the products. Conventionally products are put in the rows and skills in the columns. The cells then represent the work of the project. The hierarchy of responsibility charts defines a hierarchy of work to be done, called the *work breakdown structure* (*WBS*). Pedantically there is a difference between PBS and WBS. However, on many projects the difference is slight; each product is synonymous with the work to deliver it and so people sloppily refer to them as the same thing. Most of the time I will not draw a clear distinction between them, but occasionally I will, such as when discussing *configuration management* in Chap. 7.
3. Quality is managed using techniques including quality control, quality assurance, configuration management, procedures manuals, and audits.
4. The cost is managed through a third breakdown structure of cost types, labor, materials, overhead, and finance. This is the *cost breakdown structure* (*CBS*). The three breakdown structures combined produce what is called the *cost control cube*, and are part of a methodology invented by the US military in the 1950s called the *cost and schedule control systems criteria* (*C/SCSC*). This has now been incorporated into *earned value analysis* (*EVA*).
5. Time is managed using networks and bar charts. Networks are a mathematical tool to help calculate the time scale; bar charts are a communication tool to communicate the schedule to the project team. Networks are part of a methodology variously called *critical path analysis* (*CPA*), *critical path method* (*CPM*), or *program evaluation and review technique* (*PERT*).

The functions of project management are the first dimension of the structured approach, project-based management described in this book. They are the things that need to be managed throughout the project life cycle, together with the risk that pervades all five. They are the subject of Part Two of this book. We turn our attention now to the life cycle or management process.

Management of the Project

The second dimension of the structured approach are the management processes we follow to convert vision into reality. There are two components of the management approach:

1. *The project life cycle:* the stages we go through that take us from the initial germ of an idea that there is some change we can make to improve performance to the point where we have an operating asset providing benefit.
2. *The management process:* the management steps we follow at each stage to deliver that stage.

The Project Life Cycle. The project life cycle is the process that takes us from vision to reality, from the first idea that there is a potential for achieving performance improvement to delivering an operating facility that enables us to achieve that benefit. We cannot go straight from a germ of an idea to doing work. We need to effectively pull the project up by its boot straps, gathering data and proving viability at one stage in order to commit resources to the next. There are many versions of the life cycle, and we will discuss several in Chap. 11.

However, there is growing agreement about a basic five-step process (Fig. 1.4 and Table 1.3): concept; feasibility; design; execution; and close. Figure 1.5 overlays the life cycle on Fig. 1.1. Table 1.3 also shows tools and techniques used in the management of the stages of the project life cycle.

TABLE 1.3 The Basic Project Management Life Cycle

Stage	Name	Process	Outputs	Cost as % of project
Germination	Concept	Identify opportunity for performance improvement Diagnose problem	Initial options Benefits map Commit resources to feasibility Estimates ±50%	0.05%
Incubation	Feasibility	Develop proposals Gather information Conduct feasibility Estimate design	Functional design Commit resources to design Estimates ±20%	0.25%
Growth	Design	Develop design Estimate costs and returns Assess viability Obtain funding	Systems design Money and resources for implementation Estimates ±10%	1%
Maturity	Execution	Do detail design Baseline estimates Do work Control progress	Effective completion Facility ready for commissioning Estimates ±5%	Detail design 5%
Metamorphosis	Close-out	Finish work Commission facility Obtain benefit Disband team Review achievement	Facility delivering benefit Satisfied team Data for future projects	

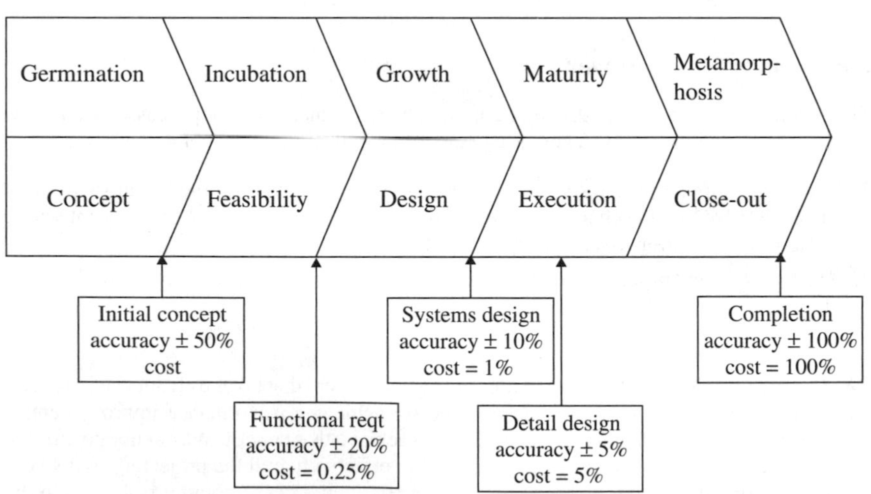

FIGURE 1.4 The project management life cycle.

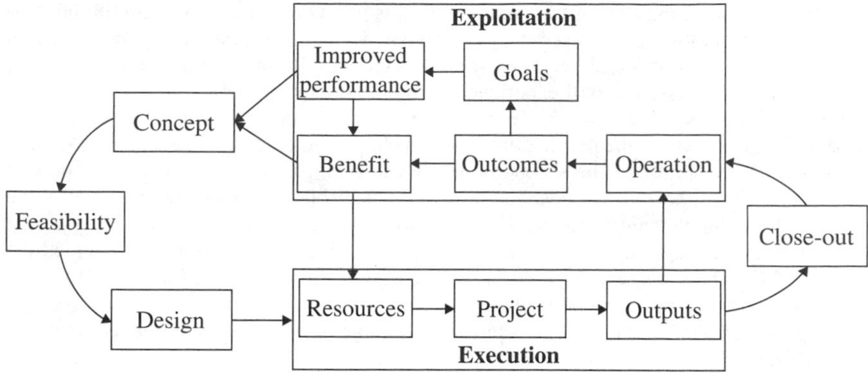

FIGURE 1.5 The project management life cycle overlaid on Fig. 1.1.

1. We start with a concept. We believe there is a problem to solve or opportunity to exploit which will help us improve performance and provide value. We do some initial problem solving, develop options, and derive very rough estimates of costs and benefit. For instance, we may think that if we spend $100, we can make $50 per year; 2-year payback; that's good business. However, at this level of accuracy, the $100 might be as little as $50 and might be as much as $150, and the $50 something between $25 and $75. To spend $50 to get $75 per year is wonderful, 8-month payback. To spend $150 to get $25 per year is awful, 6-year payback. However, at the mid-range the project seems worthwhile so we initiate the project by conducting a feasibility study. We have typically spent 0.05 percent of the cost of the project at this point.
2. During the feasibility study, you gather more information. You compare the options and choose one for further development. You develop a functional design and improve the estimates. In our example, for instance, you show the cost is more like $120, and the benefit $40, still 3-year payback—probably good business. However, the $120 may range from $100 to $140, and the $40 from $30 to $50. Best case is now 2-year payback, still excellent. Worst case is almost 5-year payback, marginal. However, the mid-range value is still worthwhile, so we commit resources to systems design, and initiate the project proper. Up to this point we have typically spent 0.2 percent of the cost.
3. In design and appraisal, we develop a fuller systems design and compose a capital expenditure proposal. We prove the viability of our project, and find a sponsor to pay for it. In our example, we may confirm the $120 cost, now accurate to $10, and the $40 per year benefit, accurate to $5. We prepare a project manual and move into implementation. Up to this point we have typically spent 1 percent of the project budget.
4. We can then move into detail design and execution. We now prepare working drawings and detail activity plans. On an engineering project, we typically spend about 5 percent of the project budget. We then do the work of the project.
5. Finally, we complete the project; we must ensure all work is finished. We then commission the facility and transfer ownership to the users. We ensure it is operated in a way that delivers the benefit expected to justify the cost. We disband the team in a way that looks after their development needs and repays any commitments made to them during the startup stages of the project. Finally we must review how we did. We cannot improve performance on this project, but we can improve performance on future projects.

There are two points I want to make. The first is that you cannot leap from initial concept to implementation with an accuracy of ±50 percent. We saw that the payback could be anywhere from 8 months to 6 years, wonderful to terrible. If you start the project and find the payback is 6 years, you will experience a loss. You must commit stepwise to the next stage in the process, allocating a bit more money on the information you have now to proceed to the next stage of the project, until you reach the end of design and are comfortable to move to full execution. The second point I want to emphasize is that the concept estimates of \$100 and \$50 for the cost and benefit of the project were not wrong, even if we discover later that they are \$120 and \$40. They were correct to the level of accuracy at that stage. In fact the range for the costs was \$50 to \$150. When at feasibility we decided the cost was \$120, the range was \$100 to \$140. The range at feasibility lies wholly within the range at concept, so the concept estimate was correct at that level of accuracy. I discuss in Chap. 8 the concept of being comfortable with a range of estimates, and indeed how it is necessary.

There are many forms of the life cycle; several are given in Chap. 11. The only other one I want to discuss here is the problem-solving cycle (Fig. 1.6 and Table 1.4). This effectively treats the project as a problem to be solved and applies standard problem-solving techniques. This also illustrates that you cannot go from recognizing you have a problem to implementing the solution in one step. If you do that, you will probably cover up the symptoms of the problem without curing the underlying malaise. Only by solving the problem in a structured way can you identify and eliminate the root cause.

TABLE 1.4 Management Process Derived from the Ten-Step Problem-Solving Cycle

Step	Management Process
Perceive the problem	Identify the opportunity for providing benefit to the organization
Gather data	Collect information relating to the opportunity
Define the problem	Determine the value of the opportunity and its potential benefits
Generate solutions	Identify ways of delivering the opportunity and associated benefits
Evaluate solutions	Identify the cost of each solution, the risk, and the expected benefit
Select a solution	Choose the solution that gives the best value for the money
Communicate	Inform all parties involved of the chosen solution
Plan implementation	Complete a detail design of the solution and plan implementation
Implement the solution	Authorize work, assign tasks to people, undertake the work, and control progress
Monitor performance	Monitor results to ensure the problem has been solved and the benefits obtained

The Management Process. At each stage of the project, it is necessary to follow a management process to deliver the work of that stage. Figure 1.7 is a four-step process delineated by Henri Fayol.[5] Fayol put the word command in the central box. I did not like this; it is too reminiscent of command and control structures. I have used *manage* and *lead*. This management process can be derived from the definition of a project. We need to plan the work to create the temporary organization that is the project by assigning resources to the project, to assign work to the resources to undertake the work of the project, and to control progress.

Those of you familiar with the PMI® PMBoK®[3] will know they use a slightly more extensive management process with

- Initiating processes
- Planning process

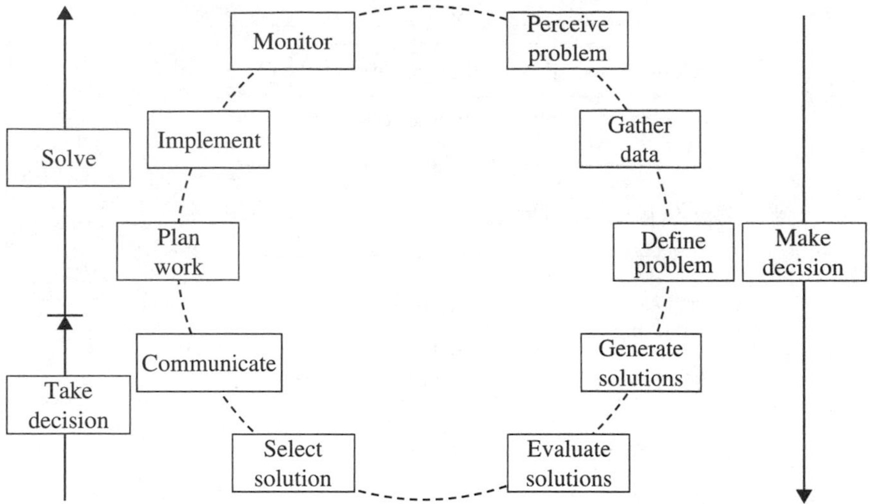

FIGURE 1.6 The ten-step problem-solving cycle.

- Organizing processes
- Implementing processes
- Controlling processes
- Closing processes

This is becoming very much like the project life cycle.

Project Management is Fractal Management. The last comment illustrates a key point. In the first edition of this book, I thought that the difference between the life cycle and the management process was so important I devoted a chapter to each. By the second edition I couldn't remember the difference. My view now is the difference is significant on large

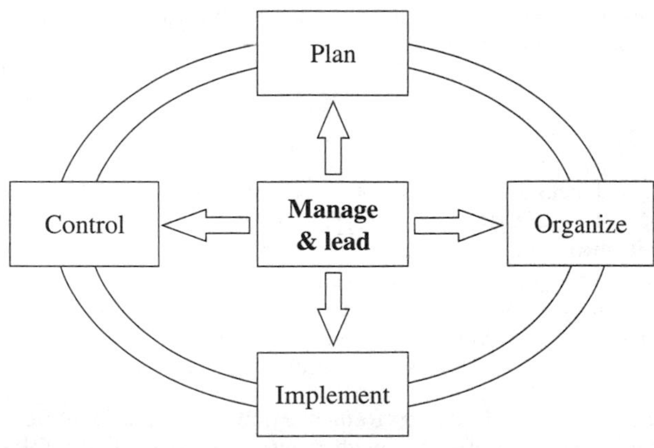

FIGURE 1.7 The management process.

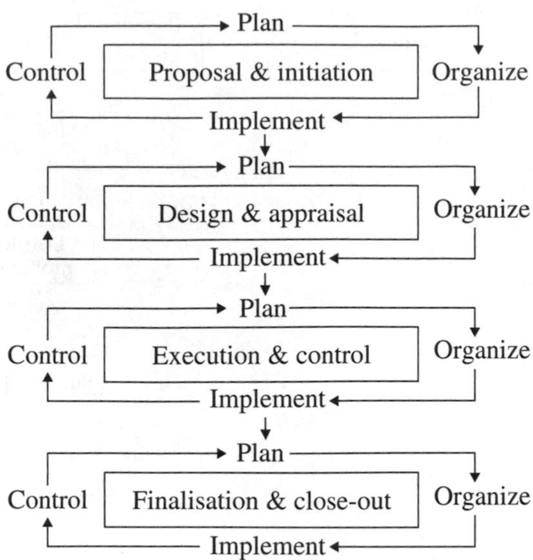

FIGURE 1.8 Project Management is fractal management.

projects but not on small projects. Large projects progress through quite distinct stages—concept, feasibility, design, execution, close—as we improve our understanding of the project. At each stage we need to plan the work, organize the resources, implement the work, and control progress, and so the management processes are repeated at each stage. On smaller projects, especially projects that are part of a program, there will be only one stage in the whole project, and so the project life cycle and the management processes may be indistinguishable. Then, in fact, using the PMI model of the management processes shows that the project life-cycle is being applied to the smaller entity.

Figure 1.8 illustrates the management process being applied at each step of the project. This shows that project management is fractal management—each stage of the project is almost a mini-project in its own right—and so the management process is the life cycle being applied at lower levels. This takes us to the third dimension of the structured approach, the different levels of the project.

The Levels

The third and final dimension of project-based management is the levels over which the project is managed. I showed earlier that a project is fractal: each component of a project is a mini-project in its own right; it is a temporary organization to which resources are assigned to deliver beneficial change, the beneficial change in that case being a component of the main project. Thus the concept of breakdown structure is an inherent part of the approach, and projects can be managed over several levels by breaking them into their component parts.

There are three fundamental levels over which a project is managed:

1. *The integrative level:* The desired performance improvement is identified, and the facility required to deliver it is defined through quantitative and qualitative objectives. Areas of work and categories of resource required to undertake them are defined, and basic

parameters or constraints determined for time scales, costs, benefits, and performance. Any risks and assumptions are stated. The *Project Definition Report* (Chaps. 5 and 12) is a tool used to record this information. A *functional design* of the facility is developed. This defines the basic features or processing steps of the facility required. For a chemical plant or computer program, this will be a *flow chart* showing inputs and outputs from each major processing element. For a training program, it will be the definition of the major elements of the program, and the learning objective of each. The definition of this level starts in the concept stage of the project and is consolidated in the feasibility stage. (Work on its definition does not end until the project ends.)

2. *The strategic or administrative level:* Intermediate goals or milestones required to achieve the objectives are defined. Each milestone is the end result of a package of work. The responsibility of organizational units, functions, and disciplines for work packages is defined. Work packages are scheduled in the project, and budgets developed. At this level the manager aims to create a stable plan which remains fixed throughout the project. This provides a framework for the management strategy and allows changes to be contained within the third level. Responsibilities are assigned to organizational units. The *milestone plan* (Chap. 5) and *responsibility chart* (Chap. 6) are tools used for this purpose. A *systems design* of the facility is developed. This shows what each of the major processing elements does to deliver its outputs, and includes a design of the processing units within each element. For a chemical plant, the systems design is based on a *piping and instrumentation diagram*, and includes specifications of all the pieces of equipment. For a computer program, it describes what each subroutine within the program achieves, how each handles the data, and the hardware architecture. For a training program, it will break each element into sessions and describe the format and learning objectives of each session. The definition of this level starts during the feasibility stage and is consolidated in the design stage.

3. *The tactical or operational level:* The activities required to achieve each milestone are defined, together with the responsibilities of named people against the activities. Changes are made at this level within the framework provided at the strategic level. The *activity schedule* (Chaps. 5 and 13) and *responsibility chart* are tools used for this purpose. A *detail design* of the facility is developed. This provides enough information to the project team to make parts of the facility and assemble them into a working whole that meets the purpose of the project. For a chemical plant, this includes piping layout and individual equipment drawings. For a computer program, it includes the design of data formats, the definition of how each subroutine achieves its objectives, and the detail specification of the hardware. For a training program, it will include the script and slides of lectures, structure of exercises, and perhaps details of testing procedures. The definition of this level starts during the design stage and is consolidated during the execution stage with the detail design.

Figure 1.9 gives a much wider view of the levels. This illustrates a cascade of objectives at different levels of management, from development objectives for the parent organization down to task objectives for individuals. At each level, the strategy for achieving the objectives at that level will imply the objectives at the next level down. I quite like this model because it gets away from hair splitting arguments about the difference between visions, missions, aims, goals, and so on (although I did use some of these words in this chapter). We just have objectives at different levels of management. This model was first shown to me by Bob Youker, who used to work for the World Bank. He illustrated it by reference to a project to develop a palm nut plantation in Malaysia (Example 1.3), a project he helped finance while with the World Bank. I show in the example how this project illustrates an important point—that often our projects do not deliver their full potential until we have completed

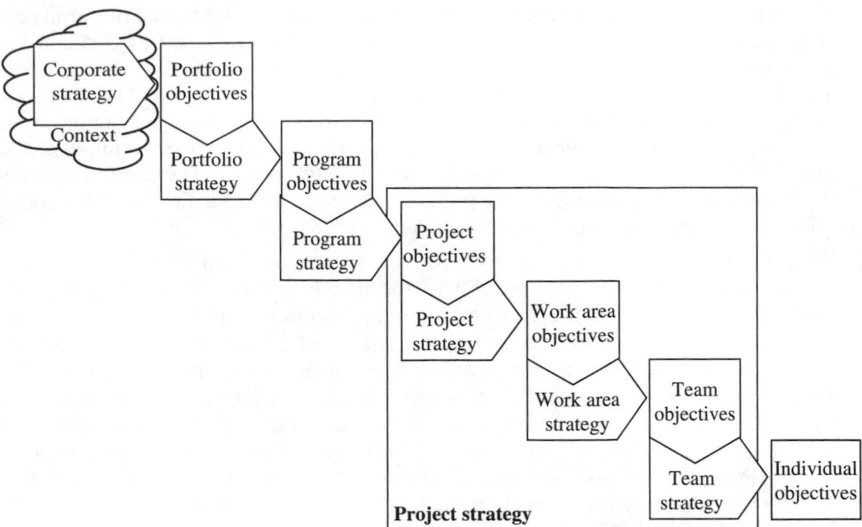

FIGURE 1.9 A cascade of objectives.

other projects in the program of projects of which they are a part. Sometimes, as in the case of the palm nut plantation, we will get no benefit at all. Table 1.5 shows the components in the PBS and the work elements in the WBS that result at different levels of the cascade in Fig. 1.9. This table also acts as something of a vocabulary for the use of these words in this book.

Example 1.3 Cascade of objectives for a project to develop a palm nut plantation

The project is a palm nut plantation. The work areas are things such as:

1. The cutting down of the jungle and the planting of trees
2. The development of an establishment to run the plantation
3. The development of systems for gathering, storing, and shipping nuts

TABLE 1.5 Standard Product and Work Breakdown Structures, PBS and WBS

Level	Product	Example 1.3	Work	Duration
	Vision	Good life		
Development	Mission	Economic growth		5 years
Program	Aim or purpose	Palm-nut oil industry		2 years
Project	Facility	Plantation	Project	9 to 18 months
Work area	Assembly	Cultivation	Work area	9 to 18 months
Team	Milestone	Orchards	Work pack	2 months
Individual	Deliverable	Planted trees	Activity	2 weeks
		Holes dug	Task	1 day

In work area 1, one team will be given the objective to plant areas of trees. On a given day, an individual will be given a bag of trees and told to plant them. (This illustrates quite nicely that the lower the level, the more the product and the work are synonymous, and the higher the level, the more the objectives have many ways of being achieved, and so are not so directly related to the work that will deliver them.)

Working upward, the program of which the project is a part is the development of a palm nut oil industry for Malaysia, and the development objectives are economic growth and employment in Malaysia. (This also illustrates that the higher in the hierarchy, the less specific the objectives.)

There is one final point: The project is part of a program to develop a palm nut oil industry. Other projects in the program might include:

- The creation of distribution systems to take nuts from plantations to factories
- The building of factories to process nuts into oil
- The creation of distribution systems to take oil from factories to customers

The palm nut plantation project will not deliver any benefit until these other projects are completed. If all we do is develop a palm nut plantation, all we will end up with is mountains of useless nuts. We can give those nuts a notional value and work out the expected return from the plantation, but we cannot realize that return until we have completed all the projects in the program. Many projects are like this; we can get the full benefit from the project only after we have completed other projects in the program.

1.2 THE PROCESS APPROACH

In the preceding discussion I emphasized two perspectives on management:

- The management of the routine versus the management of the unique, novel, and transient
- A discrete, internally focused approach versus a process-based, customer-focused approach

Together these two parameters define four types of management (Fig. 1.10) (the first of many two-by-two matrices to be introduced).

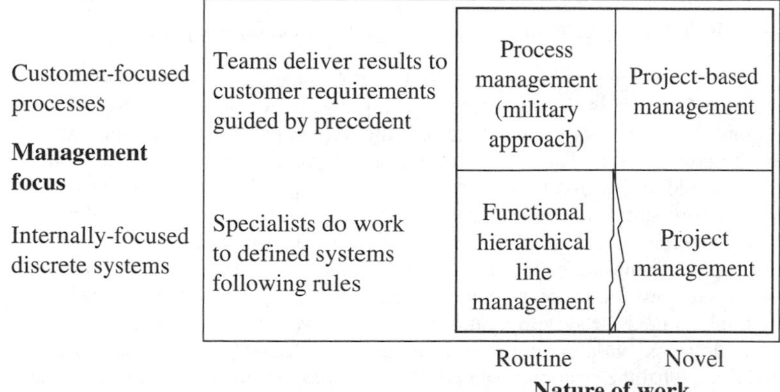

FIGURE 1.10 Four types of management.

Traditional functional, hierarchical line management, often called *classical management*, is the discrete approach to the management of the routine. The organization breaks its work into discrete steps, and creates functions to undertake the work of each step. Their products, as they move through production, are passed between the functions like batons in a relay race, except the baton is more "thrown over the wall" because little contact takes place between the functions as the product passes between each one. The idea of breaking the work of the organization into functions was the idea of Adam Smith.[6] He argued that it is much more efficient for the work of the organization to be done by specialist functions that become highly skilled at what they are doing. Frederick Taylor[7] suggested that the organization could operate like a machine, with the work processes of each function precisely defined and repeatable. (Henri Fayol's work[5] is the third component of classical management, with the functional hierarchy created to direct the functions.) Under classical management, each function takes a predefined intermediate product from the previous function, processes it, and passes it onto the next. As long as the design of the intermediate products doesn't change, the functions become decoupled, and each can focus on improving its work processes. Under a quality procedure such as the ISO 9000 series[*] for instance, each function can define its inputs and outputs, and its work processes to convert its inputs to its outputs, and then work on improving its work processes independently of the other functions as long as its inputs and outputs (the intermediate products) don't change. The organization gets better by gradual incremental improvement. However,

> If you are second best in the world you don't become best by gradual incremental improvement.[8]

When people first started embracing project management in the 1950s, they tried to adopt the functional approach with which they were familiar (the bottom right-hand box in Fig. 1.10). However, the problems associated with this approach were illustrated by the experience of a student of mine who was a quality manager with a medium-sized construction company (Example 1.4).[9] This example illustrates that:

- In a project the management approach needs to be aligned horizontally with the project and not vertically with the functions
- Every project is different so the project process needs to be tailored to the needs of the project—but be warned, the more you tailor the processes the more likely you are to make a mistake; the more you use the standard processes the more likely you are to get it right

Example 1.4 Implementing ISO 9000 in a construction company

My student was Quality Manager with a medium-sized construction company implementing ISO 9000. As a first attempt the company applied the approach described in the previous paragraph. Let's say the steps in the overall process are design, procurement, and site construction. They wrote down how each of those functions should work. Design would take instructions from the client and pass the completed designs to procurement; procurement would take the completed designs from design and pass materials to construction; and construction would take materials from procurement and pass the completed building to the client. Each function wrote down separately the work processes they would follow to convert their inputs to outputs. However, no sooner had they implemented the system than problems occurred. Difficult customers wanted the designs done and buildings constructed to their requirements. Design started saying they couldn't do what the customer wanted; it would make them noncompliant. Procurement

[*] A list of ISO procedures relating to quality is given in Table 7.1.

said they couldn't take designs according to the customer's requirements, it would make them noncompliant. They insisted doing what their ISO 9000 procedures required, not what the customer wanted. The consequence was quality fell.

As a result, they reimplemented ISO 9000, but instead of writing down what each function did, they wrote down how they processed a project from receipt of customer order to delivery of the building to the customer. Rather than aligning the procedures vertically with the functions, they aligned them horizontally with the project process. They took a process approach. They also recognized that every project is different, so at the start of every project the project manager had to develop the quality procedure for this project, defining how the standard project process would be tailored to the needs of this project.

The *Milestone Plan* introduced in Chap. 5 is the process flow diagram for the project. The process approach requires three things:

1. Functions may need to work together at some steps of the process, especially at the handover from one function to the next at each step in the process.
2. The way functions work together may vary project by project to meet the requirements of the particular customer.
3. As the project passes from one stage to the next, one function to the next, it needs to be approved against the customer's requirements and the needs of functions working further down the project process.

The concepts of stage gates, toll gates, gateway, or end-of-stage reviews are now common. At the completion of each stage of the project an assessment is made to ensure it is ready to proceed to the next stage. The business case is checked—the ratio of cost to benefit. Also it is checked that the project still meets customer requirements and the needs of functions further down the project process. End-of-stage reviews also meet another important function. I have just said that the process approach requires the management structure of the project to be aligned with the project, which means functional line managers must release authority to the project manager. Functional line managers are uncomfortable with this, but with end-of-stage reviews they can take back authority at defined intervals to check the project before it is allowed to proceed.

The process approach is that recommended by the PRINCE2(tm) methodology,[10] developed for the UK government by the Office of Government Commerce, OGC, and by ISO 10,006, the international procedure for quality in project management.[*] It is also the approach adopted in this book. Indeed, that is how I differentiate between project management and project-based management. The former is the discrete, functional approach to the management of the nonroutine, the bottom right-hand quadrant in Fig. 1.10, and the latter is the process approach, the upper half of Fig. 1.10. (Using the process approach it is much easier to move between the routine and nonroutine, being equally comfortable with runners, repeaters, strangers, and aliens.)

In Fig. 1.10, I describe the process approach to the routine as the "military approach." Some people would say that functional hierarchical line management is the military approach. It is not. The military approach is about defining process chains to support the soldier in the front line. During the battle, you cannot extend the time taken to supply him, by having functions work separately, waiting until one function is finished before the next begins. People must be empowered to support the customer within the constraints set by their orders. Functional, hierarchical line management is used in private industry and parts of the civilian civil service.

---

[*]A list of ISO procedures relating to quality is given in Table 7.1.

1.3 THE MANAGEMENT OF PROJECTS AND THIS BOOK

Figure 1.9 illustrates that at each level of management we need a strategy to achieve the objectives at that level. The project is part of the strategy by which the parent organization achieves its development objectives, but the project manager needs a strategy for undertaking the project. In Chap. 3, I present a detailed model of the strategy for undertaking a project. For now, suffice it to say that we should adopt a structured approach to the management of a project. Figure 1.11 combines the three dimensions into a single model for the management of projects. It shows that as we work through the early stages of the life cycle, we improve our understanding of the five functions—scope, organization, quality, cost, and time. It then shows that on completion of the work, in the close-out stage, we deliver first the completed work and then the commissioned facility, and then the operating benefit. The figure also shows the project taking place within a context, which itself has three components: the strategy of the parent organization, which we have already met; the project strategy; and the people involved. We revisit this again in Chap. 6. Table 1.2 lists some of the methods, tools, and techniques used in the process of managing the project and shows where in the book they are covered.

Figure 1.11 is the basis of the structure of this book.

1. In Part One I describe the context of the project: Chapter 2 describes the relationship between the project and the strategy of the parent organization, and particularly how the parent organization identifies the need for performance improvement, the desired asset to help achieve that and show how they are linked; Chap. 3 describes how we judge projects to be successful, what are the success factors that help us achieve success, and how we can develop a project strategy to deliver them; Chap. 4 discusses the people involved in the project. We look at stakeholder management and the communication with them, and project teams and leadership.
2. Part Two describes the management of the six project management functions: Chapter 4 describes the management of the scope; Chap. 5, the project organization; Chap. 6, communication with stakeholders; and Chaps. 7, 8, and 9 the quality, cost, and time, respectively. In Chap. 10, I describe the management of risk inherent in projects.

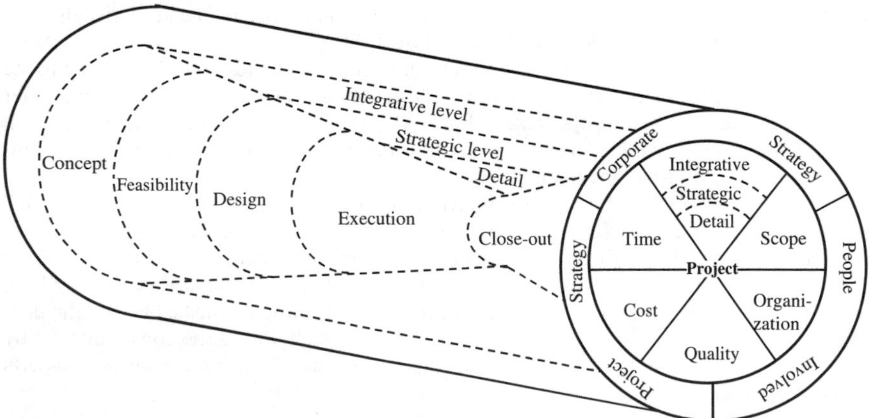

FIGURE 1.11 The structured approach to project management.

3. Part Three describes the management process: Chapter 11 gives an overview of different versions of the project life cycle and management processes. In Chap. 12, I describe the start-up processes, project definition, and feasibility. Chapter 13 covers implementation and control, and Chap. 14 discusses close-out.
4. Part Four describes the governance of projects and project management, and so gives a review of administrative techniques used in the management of projects. Chapter 15 introduces to governance the structures and associated roles. In Chap. 16, I consider the management of programs and portfolios of projects, and the role of the project support office. Chapter 17 describes corporate governance in the project-oriented organization, and the governance model recommended by the UK's Association for Project Management. This indicates the use of gateway reviews and these are further discussed. Chapter 18 describes how to develop enterprise-wide project management capability, including the development of individual project management competence. Chapter 19 describes the management of international projects (not strictly a governance issue).
5. Finally, in Chap. 20, the Epilogue, I summarise some of the principles of good project-based management introduced in the book.

1.4 IMAGES OF PROJECTS

I close this chapter by discussing two further issues relating to project management in general. The first is a classification of projects that will influence some of the thinking throughout this book. The other is a view of management that challenges some of the traditional thinking.

The Goals and Methods Matrix

It is possible to classify a project according to two dimensions—how well defined are the goals of the project, and how well defined are the methods of achieving those goals.[11] This introduces our second two-by-two matrix (Fig. 1.12), defining four types of projects. It is assumed you do not have a project until you have a clear purpose or business objective, a clear aim of the desired performance improvement. But you may not know precisely the asset that will give you that performance improvement, or how it will be constructed—part of the project may be to define the nature of the asset and how it will be delivered. The four types of project are as follows:

Type 1 Projects: These are the projects for which both the goals and methods of achieving those goals are well defined. These are typified by engineering projects. Because the goals and methods are both well defined, it is possible to move quickly into planning the work to be done, and so you will find on engineering projects an emphasis on activity-based planning. They are in the bottom right-hand quadrant of Fig. 1.10. These are earth projects built on solid foundations.

Type 2 Projects: These are the projects for which the goals are well defined, but the method of achieving them is poorly defined. These are typified by product development projects, where we know the functionality of the product, but not how it will be achieved. Indeed, the point of the project is to determine how to achieve the goals. It is not possible to plan activities, because the project will determine them. Hence we use milestone planning, where the milestones represent components of the product to be delivered. These are water projects. Water flows downhill but may cut the channel as it goes.

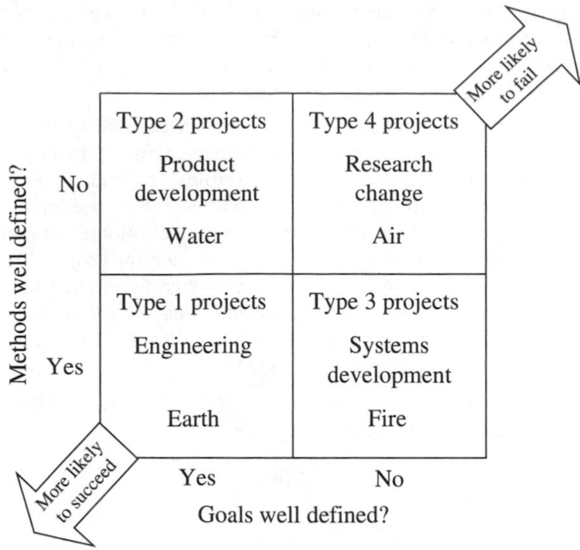

FIGURE 1.12 The goals and methods matrix.

Type 3 Projects: These are the projects for which the goals are poorly defined, but the methods well defined. These are typified by information systems projects. When I started to work as a consultant and trainer in project management, it used to amuse me that when people from the information systems industry talked about project management, all they talked about was life cycles and phases. The goals and methods matrix explains why. On an information systems project, to get the users to say what they want is difficult enough; to get them to hold their ideas constant for any length of time is impossible. All people have to hold onto is the definition of the life cycle. Hence, on information systems projects one tends to use milestone planning, where the milestones represent the completion of life-cycle stages. These are fire projects; be careful you don't get burnt.

Type 4 Projects: These are the projects for which both the goals and methods of achieving them are poorly defined. These are typified by research or organizational change projects. The planning of these may use soft systems methodologies,[12] and the plan itself will again be milestone based, but the milestones will represent gateways, go/no go decision points, through which the research project must pass or be aborted.

We see that each of the four types of project requires a different approach to its planning and management. In reality, a given project will involve more than one type of project. The example project used from Chap. 5 onward has engineering work (Type 1), product development work (Type 2), information systems work (Type 3), and organizational change work (Type 4).

Project Management as Sailing a Yacht

My analogy of project management—indeed all management—likens it to sailing a yacht. This analogy works on two levels.

Micro-Level. When yachts are sailing in a race, they sail around in a triangle, the longest leg of which is arranged to be sailing up wind. If while sailing that leg, a crew aims their boat directly at the next buoy, they will be blown backward. What they have to do is sail across the wind, called tacking, and slowly make their way upwind by tacking back and forth. Hence, they achieve the next objective, not by sailing directly toward it, but by sailing for something they can achieve, and then something else they can achieve, and eventually making the objective. There is a joke about asking an Irishman the way to Dublin station: He says, "I wouldn't start here, if I were you." You would prefer not start at this buoy to get to the next one upwind, but you have to, and you do it by taking it in steps you can achieve. All life is like that, all management is like that.

While tacking the current leg, you will choose a sail setting and a rudder setting, and plan to sail so far, say 100 yards, before tacking about. While sailing that leg, you do not say, "This is my sail setting, this is my rudder setting, good project management is adhering to my plan, come what may." You continually adjust your sail and rudder setting as the wind fluctuates. You monitor the actual conditions and respond accordingly. And if the wind comes around far enough, it may be better to be tacking in the other direction, and you will change course. You should treat your project plan as flexible. It was your best view of how to achieve the project when you developed the plan, but you must be willing to adapt it as you get new information and external conditions change.

Macro-Level. A classic yacht race is the Whitbread Round the World Race. Before the start of the race, the competing yachts will have spent months before the race pouring over weather charts, and will have chosen a strategy for the race based on the normal range of weather conditions. But while they are sailing, they must respond to the conditions they actually encounter. They will have a strategy for the race, but will determine their detail plans as they sail the race, responding to today's conditions and the forecast for tomorrow. In spite of not being able to plan the detail, there are three things that can be asserted:

1. They can predict the duration of the race to a very high degree of accuracy, a few days in nine months.
2. The boats that come first and second, after nine months, are only a few hours apart.
3. There is a large degree of luck involved.

The crew who wins is not the crew with the best detail plan to which they adhere doggedly. The people who win are the ones with the best strategic plan, who respond best to the actual conditions on the day. In spite of having to change the plan as the race progresses, the competitors are encountering the same conditions, and are very close behind. The most competent crew—the one with the best strategic plan—is the one who wins. (See Example 1.5.) Our projects are the same.

Example 1.5 The Whitbread Round the World Yacht Race

The crew first to arrive into Cape Town in October 1997 was generally regarded as the third best crew in that year's race. The two best crews arrived a day later, a couple of hours apart, having repeatedly overtaken each other over the preceding few days. The team that won took a more southerly, longer route, but picked up a stronger easterly wind. They had a better strategy, based on an assessment of the chance of achieving a stronger wind to compensate for the longer route, and their risk assessment paid off. However, they might have been unlucky and encountered lighter winds that year. They assessed the probabilities and were lucky.

SUMMARY

1. There are three dimensions to the management of projects:
   - The project
   - The management process
   - The levels
2. A project is a temporary organization to which resources are assigned to do work to achieve beneficial change. Resources from across the organization need to be integrated to work on the project. They work under a sense of urgency and uncertainty. To coordinate their efforts they must have a plan that is robust but flexible, and that means it should be goal oriented and staged.
3. The essence of project management is managing the risk and uncertainty.
4. There are seven functions of project management: managing the scope, project organization, the stakeholders, quality, cost, time, and risk.
5. The project life cycle is the process by which the project is undertaken. There are five basic stages:
   - Concept
   - Feasibility
   - Design and appraisal
   - Execution and control
   - Close-out
6. The management process is the management cycle that is followed to implement the work of each stage. There are five basic processes:
   - Planning the work
   - Organizing the resources
   - Implementing by assigning work to people
   - Controlling progress
   - Managing and lead
7. In a project, the management focus should be aligned horizontally with the process, the project, and not vertically with the functions. Every project is different, so the standard project process should be tailored to meet the needs of the project. But the greater the changes you make from the norm the more likely you are to make a mistake.
8. Projects can be categorized according to how well defined are the goals and the methods of achieving the goals. This gives four types of projects with four different approaches to planning:
   - Type 1: well-defined goals, well-defined methods, activity-based planning
   - Type 2: well-defined goals, poorly defined methods, component milestone–based planning
   - Type 3: poorly defined goals, well-defined methods, life cycle-based planning
   - Type 4: poorly defined goals, poorly defined methods, gateway-based planning
9. Project management is like sailing a yacht:
   - You cannot always achieve your objectives in one step
   - You must continually adapt your plan in response to changing circumstance
   - You cannot plan the detail, you can only plan the strategy
   - Even still it is possible to achieve an accurate forecast of the cost and duration of the project
   - The winners are the most competent team, with the best strategic plan, who respond best to the conditions actually encountered

REFERENCES

1. Gareis, R., *Happy Projects!,* Vienna: Manz Verlag, 2005.
2. Anbari, F.N., Bredillet, C.N., and Turner, J.R., "Exploring research in project management: the nine schools of project management," paper in production, 2008.
3. Project Management Institute, *A Guide to the Project Management Body of Knowledge*, 3rd ed., Newtown Square: Project Management Institute, 2004.
4. Carroll, G.R., "On the organizational ecology of Chester I. Barnard," in Williamson, O.E. (ed.), *Organization Theory: From Chester Barnard to the Present and Beyond*, New York: Oxford University Press, 1995.
5. Fayol, H., *General and Industrial Management*, London: Pitman, 1949.
6. Smith, A., *The Wealth of Nations*, London: Stratton and Cadell, 1776.
7. Taylor, F.W., *The Principles of Scientific Management,* New York: Harper & Row, 1913.
8. Johansson, H.J., McHugh, P., Pendlebury, A.J., and Wheeler, W.A., *Business Process Reengineering,* Chichester: Wiley, 1993.
9. Turner, J.R. and Peymai, R., "Organizing for change: a versatile approach," in Turner, J.R., Grude, K.V., and Thurloway, L. (eds.), *The Project Manager as Change Agent,* London: McGraw-Hill, 1996.
10. Office of Government Commerce, *Managing Successful Projects with PRINCE2,* 4th ed., London: The Stationery Office, 2005.
11. Turner, J.R. and Cochrane, R.A., "The goals and methods matrix: coping with projects for which the goals and/or methods of achieving them are ill-defined," *International Journal of Project Management,* 11(2), 1993.
12. Checkland, P.B. and Scholes, J., *Soft Systems Methodology in Action*, Chichester: Wiley, 1990.

P • A • R • T • 1

MANAGING THE CONTEXT

CHAPTER 2
PROJECTS FOR DELIVERING BENEFICIAL CHANGE

In this part I describe the project's context, the three elements in the outer ring in Fig. 1.11.

A project is a temporary organization to which resources are assigned to deliver beneficial change. The first step in the management of a project is to identify the need for performance improvement, and then diagnose the change, the new asset, most likely to deliver that performance improvement. You then need to demonstrate how the change will deliver the performance improvement, and this is done through a benefits map.

The need for performance improvement can be a positive thing; all organizations need to improve performance. You may be best at what you do, but if you want to go on being the best you have to remain ahead of the competition (Example 2.1). Or you may be second best in the world, but to become best you have to make a stepwise change in your operations. Or you may have an area of business that is not performing as you would like, and you have to either radically improve it or exit that business.

In this chapter, I discuss how to identify the need for change, diagnose the change required, and draw the benefits map that shows how the new asset will deliver the required benefit. I also consider how projects should be linked to corporate strategy.

Example 2.1 The need for performance improvement

I worked with an insurance company which had had more than 50 percent market share for insurance in the country. They had been very profitable. They were being threatened by low-cost entrants to the market, so they were losing market share. They still had the largest market share, but no longer over half. They had also lost profitability. They were best in their market at what they did, but they were losing that position. They had to improve performance to remain dominant.

2.1 IDENTIFYING THE NEED FOR PERFORMANCE IMPROVEMENT

There is a simple tool called the *performance gap* for identifying the need for performance improvement (Fig. 2.1). You have a measure of performance of an area of your business. This may be a quantitative measure, such as level of sales, return on sales, or return on assets. Or it may be a qualitative measure, such as motivation of your employees, environmental performance, or the satisfaction of your customers. Or it may be partially quantitative and partially qualitative, such as rates of absenteeism, noise levels, or number of complaints.

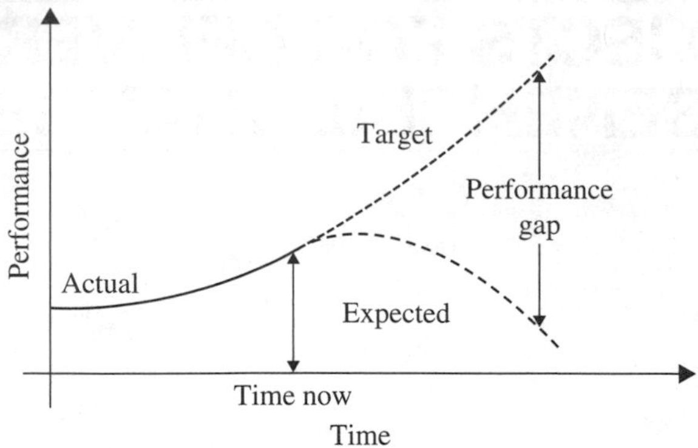

FIGURE 2.1 The performance gap.

For sometime now you have been improving and you want to go on improving to remain ahead of the competition. But you identify that if you remain where you are, performance into the future will actually fall. So the difference between where you are and where you would like to be identifies a performance gap, and you need to fill this gap by undertaking projects to introduce change. If you have a quantitative measure of performance, the predicted size of the gap will indicate the benefit of the change from undertaking the project.

Examples of desired performance improvement have been presented:

- In Example 1.2 the Chinese government wanted to improve the economic performance of the costal region on the north side of the Yangtze River. To do that they believed they needed to improve speed of traffic flow across the river at that point.
- In Example 1.4 the company wanted to improve its quality performance and improve customer satisfaction with its performance.
- In Example 2.1 the company wanted to maintain market share and maintain profitability in its product portfolio.

This last example illustrates that performance improvement may mean maintaining current levels. That will be an improvement over the future where current levels are threatened.

The shortfall in performance can be caused by internal or external pressures. Internal pressures can include an aging workforce, changing technology, or strategic initiatives. External pressures can include government legislation, new products introduced by competitors, or changing customer preferences. External pressures driving change are often categorized as being political, economic, social, technical, legal, and environmental, forming the well-known acronym *PESTLE*. The use of diagnostics or benchmarking can help to identify the need for performance improvement. You can benchmark your performance against other departments from the same company or against other companies. You can make quite detailed and direct comparisons with other parts of the same company, comparing, for instance, productivity, absenteeism, or profit on sales. However, you need to beware that differences in performance are not caused by some fundamental differences in the nature of the work done by the different departments (Example 2.2).

Example 2.2 Differences in performance of different businesses

I worked for a company called Imperial Chemical Industries, ICI, and I was involved in doing investment appraisal for a new process plant for my division making bulk chemicals. The internal rate of return (IRR) was calculated as 15 percent. The main board said that wasn't good enough; they could get 25 percent from speciality chemicals. But if you looked at the end-of-year profits my division was making 15 percent while speciality chemicals was making 1 percent. That was because in bulk chemicals all the costs were in building and operating the plants, whereas in speciality chemicals, all costs were in research and development, spent before the decision to build a plant was taken. Once the decision to build a plant was reached, the returns were high, but there were also high sunk costs to reach that point. However, persuaded by the high IRR from building plants, ICI concentrated on speciality chemicals and withdrew from bulk chemicals. It fell from being the fifth largest company in the United Kingdom and was eventually broken into small businesses. It was wrong to compare the performance of projects in one division with those from another because of the fundamentally different nature of the cost base.

You can also make comparisons with competitors, but the data may be more difficult to come by—your competitors are not going to open their books for you. However, there is a lot of published data. From published company accounts, for instance, you can work out return on sales, return on assets, and levels of working capital. Also it is sometimes possible to compare your performance to published data for the industry, so although you cannot compare your performance directly to a competing company, you can compare to industry norms. For instance, the Construction Industry Institute based in Austin, Texas maintains a database of project performance statistics that can be used to benchmark project performance. A range of diagnostic tools also exists that can help benchmark performance. The CMM model developed by the Software Engineering Institute of Carnegie Mellon University[1] can be used to benchmark software development performance, or the OPM3 model developed by Project Management Institute (PMI)[2] can be used to benchmark project performance.

Case Study

Throughout the book I use a case study to illustrate the concepts, so we see the project plans and control mechanisms developing as the book progresses. The project is taking place in a company called TriMagi Communications, which supply visual, data, and voice networks, including cable television. Table 2.1 introduces the company and sets out the background to a particular area of performance improvement they hope to achieve. They are seeking two areas of performance improvement:

1. Improve customer quality by:
   - Never having an engaged line
   - Reducing the wait time for call to be answered
   - Reducing the time for a repair engineer to reach customer's premises
2. Improve productivity and flexibility of staff

2.2 DIAGNOSING THE CHANGE REQUIRED

Having identified the need for performance improvement, we now need to work out what changes we need to make to achieve them. Figure 2.2 illustrates that performance improvement can come from several sources, including:

1. *Operational efficiency:* changes to work flow, continuity of production, production machinery, the use of automation, and supply chain
2. *Organization effectiveness:* changes to management processes, information systems, management style, personnel competence, and rewards

TABLE 2.1 Performance Improvement at the TriMagi Communications Customer Repair and Maintenance Offices (CRMOs)

TriMagi Background
TriMagi Communications is in business to supply visual, voice, and data communication networks based on its leading edge in glass fibre and laser technology. It will supply two-way cable television services to domestic and educational customers, data communication networks to these and commercial customers, and telecommunication services through its cable and data networks. It will be the first choice provider in the European countries within which it operates. It currently operates in its home base of the Benelux countries (Belgium, the Netherlands and Luxembourg), but plans to expand into other European countries. With its expansion in Europe, TriMagi Communications intends to rationalise its Customer Repair and Maintenance Offices, CRMOs, in the Benelux countries, starting in its home base in Holland. There are currently 18 CRMOs in the region. Each office is dedicated to an area within the region. An area office receives all calls from customers within the area reporting faults. The fault is diagnosed either electronically from within the office, or by sending an engineer to the customer's premises. Once diagnosed the fault is logged with the field staff within the office, and repaired in rotation. Each area office must cope with its own peaks and troughs in demand. This means that the incoming telephone lines may be engaged when a customer first calls, and it can take up to two days to diagnose the fault. To improve customer services the company plans to rationalise the CRMO organization within the region, with three objectives: – never have engaged call receipt lines within office hours – achieve an average time of two hours from call receipt to arrival of the engineer at the customer's premises – create a more flexible structure able to cope with future growth both in the region and throughout Europe, and the move to "Enquiry Desks," dealing with all customer contacts.

3. *Business portfolio:* changes to the product, price, and place of same, technology, quality
4. *Higher order strategic issues:* changes to technology, culture, or overall business strategy

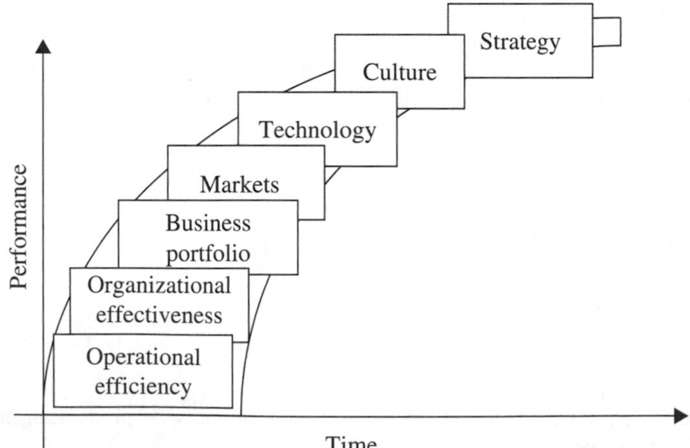

FIGURE 2.2 Achieving performance improvement.

Figure 2.2 illustrates that the time it takes to achieve performance improvement takes longer the further you move down this list, but so too the levels of performance improvement are commensurately greater. So you can achieve "quick and nasty" levels of improvement with operational efficiency, but not very great levels, whereas you can achieve much greater levels of improvement by changing your technology, culture, or strategy, but it will take longer. That means whichever you choose will depend on the levels of performance improvement required, and how long you have got. If you are best in the world and want to maintain that position, or are second best and want to become best, you will probably start by changing your strategy first. If on the other hand you are going to go bankrupt tomorrow if you don't do something quick, you will look to make quick and nasty improvements in operational efficiency.

Diagnostic Tools

There are a number of diagnostic tools that can help identify where the problems preventing performance improvement lie, or what changes can help achieve performance improvement.

Diagnostic Questionnaire. The first type of tool is a diagnostic questionnaire. In effect the maturity models[1,2] mentioned previously are diagnostic questionnaires. Or you may have or develop your own questionnaires. Table 2.2 is a questionnaire[3] I have used with client organizations to help them pinpoint where their problems lie. I applied this with the insurance company in Example 2.1. The results are given in Example 2.3.

Example 2.3 The cause of falling performance

I applied Table 2.1 with the insurance company in Example 2.1. When we scored them for managerial and financial factors, they scored average plus to good. There were a couple areas of weakness but overall they were good. It indicated that perhaps there was a possibility for some improvement but it was not the cause of their problems. But when we applied the competitive and technical factors a different picture emerged. They were scoring poorly against many of them. They were selling their products through agents, based in shops in the town centre whose average age was over 60. Many low-cost players were entering the market and selling their products over the Internet. Some products such as house insurance and car insurance can be sold easily that way. For others, such as life insurance and health insurance, local laws (PESTLE) make that more difficult. Some low-cost entrants were also selling house insurance with premiums one-third those of my client. Their products were more comprehensive, but it was difficult to make people buying over the Internet aware of this. The company recognized they needed to change their distribution channels for some products and improve the effectiveness of their agent network.

Boston Consulting Matrix. The Boston consulting matrix (Fig. 2.3) can help pinpoint where a company's products are in the product life cycle. (We return to the product life cycle in Chap. 11.) The Boston matrix views products in terms of their market competitiveness and growth, and identifies products at four stages of development: cash cows, rising stars, dogs, and problem children. Cash cows are generating large profits but absorbing little cash for further growth and so are cash positive. On the other hand, problem children, newly introduced products, are not generating significant profits yet but are absorbing money to fund their growth. Cash from the cash cows needs to be used to fund the development of new products to maintain the company's portfolio. All cash cows will eventually become dogs, which are cash neutral, not generating profits but not absorbing any money to fund their

TABLE 2.2 Diagnostic Questionnaire for Organizational Capability

Capability factors	Poor	Weak	Avg−	Avg+	Good	Excl
Managerial factors						
1. Corporate image, social responsibility						
2. Use of strategic plans						
3. Environmental assessment						
4. Speed of response						
5. Flexibility of organization						
6. Communication and control						
7. Entrepreneurial orientation						
8. Ability to attract and retain good people						
9. Response to changing technology						
10. Aggressiveness in meeting competition						
Competitive factors						
1. Product strength, uniqueness						
2. Customer loyalty, satisfaction						
3. Market share						
4. Selling and distribution costs						
5. Use of experience curve in pricing						
6. Use of product replacement life cycle						
7. Investment in new products						
8. Barriers to entry						
9. Takes of market growth potential						
10. Supplier strength						
Financial factors						
1. Availability of capital						
2. Capacity utilization						
3. Ease of exit from market						
4. Profitability, return on investment						
5. Liquidity						
6. Leverage, financial stability						
7. Capital investment, to meet demand						
8. Stability of costs						
9. Ability to sustain effort						
10. Elasticity of demand						
Technical factors						
1. Technical skills						
2. Resource and people utilization						
3. Level of technology in products						
4. Strengths of patents and processes						
5. Production effectiveness						
6. Value added to product						
7. Intensity of labour to make product						
8. Economies of scale						
9. Newness of plant and equipment						
10. Level of coordination and integration						

growth. If problem children and cash rising stars are not being nurtured there will be no new cash cows to replace the old ones as they wane. If a product is losing performance it may be because it is on the wane. The insurance company in Example 2.3 is finding several of its products turning to dogs, especially home and car insurance. With dogs, you can decide either to shed that product, or to relaunch it and turn it back into a problem child.

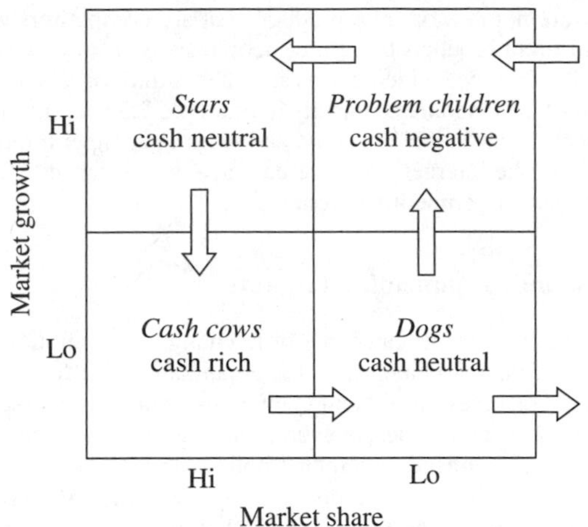

FIGURE 2.3 Boston consulting matrix.

Ansoff's Matrix. Ansoff's[4] matrix (Fig. 2.4) is a tool that has stood the test of time. An organization that is looking to expand its business can move toward introducing new products or entering new markets. That gives four growth strategies: growth, development, penetration, and diversification. Growth is low-risk but low-potential performance improvement. Diversification is high-risk but high-potential performance improvement.

Porter's Five Forces. Porter's five forces model is also a tool that has stood the test of time. It suggests that an organization operating in an industry faces pressures from five sources:

1. Competitors within the industry
2. Suppliers
3. Buyers
4. New entrants to the industry
5. Substitute products

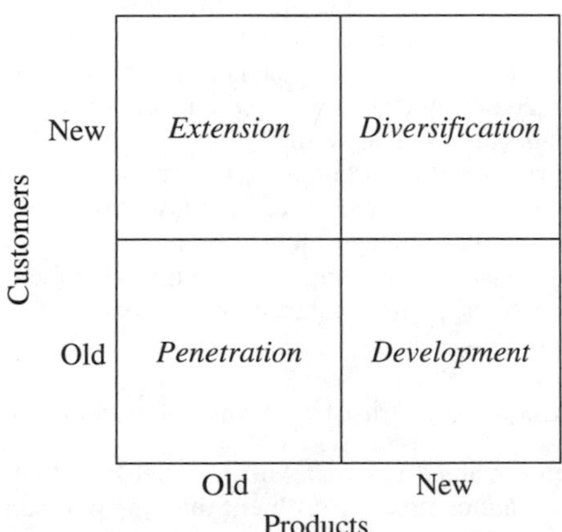

FIGURE 2.4 Ansoff's matrix.

Substitute products are not new, similar products that are competitors within the industry. They are entirely different products that entice people away from your product, offering an alternative benefit. For instance, video games are a substitute for television, which is a substitute for the cinema. The insurance company in Example 2.3 was suffering pressures from buyers finding new channels in which to buy, and seeking cheaper products, and from new entrants to the market. The Internet had lowered barriers to entry enabling new companies to enter the market to offer competitive products.

People, Systems, and Organization Projects

Figure 2.2 suggests improvements can come from changes in operational efficiency, often achieved by making technical changes, or by organizational effectiveness, achieved by making organizational changes, introducing new management processes, developing people with new competencies and values, or even changing the organization structure. In reality, on most projects you will need to combine both technical and organizational changes. This leads to the concept of people, systems, and organization (PSO) projects, projects that involve a mixture of technical and organization changes, the latter requiring changes to PSO. Often, however, the technical changes are easy and well understood, and people focus on those. In Chap. 5, I introduce a planning process that forces one to address the organizational changes as well, to make sure that not only do you introduce the new technology to make the operational improvements, but that you also introduce the new management processes, that you train and educate the people, and that you make any structural changes to achieve the organizational improvements. Through this technique you can also ensure that you make necessary changes to the business portfolio, markets, technology, culture, and strategy.

During the 1980s and 1990s, there was something of an evolution in people's focus in addressing the need for the organizational changes. In the mid 1980s, people viewed their projects only in terms of the technical changes. They focused on managing the technology, and assumed that the changes to PSO would happen automatically, or worse, weren't required at all. Then, in the late 1980s people began to recognize that the organizational changes were necessary. You have to introduce new management processes to operate the new technology, people need new competencies to operate the new technology and the new management processes, and sometimes you need new organizational structures. However, they behaved as if the technology was the main focus of the project, and the organizational changes would just piggyback on those. By the early 1990s, people realized that often the organizational changes are in fact the drag on the rate at which you can introduce change. Sometimes you need to introduce the change in two steps (tack like the yacht in Sec. 1.4). But you can't introduce all the technical changes and only some of the organizational changes. You have to make technical changes which support the organizational changes. So if you make the organizational changes in two steps you probably also need to make the technical changes in two steps (see Example 2.4). By the late 1990s the pendulum had swung completely the other way. As a result of Business Process Engineering, people now behaved as if the organizational, strategic, and cultural changes were the main focus of the project and the technical changes were merely something that facilitated that. The pendulum has now swung back to the middle; but the risk is it goes too far and people start focusing on just the technical changes again.

Example 2.4 Introducing Internet-based ways of working

Often when replacing manual systems with Internet-based ones, the ideal, Internet-based, user interfaces are completely different than the old manual-based, legacy systems. However, people can be resistant to adopting both Internet-based data entry and new interfaces simultaneously. They want to continue entering data into forms with which they are familiar, while they adjust to using the Internet or Intranet as the medium. Thus the transition should be made in two steps, first making the online user interface look like the old manual forms allowing users to become familiar with Internet-based data entry, and then introducing the new user interface later.

Case Study

After applying the diagnostic in Table 2.1, TriMagi determines that they can use new technology to change the organizational structure of their CRMOs (Table 2.3). It is not that their old technology and structure were wrong; it is just that new technology now exists that makes a different structure better able to meet their quality targets.

TABLE 2.3 Proposed Changes at the TriMagi CRMOs

TriMagi Project outputs
The proposed improvement can be achieved by changing the CRMO structure using new technology recently developed by the R&D department. In the new structure there will be three call receipt offices, two diagnostic offices, and four field offices servicing the entire region. It would be possible to have just one office for each of call receipt and diagnosis, but that would make the service exposed to technical failure. Incoming calls would be switched to a free line in any call receipt offices. It will be logged automatically, and passed on to a diagnostic office. The diagnostic office will try to diagnose the fault electronically, which should be possible in 90% of cases. The diagnostic offices are also able to discover faults before the customer notices them. The diagnostic offices will pass the faults to the field offices to repair the faults, and diagnose the remaining 10%. The field offices will be nominally assigned to an area within the region, but will share cases to balance their workload. With time the call receipt and diagnostic offices can be off-shored to achieve further savings.

2.3 THE BENEFITS MAP

In the old days of project management, 50 years ago when projects mainly came from engineering and construction, when the new asset was switched on it immediately gave the desired performance improvement. For instance, if the project is to build a new electricity-generating station, the desired performance improvement was more electricity to meet industrial and domestic demands, and as soon as the new station is switched on you have that. However, with modern change projects, especially PSO projects, including the delivery of new computer systems, how the asset will be used to deliver the performance improvement is not always quite so simple. Some additional steps are required to bridge the gap between commissioning the new facility to achieving the desired performance improvement. For instance, it may be necessary to:

- Allow time for people to gain experience to convert new skills into competencies
- Bed down new management systems and processes to gain experience in their use
- Wait for customers to become aware of a new service and so start to use it more frequently
- Give customers time to gain experience with the new service and so use it more effectively

It is important to understand how the new facility will deliver the desired performance improvement for at least two reasons:

1. It will influence the design of the new facility. A standard computer system may be used in many ways, and so its design has to be tailored to meet the exact requirements (Example 2.5).
2. After the system is implemented, the users need to know how they should use it to achieve the desired performance improvement. Further, because it takes sometime for the benefits to work through, they need to be tracked. It can take several months for full benefits to be realized, and it is the responsibility of the users to monitor that the new facility is being used to deliver the desired benefits (Example 2.6).

Example 2.5 Designing the system for the desired performance improvement

A client of mine was implementing SAP. The strategic objective set by the parent company was that my client should aim to maximise annual profits, so the system and associated management systems were designed to achieve that. Before the system was implemented they were bought by another company. The strategic objective set by the new parent company was that they should maximise annual cash flow. This required the SAP system and associated management systems to be redesigned.

To meet these twin requirements it is suggested that you draw a benefits map (Fig. 2.5), to show how the new system will be used to deliver the desired performance improvement, and to track achievement of the performance improvement post project. The right-hand side of Fig. 2.5 shows that there are several problems stopping the organization achieving the desired performance improvement. The left-hand side shows that the new asset delivers several new capabilities to the organization. But they do not themselves immediately solve the problems to deliver the desired performance improvement. Several steps are required. After the project, each of these steps may take several months to achieve, and so the users,

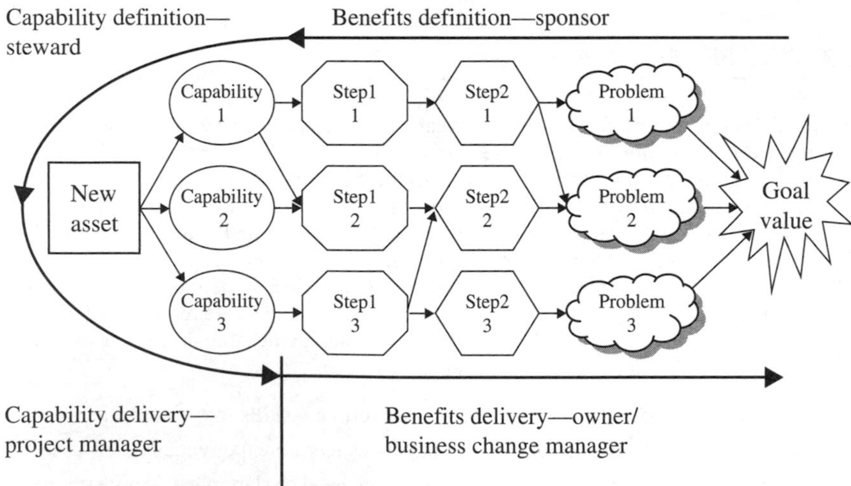

FIGURE 2.5 A benefits map.

as part of the control process, need to track that each of these steps is being achieved, and that with time the desired performance improvement is achieved.

Example 2.6 Monitoring achievement of the benefits

Another client was implementing a Customer Requirements Management System (CRMS). They wanted to improve return on sales, and to do this they felt that they needed to better understand their customers' buying habits, better segment the market, and better communicate with customers. The marketing department decided a CRMS was the solution, and commissioned the information systems department (ISD) to deliver one. At that point, all communication between marketing and ISD ceased. So ISD delivered a system in accordance with best practice. A year after the system was commissioned there had been no improvement. "Yet again ISD have failed," said marketing. No! marketing are not using the system to achieve the desired performance improvement. They are not sure that the system has actually been designed to do what they want (Example 2.5). But even if it has, they are not tracking the benefits realization. It is marketing's responsibility to use the system to achieve the benefit, not ISD's.

Figure 2.5 shows four governance roles associated with the project: the sponsor, the steward, the project manager, and the owner or business change manager.

The sponsor: He or she is somebody from the business who identifies the need for performance improvement and the possible change that will deliver it. He or she does the initial project definition and draws the draft of the benefits map. He or she also wins resources for the project. But the sponsor is not a technical expert and so needs support from the steward.

The steward: He or she is a senior technical manager who advises the sponsor about what the technology can do. The sponsor and the steward conduct the feasibility study and finalize the project definition and the benefits map.

The project manager: He or she defines the project process to deliver the change, and is responsible for managing its delivery. But it is not the project manager's responsibility to embed the change and ensure it is used to actually achieve the performance improvement.

The owner: He or she owns and operates the new asset to achieve the performance improvement and receive the benefit. Either he or she, or a subordinate, sometimes called the business change manager, is responsible for embedding the change and ensures that it is actually used and works to deliver the desired performance improvement. This is done by tracking progress through the benefits map.

I return to these roles again in Chap. 15, where they are fully described.

Case Study

Figure 2.6 shows the benefits map for the case study project.

2.4 PROJECTS FOR IMPLEMENTING CORPORATE STRATEGY

I spoke above as if the user department acts fairly independently to identify the project opportunity. However, it is important that the project should be aligned with the company's strategic objectives. Example 2.7 describes what can happen when it is not. In reality, the user department will be operating within the corporate planning process, and most identified

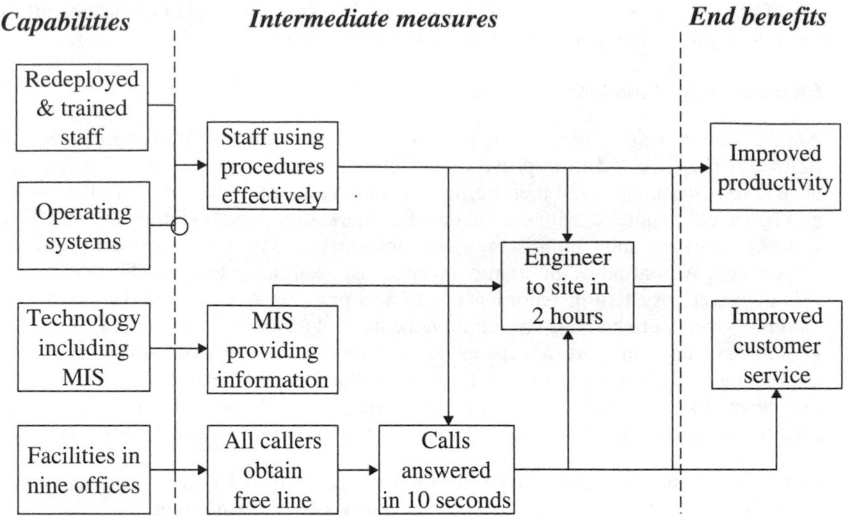

FIGURE 2.6 Benefits map for the case study project.

project opportunities will almost automatically fall under the company's strategic objectives. Sometimes the project will be a direct result of the corporate planning process. A new business opportunity will be identified; the performance improvement will come from exploiting that opportunity. Other times, the link will be a step removed. The department will not be achieving its strategic objectives as well as it would like, and the desired performance improvement will be to raise its game to meet its objectives. In both those cases, the project will be aligned with the corporate planning process. Sometimes a department may identify a project that is not immediately linked to the corporate objectives (like Example 2.7). My view is that either the department should change the corporate objectives or drop the project. Example 2.7 is a case where they did not change corporate objectives and that was the right outcome. Example 2.8 is a case where the company did not change the strategic objectives and that was the wrong outcome. Example 2.9 is a case where the company did, and that was the right outcome. In this section, I give a brief overview of the business planning process and indicate how the project should be aligned with it.

Example 2.7 A project not aligned with corporate strategy

I worked with a company in the computer industry running a series of project launch workshops in the research and development department. One project was to develop an accountancy package, which a salesman had suggested as a result of several client requests. However, this was at a time when the senior management of the company was trying to focus on software more orientated towards the requirements of managers (such as estates management and manufacturing planning), rather than functionally orientated packages. When we came to assign resources, the only person available was the project manager, and the project quickly died.

Example 2.8 The graphical user interface

The man who invented the graphical user interface for computers worked for a company not directly involved in the computer industry. "Not part of our strategic objectives," they said. So he left to join Apple.

Similarly, the company first offered the opportunity to fund the development of the telephone turned it down. They are reputed to have said that there would be a need for at most one of those in every town.

Example 2.9 The glue that wouldn't stick

There is the story of the man, who worked for the multinational company, 3M, who invented that glue that would not stick. Innovation is so important to 3M they allow people who work in their research department a day a week to work on their own projects. However, the story goes that the company tried to stop that research project because selling glue that did not stick was not part of their development objectives. Undeterred the man pressed on, and found a use for the glue. He sang in his local church choir and wanted to mark the hymns in his hymnal. Bits of paper often fell out, but if he pasted them with his glue, he could securely mark his place, and remove them at the end of the service. He went back to his organization and said he had found a use for his glue. His organization said that not many people sang in choirs. Undeterred the man made some sample pads of paper pasted with the glue (in blocks of yellow paper), and gave it to the secretaries to try out. Soon, bits of yellow paper were everywhere. The organization decided that perhaps there was a market for this product after all, and the rest, as they say, is history.

The Business Planning Process

There are four essential steps in the business planning process (Fig. 2.7).

1. Define the mission of the business.
2. Set long-term objectives for achieving the mission.

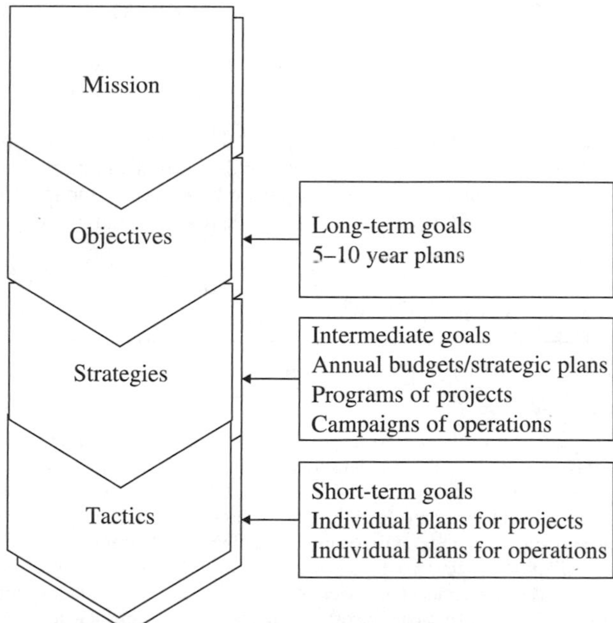

FIGURE 2.7 The business planning process.

3. Develop strategies for achieving the objectives.
4. Develop tactical plans for achieving each element of strategy.

Define the Mission of the Business. The mission is the axiom which initiates the business planning process. It is a statement of the reason for the organization's existence; it's purpose for being in business. It may be a statement as simple as to make profit for the shareholders. However, it is more common to include statements on:

- The type of products
- The positioning of the products in the marketplace
- The relationship with the employees
- Other hygiene factors
- Relationships with other stakeholders, especially local communities

A mission statement for TriMagi is given in Table 2.4.

Set Long-Term Objectives for Achieving the Mission. Having defined the mission, the company sets objectives for the next 5 to 10 years to deliver it. These are statements of the position the organization will reach in the relevant timescale covering:

- The types and ranges of products, and turnover from each
- Return on sales and assets, and growth of dividends
- Type, number, skills, and remuneration of employees
- Environmental impact
- Social and community activities

A set of objectives for TriMagi is given in Table 2.5. The CRMO supports these objectives.

Develop Strategies for Achieving the Objectives. Having set objectives, the organization can then develop strategic plans for achieving the objectives. These can take several forms.

Annual budgets: These show, year by year, how the business will develop towards the position envisaged in the objectives. Budgets for the current and immediately following year are the plans to which the business is presently working. Budgets for future years become increasingly more speculative, and will be revised annually. For example, each

TABLE 2.4 A Mission Statement for TriMagi

TriMagi Mission Statement
TriMagi Communications is in business to supply visual, voice, and data communication networks based on its leading edge in glass fibre and laser technology. It will supply two-way cable television services to domestic and educational customers, data communication networks to these and commercial, customers and telecommunication services through its cable and data networks. It will be the first choice provider in the European countries within which it operates. The company will provide secure, competitive employment for its staff. All its services will be provided in a way which has no impact on the environment. Above all, TriMagi Communications will supply its shareholders with a secure investment which increases in value annually.

TABLE 2.5 A Corporate Objective Statement for TriMagi

TriMagi
Corporate Objectives

From its current domination of the market in the Benelux countries, TriMagi Communications will establish operating subsidiaries in the following regions:

Year 1: France
Year 3: Germany and British Isles
Year 5: Iberian Peninsula, Italy, Austria, and Switzerland
Year 7: Scandinavia and Baltic States

Each new subsidiary will break even within two years, with a turnover of at least 100 million Euro, and from there will achieve a growth of 50% per annum for the next three years. By the fifth year, it will have achieved a return on assets of 20%, and will contribute 10% of turnover to the parent company to fund further product development.

Each subsidiary will employ operating personnel, and sufficient technical staff to install and maintain the networks. They may maintain a small marketing effort to develop local opportunities for using the network. These local opportunities will contribute at least 15% of turnover.

The parent company will employ technical staff to maintain the company's leading technical edge, and to develop new products and opportunities for using the networks. New products and opportunities will enable established subsidiaries to maintain a growth of at least 20% over and above that available from increased market, or increased market share, beyond their initial five years.

The objectives which TriMagi has set indicate that it will maintain its existing operation, in the Benelux countries. It will fund further growth by using the income from those operations to expand into new markets, then achieving further growth as each new market becomes established. Initially, it will sell existing products into the new markets, but, as they become established, develop new products for them. It will also try to use those new products in its old markets, where possible, to achieve further growth. The objectives also imply that the operation in the Benelux countries will split into an operating company, and a parent company undertaking new product development.

of TriMagi's subsidiaries would have annual budgets for capital expenditure, income, and revenue costs.

Subsidiary goals and milestones against each objective: The annual budgets show where the business is expected at each year end against the objectives. These can be summarized into a plan against each objective, showing intermediate milestones for achieving each one. These are sometimes called the goals of the business, and may be drawn as one or more milestone plans for the development of the objectives (Chap. 5).

Campaigns or programs for functions, operations, or projects: The annual budgets are set, or are based on, campaigns or programs for individual departments or functions within the organization. These may be campaigns for continuing operations, or programs for new projects. The business planning process is iterative, and so these programs are developed in parallel with the annual budgets, through negotiation and compromise. However, all but the first of the programs below tend to be set within constraints of the annual budgets. The first sets the basis from which the budgets and goals are derived. There are several types of campaign or program including:

1. *Programs of corporate planning or marketing:* They describe the evolution of technologies, products, or markets of the business (the upper elements in Fig. 2.2). The term

strategies is sometimes reserved for the corporate planning program, because that sets the basis for deriving the goals and annual budgets. The marketing campaign is shorter term, and deals more with the balance between products, pricing, distribution channels, and promotional campaigns to achieve annual budgets.

2. *Campaigns for existing operations:* These are undertaken when a business decides to maintain its existing products, markets, or cash cows. This may be for production, sales, or services.
3. *Programs for new projects:* These are undertaken when the analysis suggests the business should adopt new products, markets, or technologies, or undertake some other improvement to its existing operations. The projects will deliver new facilities, in the form of product designs, factories, or technologies to produce them, computer systems to manage their production, or new organization structures with trained staff and managers to undertake the production.

The last paragraph of Table 2.5 indicates how TriMagi plans to achieve its objectives.

Develop Tactical Plans for Achieving Each Element of Strategy. Plans for individual campaigns or programs, or for functions, operations, or projects are the tactical level plans. They describe how the organization will achieve each element of its strategic plans. These tactical plans may be marketing plans, production plans, or milestone plans for projects.

The Role of Projects and Operations

I have just shown how the business planning process can identify a need for routine operations and projects. These are the vehicles through which organizations achieve competitive advantage. Either they do more of the same, though always striving to improve efficiency through habitual increment improvement; or they do new things with novel organizations, that is, projects. Until the 1980s, the former dominated. However, with the development of more sophisticated corporate planning techniques, and with the explosion of technical innovation and communication, the second is beginning to dominate. Thus management by projects is becoming the way in which organizations fulfil their business plans. Just like the business as a whole, each operation and project has three levels of planning (Fig. 2.8): the integrative level, the strategic level, and the tactical level. There may be lower, more detailed levels of planning. For particularly large projects, there can be up to seven levels of work breakdown, and we shall return to this concept in Chap. 5.

The integrative level defines the purpose of the campaign or program, as defined by the corporate objectives, and the objectives it must achieve in order to satisfy the annual budgets:

1. *For sales and marketing:* This will be objectives for turnover expected from each product, budgets for distribution, and promotion and overheads for the sales department.
2. *For operations:* This will be production targets and budgets for cost of sales.
3. *For projects:* This will be a definition and specification of what the project is to produce and constraints of time and cost.

The strategic level defines subsidiary objectives each function must achieve to satisfy its overall objectives.

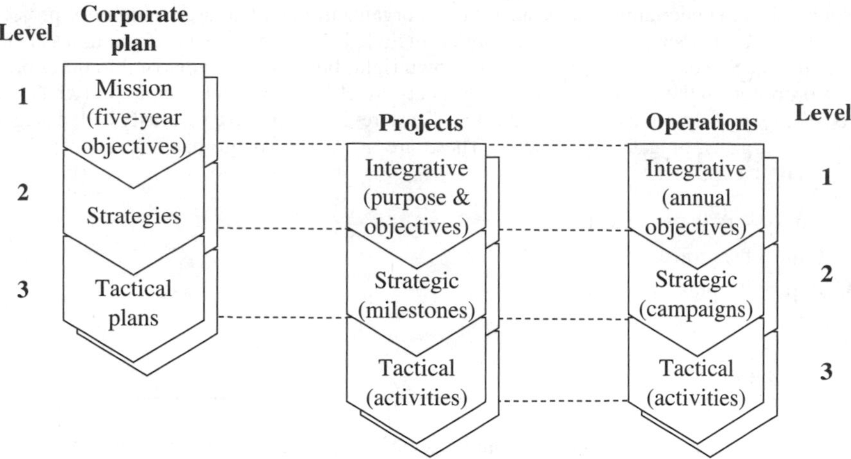

FIGURE 2.8 The projects and operations hierarchy—product breakdown.

1. *For sales and marketing:* This may be individual campaigns for selected products, product launches, advertising campaigns, or testing of new outlets, each of which may result in a project.
2. *For operations:* This will be targets for each product, or for efficiency improvements.
3. *For projects:* This will be a milestone plan or work package plan for the project.

The tactical level defines the detail of how the work to achieve each of the subsidiary objectives is to be achieved.

Selecting Projects

The business planning process may identify several possible projects. Usually, there will be insufficient resources, money, people, and materials to fund them all, and so the organization must assign priorities to select projects which are most beneficial. I will discuss the project prioritization process further in Chap. 16, when I describe portfolio management. There are several quantitative and qualitative techniques for appraising the value of projects, and making this selection. It is not my intention to give a detailed description of them here; that is more appropriate for a book on project appraisal and finance.

What I would like to reinforce is the appraisal processes is repeated at several stages of the project life cycle, using increasingly accurate data. In Sec. 1.2, I suggested that at the end of the concept stage the accuracy of the estimates may be ±50 percent, giving a wide range for the estimate of the potential value of the project (benefit divided by cost), or inversely potential payback (cost divided by benefit). (These are very crude methods of investment appraisal, but as I say I don't intend to go into more sophisticated methods.) However, based on the appraisal at the concept stage a small amount of resource is committed to the feasibility study. At the end of feasibility the appraisal process is repeated, but now the accuracy of the estimates of cost and benefit are typically ±20 percent, giving a narrow range for the estimates of value. If the project still looks worth, a larger amount of resources is committed to design. At the end of design, the estimates are typically accurate to ±10 percent, giving an acceptable range for the estimate of the value that the organization can commit to undertaking the project. For an organization with a large number of project proposals, at this stage, the project may be included in the project portfolio assessment (Chap. 16). So not only is it assessed in its own right, but also in comparison to other project proposals in the portfolio. The best projects are chosen for implementation (whatever "best" means as I will discuss in Chap. 16). These review points are variously called stage-gates, toll-gates, or gateway reviews. These are go, no-go, or go-back decision points. Based on the assessment you either:

1. Go forward to the next stage
2. Cancel the project
3. Repeat the previous stage

SUMMARY

1. The performance gap can help identify the need for performance improvement.
2. A shortfall in performance can be caused by internal or external pressures, and can be identified by using benchmarking and diagnostic techniques.
3. In order to achieve the desired performance improvement, the organization will introduce a change, delivering a new facility or asset which will give it new capabilities, and those will enable it to solve the problems or exploit the opportunities which will lead to the performance improvement.
4. Many diagnostic techniques exist to help the organization to identify the change that will enable it to achieve the performance improvement
5. Usually the new capabilities will not solve the problems blocking performance improvement directly. It is necessary to achieve several intermediate benefits before realizing the main benefit. The benefits map shows the link between the new capabilities and the desired performance improvement via the intermediate benefits.
6. The benefits map may be used to change the definition of the desired asset, and the definition of the desired performance improvement. It should also be used to ensure realization of the benefit after the project.
7. There are four governance roles associated with the management of the project and realization of the benefit: the sponsor, steward, project manager, and owner.
8. The project should be linked to the corporate planning process to ensure it will help the organization achieve its long-term goals.
9. The project needs to be appraised repeatedly throughout the life cycle to ensure it will deliver value to the organization.

REFERENCES

1. Software Engineering Institute, *CMMI® for Development, Version 1.2,* Pittsburg, Pa.: Carnegie Mellon University, 2006.
2. Project Management Institute, *Organizational Project Management Maturity Model,* Newtown Square, Pa.: PMI, 2003.
3. Rowe, A.J., Dickel, K.E., Mason, R.O., and Snyder, N.H., *Strategic Management: A Methodological Approach,* 3rd ed., Reading, MA: Addison-Wesley, 1989.
4. Ansoff, H.I., "Strategies for diversification," in *Harvard Business Review*, May, 113–124, 1957.

CHAPTER 3

PROJECT SUCCESS AND STRATEGY

In Chap. 2, we defined the project objectives: the desired project outcome, the desired performance improvement, and the problems or opportunities that the new asset will solve to help achieve that performance improvement. We then defined the project output, the new asset, and the new capabilities it will give the firm to enable it to solve the problems or exploit the opportunities to achieve the desired benefit. Figure 1.9 suggests that as we cascade down the product breakdown structure, before defining the objectives at the next level, we should define a strategy for how to achieve the objectives at the current level. So before we begin to plan the project we should derive a strategy for achieving the overall project objective. The first step of that is to round off the definition of the objectives by determining the criteria by which we will judge their successful achievement. Then you can determine what factors will increase the chance of achieving, and from them derive a strategy for implementing the project.

There are two components of project success:

1. *Success criteria*: The dependent variables by which we will judge the successful outcome of the project.
2. *Success factors*: The independent variables which will influence the successful achievement of the success criteria.

A doctoral student of mine, John Wateridge, identified what I consider to be a necessary condition for project success.[1] In order for a project to be successful, you must agree the success criteria with all the key stakeholders before you start. This is a necessary condition for project success, not a sufficient condition; unfortunately there is nothing that will guarantee project success. To meet this condition you must make an attempt to identify who most of the key stakeholders are. There are several reasons why it is important to agree the success criteria before you start, including:

- You want everybody to have the same vision of the end point of the project. If people have been working towards different end points, even inadvertently, it is impossible to pull them all together at the end.
- You want everybody to be applying the same success factors, following the same project strategy, and following the same road to its successful achievement. You don't want the project team members all chasing off in different directions.
- Even quite small differences in interpretation of the success criteria can lead to quite different outcomes, even down to whether you treat time, cost, or quality as more important (see Example 3.1).

Example 3.1 Different interpretations of the importance of time, cost, and quality

A colleague of mine was working with a shipbuilding company that had traditionally built submarines. They wanted to move into surface ships. The Ministry of Defence (MOD) issued an invitation to tender for a frigate, so the company decided to win the work to demonstrate to the MOD that they could successfully build frigates. Their strategy was to bid the job at no profit and then complete the job on time and to quality to demonstrate their competence in this area. They successfully won the bid, but nobody thought to tell the project manager the company's strategy. He saw that the project was likely to make a loss, and went all out to reduce cost. As a result quality suffered and the project went late. (Actually the parent company also changed part way through the project, and as in Example 2.7 the new parent company set different objectives for the subsidiary.)

We judge the success criteria at the end of the project, and in the months and years following. But we don't want to wait until the end of the project and find we have gone off target at the start. There are also key performance indicators, measures of the success criteria which we can track throughout the project to ensure we are on course to achieve a successful outcome.

In this chapter, I consider the issue of project success. I identify potential success criteria for projects. We see different stakeholders judge success in different ways, and it is important to achieve a compromise between their different views, to achieve an overall balanced view of success. I then describe key performance indicators, and indicate ways in which they can be simply and visually tracked through the project. Having identified the success criteria, we then need to identify the success factors which will help us achieve those criteria. I used to talk about pitfalls, things that will trip us up on the project. Now I like to take a more positive view and talk about active things we can do to increase the chance of success. In the process I will identify four necessary conditions for project success. I will then describe two models for developing a project strategy: the seven forces model and the project excellence model. Finally I will describe five principles for project success which pervade the ideas in this book.

3.1 PROJECT SUCCESS CRITERIA

The standard mantra for how we judge project success is that it should be completed to time, cost, and quality. However, this is simplistic in the extreme, and can be positively dangerous. There is an apocryphal story of research done in Australia which looked at how people viewed the success of software projects 5 years after implementation. It is said that every project that was finished on cost and time was judged 5 years later to be a failure. The point is that in striving to finish on cost and time the project manager sacrificed functionality, and the users had to live with poor functionality for 5 years. Even the project team may get satisfaction from other things (Example 3.2). I also said in Sec. 1.2 that by focusing on time, cost, and quality, project managers are distracting their attention from what is important on projects: the need to manage the uniqueness, novelty, and transience, and the inherent risk and need for integration that those create.

Example 3.2 Finishing the project on time

I worked as a maintenance engineer on four ammonia plants in the northeast of England. Every 6 months we closed a plant for biennial refit. Over a period of 4 weeks we did 100,000 man-hours of work. We planned the overhauls to within 4 hours, but we were usually 2 days late. But we were *only* 2 days late. We pulled out all the stops, and managed our way through all the problems to deliver the project within 2 days of target.

Once we coasted in 4 hours early, and felt we had failed. If we had been given a tighter target we could have really proved ourselves and achieved a shorter duration!!! That overhaul did not fulfil our need to prove ourselves as managers.

In his research, John Wateridge asked people working on information systems projects to think of two projects they had worked on, one a success and one a failure, to say what their role had been (sponsor, user, designer, or project manager), and to say how they judged each project to be a success or failure. On almost all the successful projects, all four types of stakeholder said that the project had been successful because it provided value for the sponsor. On unsuccessful projects they gave different responses as to why it had failed:

- The sponsors said the projects had failed because they hadn't provided value.
- The users said they failed because they hadn't provided the functionality they wanted.
- The designers said they failed because they were not a good design.
- The project managers said they failed because they finished late and were over budget.

What a surprise! If all the project stakeholders are working towards the same objective, to provide value for the sponsor, the project is a success, but if they focus on different things they tear themselves apart and the project is a failure. Yes, the users are interested in functionality, the designers in the design, the project managers in cost and time. But on successful projects those stakeholders bring what is important to them and balance it against the needs of others to come up with an overall compromise that meets the overall need of delivering a beneficial change that provides the sponsor with value. On unsuccessful projects, people are focusing on what is important to them to the detriment of others, and tear the project team apart.

The relative importance of the different criteria also differs project by project. The team needs to understand what is important for their project, and agrees it before they start:

- In Example 3.1, time and quality were important.
- In the Olympic games, after 6 years of preparation they have to be ready to the nearest minute—the time of the starting ceremony has been set, the television companies have sold their advertising. It has to start exactly on time.
- Work done by the consultants McKinsey in the late 1980s showed that on product development projects the functionality of the new product has the greatest impact on value, time to market is very important, and cost is of almost no importance.[2]

When I describe Wateridge's results, some project managers say that they hear what I say, but in their company, in their annual appraisal, they are judged on how many of their projects were finished on cost and time. That is what determines their annual bonus, not the value of the projects to the sponsor. They ask me what should they focus on, cost and time, or value to the sponsor. I say they should focus on changing the appraisal system so that it is supportive of good project management.

Table 3.1 gives a wider range of success criteria than Wateridge's basic four. This table shows the primary stakeholder interested in each of the success criteria. As I have said, these criteria are potentially incompatible. If, at the start of the project, you work on achieving a negotiated compromise, you can achieve an overall balance which meets the needs of everybody. If you wait until the end of the project you will be trying to reconcile the irreconcilable. Table 3.1 also shows that the final assessment is made at different times. The bottom three items relate to the work of the project and the project's output. They are assessed as the project is completed. The middle three relate to the project's outcome: does the project perform as expected and produce the desired benefit. That becomes obvious in the months following the project. The top three relate to the higher level strategic goals, and

TABLE 3.1 Project Success Criteria

Measure of success	Stakeholder	Timescale
The project increases the shareholder value of the parent organization	Shareholders	End plus years
The project generates a profit	Board	End plus years
The project provides the desired performance improvement	Sponsor	End plus years
The new asset works as expected	Owner	End plus months
The new asset produces a product or provides a service that consumers want to buy	Consumers	End plus months
The new asset is easy to operate	Users	End plus months
The projects is finished on time, to budget, and with the desired quality	All	End
The project team had a satisfactory experience working on the project and it met their needs	Project team	End
The contractors made a profit	Contractors	End

can only be determined 1 or 2 years into the future. (Table 3.1 is phrased in terms for the private sector, but similar criteria can be determined for the public sector.)

Eddie Westerveld in his project excellence model used a much simpler set of success criteria.[3] He suggested a project was successful if it satisfied the needs of various stakeholders, without specifying what their needs are, as Table 3.1 does. The five groups of stakeholders he focused on are (Fig. 3.4):

- The client
- The project team
- Users
- Contractors
- Others

Ralf Müller and I, in our research of the leadership style of project managers[4], extended this list (Table 3.2). We investigated how different leadership styles are appropriate on different types of projects (see Sec. 4.5). Finally, two clients of mine had very simple success criteria for their product development projects: The project should meet its first year revenue targets and provide increasing revenue in subsequent years. To achieve this the project must achieve quite a few of the requirements in the middle three rows of Table 3.1: It must work, the customers must like it, want to go on buying it, and new customers must want to buy it. It is then also likely to make a profit and increase shareholder value. The essence of Table 3.1 is captured by this simple statement.

Hartman's Three Questions

Francis Hartman[5] suggests that during the start-up process you ask the project team three questions to help identify the success criteria and the stakeholders for the project:

Q1: On the last day of the project what will the project team deliver to the operations team?

Q2: How will the successful achievement of that be judged?

Q3: Who has an opinion on questions 1 and 2?

TABLE 3.2 Project Success Criteria Used by Turner and Müller[4]

Project success criteria
End-user satisfaction
Supplier satisfaction
Team satisfaction
Other stakeholders' satisfaction
Performance in terms of time, cost, and quality
Meeting user requirements
Project achieves its purpose
Customer satisfaction
Reoccurring business

The first question ensures that the team has a common understanding of the project deliverables. Francis Hartman reports examples where teams had a quite fundamental misunderstanding of the project deliverables (Example 3.3). The second question identifies differing opinions about the success criteria. You are not only looking at the end of the process to get agreement on the criteria: during the process you are looking to identify where differences exist, so they can be discussed and a compromise reached. The last question identifies key stakeholders.

Example 3.3 Differing interpretations of success criteria

Francis Hartman describes running start-up workshops with each of two companies, where the project teams gave contrary answers to his three questions.

The first project was the construction of a petrochemical complex in Alberta. There were two project managers, one for the design stage of the project and the other for construction. In response to question 1, one said the project was over at mechanical and electrical completion, and the other said that it would be over when the plant delivered 60 percent of its design capacity, two dates at least 3 months apart, and yet both gave the same completion date.

On the other project, the team was replacing the accounting software for their organization. About 30 people attended the workshop, and responses to the first question ranged from

- Beta test successfully completed
- The system has run for 12 months without fault
- Thirty people have been made redundant

The first two of these were now at least 15 months apart. The third was unfortunate because some of the people in the room were those to be made redundant and this was the first they had heard of it.

The teams probably blamed failure of their projects on circumstances beyond their control, saying "We were unlucky."

In Chap. 11, I describe how to build these questions into the start-up process, and in Chap. 18, I give a Project Health Check, which asks for checks if agreement has been reached.

Case study. Table 3.3 shows the answers to the three questions for the CRMO Rationalization Project.

TABLE 3.3 Hartman's Three Questions for the CRMO Rationalization Project

	TriMagi Project success
Deliverables	The project will deliver to the parent organization: – Three call receipt offices, two diagnostic offices, and four filed offices – The technology to support the operation of the new system – Operational procedures to operation of the new system – Working methods to support the new system – Adequate numbers of competent people to support the new system
Success criteria	The project will be judged successful if – There are never any engaged telephones in call receipt – An engineer always arrives on site within 2 hours of a call being logged – There are improvements in flexible working and productivity – There are fewer customer complaints – The new structure supports the company's expansion plans
Stakeholders	Relevant stakeholders include – The board of the parent company – Managers in the CRMO organization – Staff in the CRMOs – Customers – Managers of the new regions being established – Etc

3.2 KEY PERFORMANCE INDICATORS

The success criteria should be agreed with the stakeholders before you start, but you don't want to get to the end of the project and find you are well off target, and have been so since the early days of the project. In order to avoid that happening, it is necessary to track control parameters which measure progress towards achievement of the success criteria. Throughout the book I will give guidance on how to measure the key control parameters. In modern jargon, these key control parameters are called key performance indicators (KPIs). They give a measure of the performance of the project.

It is important in project reports to give a clear and visual representation of these control parameters. There are a number of tools for doing this. The first is the project dashboard (Fig. 3.1). For any quantitative KPI, it is possible to give an indication of the current performance of that KPI against the target. In the figure, the cross where the first and second box meet represents the planned out-turn for that KPI. The arrow underneath shows the current prediction, the way the needle on your car dashboard shows the speed or level of fuel, for instance. In case a colour version of the diagram, the first box would be green, the second yellow, and the third red. This colour scheme was introduced by the Lockheed Aircraft Corporation. Green means at or ahead of plan; yellow means just behind plan, but controllable; red means well behind plan and in crisis. So you have a visual representation of the current status of that KPI. Quantitatives that you may want to track can include:

- Time
- Cost
- Forecast first-year revenue
- Safety
- Variations in design
- Productivity

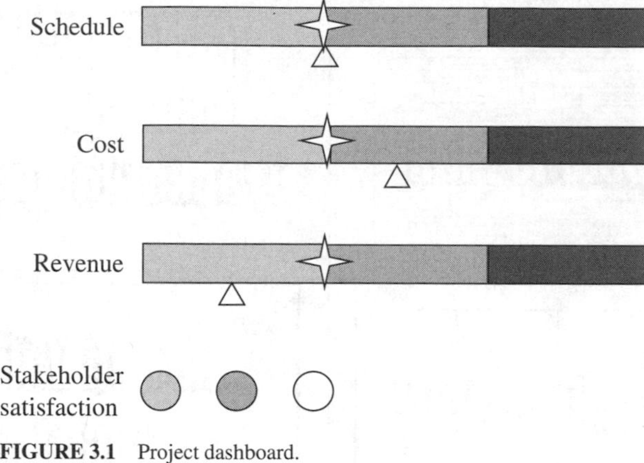

FIGURE 3.1 Project dashboard.

We see the project manager can be made responsible for forecasting first-year revenue. So, although it is probably not appropriate to make their annual bonus dependent on the total value of the project to the sponsor, which may be dependent on revenues from 5 or even 10 years, it is appropriate to make it reflect revenue from the first year—appraisal systems can be made compatible with effective project management. Figure 3.1 also shows that you can use the traffic light system to represent performance against qualitative criteria. Now you would just show the traffic light indicating red, amber, or green depending on your assessment of that criterion. I have seen people representing stakeholder satisfaction in this way.

The project dashboard provides a very effective visual representation of project progress today. However, the weakness is it does not show how progress has changed from the previous report. The project may be getting worse, it may be getting better, or there may be no change. We simply don't know. It would of course be very simple to produce a moving marker, like a seismograph. That would be very easy. However, other tools have been developed that show how the KPI is changing with time. Earned value reports (Chap. 8) show how cost performance is changing, and milestone tracker charts (Chap. 9) show how time performance is changing. These can be combined with a report against the milestone plan (Chap. 5) and a risk report (Chap. 10) to provide a complete overview of how the project is progressing (Fig. 3.2). I will return to traffic light reporting when I describe portfolio management in Chap. 16.

3.3 PROJECT SUCCESS FACTORS

Project success factors are elements of the project or its management that can be influenced to increase the chance of achieving a successful outcome. The reverse, pitfalls, are management mistakes which increase the chance of failure.

The earliest work on project success factors was done by Kristoffer Grude in Norway. This was reported in the first Norwegian edition of the book *Goal Directed Project Management.*[6] Kristoffer Grude, in his work as managing director of a Norwegian software company, identified a number of pitfalls. At the end of every project his staff had to record what went well or badly on their projects, and from this they compiled the list of pitfalls. I present the reverse as a list of project success factors below. The most often cited work is the list compiled by Jeffrey Pinto in his Ph.D.[7] Jeffrey Pinto identified ten success factors,

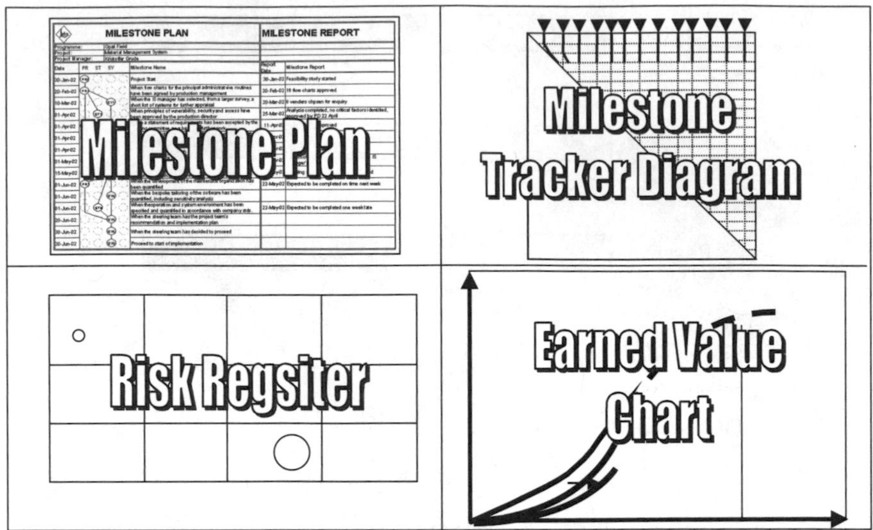

FIGURE 3.2 A single page progress report for the project.

listed in order of importance in Table 3.4. During the 1990s, work on project success focused on success criteria, but returned to consider success factors in this decade. Terry Cooke-Davies differentiated between the success of the project and the success of project management[8] (see Table 3.5). Jim Johnson, the managing director of the Standish Groups, has identified 100 pitfalls in information systems projects, which he describes as 10 items within each of 10 areas. The 10 areas are given in Table 3.6.

Up to this point, the literature almost studiously ignored the project manager's competence as a success factor on projects. It was implied that as long as the project manager used the right tools, the project would be successful. Terry Cooke-Davies identified organizational project management capability as a success factor (I consider this in Chap. 18). One of Jim Johnson's areas covers project management competence. Ralf Müller and I looked at the project manager's leadership style as a success factor on projects.[4] We found across the board that the project manager should exhibit high emotional intelligence. We also identified specific leadership competencies that contributed to project success for different types of project. I describe these results further in the next chapter.

Success Factors

I would now like to review, the success factors from the book *Goal Directed Project Management.*[6] In that book, they are presented as pitfalls, but I present them here as success factors. We identified success in four stages of the management process:

1. Establishing the project
2. Planning the project
3. Organizing and implementing the project
4. Controlling the project

TABLE 3.4 Pinto and Slevin's List of Success Factors

Success factor	Description
Project mission	Clearly defined goals and direction
Top management support	Resources, authority, and power for implementation
Schedule and plans	Detailed specification of implementation process
Client consultation	Communication with and consultation of all stakeholders
Personnel	Recruitment, selection, and training of competent personnel
Technical tasks	Ability of the required technology and expertise
Client acceptance	Selling of the final product to the end users
Monitoring and feedback	Timely and comprehensive control
Communication	Provision of timely data to key players
Troubleshooting	Ability to handle unexpected problems

Establishing the Project. These are factors in the way the project is set up within the parent organization.

Align Project Plans with Business Plans. Project plans must be derived from the business plans (see Sec. 2.4 and Example 2.1). A mistake often made is to start with detail planning, and then finding it difficult to link the project back to corporate plans. Start at the top and work down (Figs. 1.9, 2.7, and 2.8).

TABLE 3.5 Terry Cooke-Davies' List of Success Factors

	Project management success factors contributing to time completion
F1	Adequacy of company-wide education on risk management
F2	Maturity of organization's processes for assigning ownership of risk
F3	Adequacy with which a visible risk register is maintained
F4	Adequacy of an up-to-date risk management plan
F5	Adequacy of documentation of organizational responsibilities on the project
F6	Project or stage duration as far below 3 years as possible, preferably below 1 year
	Project management success factors contributing to budget completion
F7	Changes to scope only made through a mature scope change control process
F8	Integrity of the performance measurement baseline
	Additional project success factors contributing to successful benefits realization
F9	Existence of an effective benefits delivery and management process that involves the mutual cooperation of project management and line management functions
F10	Portfolio and program management practices that allow the enterprise to resource fully a suite of projects that are thoughtfully and dynamically matched to the corporate strategy and business objectives
F11	A site of project, program, and portfolio management metrics that provide direct line-of-sight feedback on current project performance and anticipated future success, so that project, program, portfolio, and corporate decisions can be aligned
F12	An effective means of learning from experience on projects that combine explicit and tacit knowledge in a way that encourages people to learn and to embed that learning into continuous improvement of project management processes and practices

TABLE 3.6 The Standish Group's Ten Areas of Success Factors

Ten areas of project success factors
User involvement
Executive support
Clear business objectives
Scope optimization (lean)
Agile processes (iterative)
Project management expertise
Financial management
Skilled resources
Formal methodology
Tools

Define Procedures for Managing Projects. Projects use transient teams to undertake novel assignments. The teams form quickly in order to undertake the task successfully. A properly structured start-up process is therefore important (Chap. 11). A consistent, company-wide approach to project management can also help (Chap. 17). However, it is necessary to obtain a balance between the need for a company-wide approach and the need to respect the individuality of project types.

Communicate Priorities to the Parties Involved. Example 3.1 shows what can happen when priorities are not communicated. People assign their own, usually different, priorities, with the result that there is no coordination, and no work is done. Agree the success criteria with the stakeholders before you start.

Planning the Project. The following factors are among those that determine how the work is defined and, the time and cost schedules calculated and communicated to the project team.

Develop Project Plans Developed on Multiple Levels. The use of breakdown structure is how we ensure the work delivers the required benefit. The usual pitfall is to plan at a detailed level only; computer software unfortunately encourages this. Sometimes work is planned only at a very high level, and there is no coordination. The following Chinese proverb illustrates that in almost every area of human endeavour work is planned on many levels. Projects should be no different:

A journey of a thousand miles begins with a single step (Lao Tsu).

On a journey there are at least two levels of planning between the end objective and the steps: the milestones (towns and villages), and the route map (roads). The former is the strategic plan, comprising intermediate goals or products, and the latter the tactical plan. At the milestone level, we make our plan robust but flexible, providing key fixed points for measuring progress towards our objective but able to incorporate changes at a lower level without changing the milestone definition. The road map we also try to keep fixed. However, there are two ways we can build in flexibility. If we find the route blocked, we can make a detour, but still aim to reach the next milestone. Sometimes the detour is better than our original route, but changes are contained at a low level. We can also adopt rolling-wave planning. We do not need to define the route between the last two towns until we reach the penultimate town. Sometimes we cannot get that information until we get there. All we need to estimate is the distance between the towns to plan the time and cost of the journey. The single steps are planned as we progress.

Use Simple Planning Tools. The complexity of project planning tools has grown over the last 40 years, due to the increasing power of software. However, at best complex plans achieve nothing; at worst they confuse the situation (see Example 3.4). The plans and progress reports should be cascaded through work breakdown structure (WBS) (Fig. 1.10). This can help build the vision for the project.

Example 3.4 Cumbersome, unfriendly tools

A delegate on a project management course said that he had 3 people on his project team of 20 who spent all day every day developing plans on a well-known PC-based package, and he got no useful information out. Thus 15% of his team was contributing nothing!!!

One reason why detail planning tools have developed is they were used so successfully on the Polaris Project in the United States in the 1950s. There is no doubt that Program Evaluation and Review Technique (PERT), which was first developed on the project, was a powerful analytical tool which helped identify and eliminate risk, which removed 2 years from an 8-year schedule. The project manager was also very charismatic and used the technique to help build the vision for the project. However, the following quotation illustrates a covert use of the technique:[10]

> These procedures were valuable in selling the importance of the mission. More importantly, the PERT charts and the rest of the gibberish let us build a fence to keep the rest of the Navy out and get across the message that we were the top managers.[6]

Complex plans were deliberately used to confuse outsiders and discourage them from getting too closely involved in the project, thereby protecting the project team from interference. This is a valid use of complex plans, but you also need to maintain the simple plans, or you will confuse yourself.

Encourage Creativity. It is the reality of modern projects that the project manager cannot be an expert in all areas of a project. Yet it is not uncommon to see project managers dictating to people more expert than themselves through the plan, telling them how to do their jobs. This can demotivate the experts, and isolate them from the project. What the project manager should do is delegate elements of the strategic plan to the experts, telling them which milestones they are responsible for, by when and at what cost, but allowing them to determine the best method of achieving that. In this way they can retain their integrity, while meeting the project's goals.

Estimate Realistically. There are several causes of unrealistic estimates.[11] It is common when preparing an estimate to believe the owner may not accept it and reduce it, or not accept the project. So people play the project management game and shave the estimates. Inevitably the work turns out as originally estimated, resulting in perceived failure. This is called strategic misrepresentation (see Example 3.5). Secondly, people may be overoptimistic about how the project will turn out; they just see things in a rosy light. This is called optimism bias. Thirdly, there may be inadequate historical data to estimate the work accurately. In that case, the risk must be identified and an appropriate contingency added. Flyvberg[11] suggests that if this were the cause estimates would improve with time, which they don't. Fourthly, people have different abilities. You must plan for the people you have, not some unobtainable ideal. Finally, it is sometimes assumed that project personnel are able to work 260 days (2080 man-hours) a year. A person working full time on a project is available much less than that. Lost time is caused by holidays, bank holidays, sickness, training, group meetings, and the like. When planning, this lost time must be accounted for (Chap 9).

Example 3.5 Strategic misrepresentation and the project management game

I know somebody who was on the French team evaluating the proposals for the Channel Tunnel in the mid-1980s. He says they knew the estimates of capital cost had been halved and the estimates of revenue had been doubled, making the project look four times better than it was. But they played the project management game because they all wanted the project.

Organizing and Implementing the Project. These are factors in building the project organization and assigning work to people.

Obtain Cooperation. It is not uncommon on projects to wonder if you all work for the same organization, as covert objectives get in the way of the overt objectives. Cooperation is achieved in two ways: by building a clear vision for the project; and by negotiating agreement to the plans (Chap. 4).

Obtain Commitment of the Resource Providers. Project managers often use resources on secondment from other managers. They will not willingly release their resources if they are not committed to the project.

Ensure Resources are Available When Required. It is not adequate just to send the resource providers a plan and expect their people to be available at some point. Even if they are committed, you must ensure they understand the requirements. This is helped by using simple plans, by discussing the requirements of the plan with the resource provider, and by negotiating their release. They must also plan to release their resources at the required time.

Define Management Responsibility. When defining roles on projects, it is common to consider only those people who do the work: cutting metal or writing code. However, people have other roles which consume time or can delay the project. These tend to be management roles, especially those which cause delay. These roles include taking decisions, managing information, and managing progress.

Ensure Good Communication. Surprisingly, poor communication on projects is sometimes caused by too much rather than too little. Communication out of a project is often achieved by sending every piece of information to everyone involved. People soon learn only a few documents are relevant to them, so all go straight in the bin. The project manager must define those who need information, so that when people receive something they know they ought to read it. If some other person wishes to be included in the circulation, they must negotiate inclusion on the responsibility chart. Similarly, committees are often used for communication into a project. Once invited people tend to stay on the committee, even if they are no longer required. Committees grow organically. Worse still, it is those people who have least to contribute who do most of the talking at meetings, as they talk to justify their presence. Channels of communication into a project must be clearly defined and limited, and any additions discussed and negotiated.

Differentiate between Technical Management and Project Management. It is still common to hear design managers refer to themselves as project managers, especially on information systems projects. Often, these "project managers" are not good at delegating work. They believe, quite rightly, they can do the work better than anyone else, and so surround themselves with idle people while they work themselves into an early grave. It is my view that an industry has truly matured in the management of projects when they stop calling design managers project managers, and stop using design engineers as such. Project management is an integrative function and design management is a specialist function.

Controlling the Project. Finally, factors in monitoring and controlling progress are illustrated by Example 3.6.

Example 3.6 Losing control

I once audited a project where the manager felt he had lost control, but was unsure why. The project was to put on a trade exhibition, held in Birmingham in December. There were 15 syndicates of 4 companies collaborating in this exhibition. Work started in June. Each syndicate prepared their own material, bringing it to a test site in September, moving it to Birmingham in late November. The project manager was a contractor. In June he had a meeting with the representative of each syndicate, showed them his plan, and said if the syndicate had any problems with the plan, to let him know. That was his first and his second mistake: First, he dictated to the experts by telling them his plan, not developing a plan with them; second, lack of comment was interpreted as agreement. The project manager then held weekly meetings attended by the representatives at which they gave verbal progress reports. Each person spoke for about 15 minutes, resulting in a 4-hour meeting; but the project had been set up in such a way that no one was interested in what the others were saying. The whole point of dividing the project into 15 syndicates was each syndicate could work on its own in the early stages. Each meeting therefore consumed 64 man-hours to no effect. At each meeting the representatives usually reported that everything was going to plan. I was called in mid-September because in spite of that, materials were not arriving at the test site at the due time. The manager wondered what was going on. What had happened was that after the first meeting most of the syndicates had ignored the project manager's plan and worked on their own. When they said things were going according to plan, they meant their own, but the project manager assumed they meant his, and the two bore no relation.

Understand the Purpose of Control. The purpose of control is not to hold meetings. It is also not to punish people for failing to achieve the plan. If people believe that is the purpose of control they will withhold information. The purpose is to monitor progress, to compare progress to the plan, and to take necessary action to achieve the project's goals. That requires people to be open and honest about progress on the project. If people know they are reporting progress because it is time to report progress, and the information will be used to help and support them, they will be more willing to give a true picture of progress.

Monitor Progress against the Plan. Control was lost in Example 3.6 because people were not reporting progress against the plan. Control will only be effective if there is a common basis for control, which means a common plan. This is achieved most effectively by reporting progress on a copy of the plan.

Hold Effective Review Meetings. To be effective formal review meetings must be held, with controlled attendance, fixed criteria for reporting, and at fixed intervals. Discussing progress at the coffee machine may be part of good leadership, but it is not of good control. At the other extreme, large meetings where most people are not interested in what others are saying waste time. People must only be invited if they have something to contribute. Holding review meetings at two or more levels of the planning hierarchy can aid this. (The manager in Example 3.6 should have had weekly meetings with the representatives individually, and less frequent meetings with the whole group to discuss common issues). The meetings must have a fixed agenda, which means reporting against fixed criteria, including the plan. Without a structure people will report progress in a way which puts them in the best light. Finally, people sometimes hold meetings only when they have something to discuss. By then control is reduced to damage limitation. Meetings must be held at fixed intervals, although the frequency may vary depending on the risk, and the point in the project life cycle.

Combine Responsibility with Authority. The manager in Example 3.6 had no direct authority over the syndicates, and was not able to use other sources, including that obtained by negotiating agreements. Without authority for control, the manager cannot take action to achieve the project's goals. I describe in Chap. 6 sources of authority available to the project manager.

Five Necessary Conditions for Project Success

Two Ph.D. students of mine, John Wateridge[1] and Ralf Müller[12], have between them identified what I believe to be five necessary conditions of project success.

Key Stakeholders Should Agree on the Success Criteria before You Start. I started this chapter by explaining the importance of this. It will repeatedly recur throughout the book.

Continue to Confirm Agreement at Configuration Review Points throughout the Project. It is not enough just to agree the project goals once at the start of the project; you need to ensure that people maintain a common vision of the project's outcomes throughout. This can be done at configuration review points (Chap. 7), and project gateway reviews (Chap. 18).

Maintain a Collaborative Working Relationship between the Project Owner and Project Manager, with Both Viewing the Project as a Partnership. There is increasing evidence that in order to have a successful outcome, the project owner and project sponsor must work together in partnership towards mutually beneficial goals. They must play a win-win game. Unfortunately they so often try to outdo each other, viewing the project as a fixed cake, and each tries to benefit at the others expense. They play a win-lose game. But there are no win-lose games on projects; it is either win-win or lose-lose. If the owner and manager try to play a win-lose game, they will both lose; one will just lose more than the other.

Empower the Project Manager, Setting Medium Levels of Structure. Unfortunately the owner often tries to impose rigid structures on the project manager to maintain control. The result is the project manager has no flexibility to deal with risk. But the other extreme doesn't work either. If the owner gives no guidance, *laissez-faire* management and anarchy reigns. In fact the owner should impose medium levels of structure. Agree the goals with the project manager and set parameters within which the project manager should operate to achieve those goals, but allow the project manager flexibility to deal with risk. Also, as I mentioned in Sec. 1.2, the owner can release authority to the project manager between stage-gate reviews, knowing they can take it back at those times.

The Owner Should Take an Interest in Project Performance. Ralf Müller observed that where the owner took an interest in progress the project performed well, but the owner usually thought the project was doing less well than it was. Where the owner didn't take an interest in progress, the project didn't perform well, and the owner had a rosy picture of progress. In Chap. 15, I will describe communication between the project manager and sponsor to satisfy the sponsor's needs for comfort.

3.4 THE STRATEGIC MANAGEMENT OF PROJECTS

Having identified the success criteria for your project, and the relevant success factors, the next step is to develop a project strategy. Several models have been developed for this, and I present two here.

The Seven Forces Model

The seven forces model (Fig. 3.3) is a model I developed from the work of Peter Morris.[13] It shows that there are seven forces acting on the project.

External context: Two forces are imposed by the external context, as described in Chaps. 1 and 2:

- *External influences*: The political, economic, social, technical, legal, and environmental (PESTLE) influences of and on the parties involved.
- *Sponsorship and schedule*: The finance provided by the owner, the benefit expected in return, and the timescale which makes that benefit worthwhile, and will repay the finance.

Project strategy: Two forces arise from within the parent organization, from the strategic importance given to the project, and the strategy for undertaking it:

- *Definition*: What the project is required to do, the approach to its design and technology expected to deliver it.
- *Attitudes*: Representing the importance attached to the project and the support given from all strata of management, from the leaders to the followers.

Internal implementation: Three driving forces come from within the project:

- *People*: Their management, leadership, teamwork, and industrial relations.
- *Systems*: Planning, reporting, and control are the systems by which progress will be measured and managed.
- *Organization*: The roles, responsibilities, and contractual relationships between the parties involved.

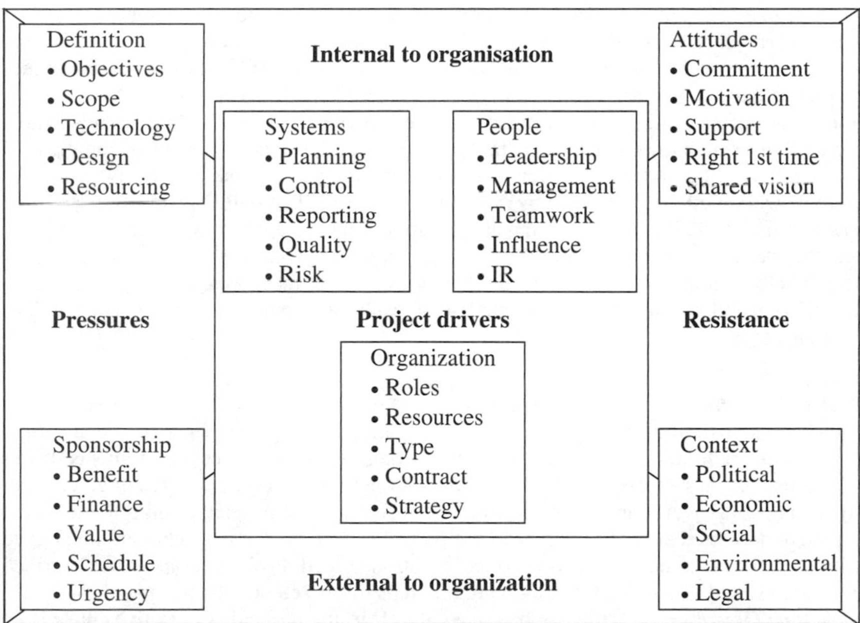

FIGURE 3.3 The seven forces model of project management.

External Influences. As well as being the primary influence of corporate strategy, external influences are a primary cause of many project overruns.[9, 13] In Sec. 2.2, I introduced the analysis of these factors as political, economical, sociological, technical, legal, and environmental factors *(PESTLE) Analysis.* We might ask how much the project manager can influence these factors. Often some influences can be exerted, if only to provide protective action or contingency.

Most projects raise political issues, and hence require political support. These issues must be considered from the outset. People working on a project must be attuned to them and be ready to manage them. To be successful, project managers must manage upwards and outwards, as well as downwards and inwards. The project manager should court the politicians and influential managers, helping allies by providing information needed to champion their program. Adversaries should be co-opted, not ignored.

Stakeholders, especially the local community, are an important external influence. The management of change must take account of this influence (Sec. 4.2).

Sponsorship and Schedule. The project cannot begin without finance, and that will only be forthcoming if the owner expects adequate benefit from the project (Chap. 2). Much of the project definition will be driven by the available sources of finance, the financiers wishing to minimise risk, especially in the choice of technology.

A key parameter in a project's viability is the completion date, with even a small slippage leading to a significant loss of revenue and increased financing charges. Determining the timing of the project is crucial to calculating the risks and dynamics of its management. How much time is available for each stage, together with the amount and difficulty of the work to be accomplished, influence the nature of the task to be managed. Therefore, in specifying the project, the manager should ensure the right amount of time is spent on the overall duration. Milestone scheduling is crucial. It is important that the development stage is not rushed or glossed over (a fault that has caused many project catastrophes in the past).

A degree of urgency should be built into a project, but too much can create instability. The manager should avoid beginning implementation before technology development and testing are complete. This situation is known as *concurrency*. (Concurrency is sometimes employed quite deliberately to get a project completed under exceptionally urgent conditions, but it often brings major problems in redesign and reworking.) Concurrency is now increasingly synonymous with *fast track*, that is, building before design is complete. If faced with this, be under no illusion as to the risk. Analyse the risk rigorously, work element by work element, milestone by milestone. The term "fast build" is now being used to distinguish a different form of design and construction overlap: that where the concept, or scheme, design is completed but the work packages are priced, programmed, and built sequentially, within the overall design parameters, with strict change (configuration) control being exercised throughout. With the use of fast build, the design is secure and the risks are much less.

Project Definition. The development of the project's definition is vital to its success. A comprehensive definition should be developed, stating its purpose, ownership, technology, cost, schedule, duration, financing, sales and marketing, and resource requirements. If this is not done, key issues essential to the viability of the project may be omitted or given inadequate attention, resulting in poor performance. Through the project definition, the vision for the project is created, the purpose of the project is defined, the project plans are aligned with the business plans and the basis of cooperation agreed. Project definition is described in Chap. 11, and is achieved by following the steps discussed next.

Setting Objectives. Little can be done until clear, unambiguous objectives have been set for the project. The project's success can be compromised by objectives that are unclear—do not mesh with organizational strategy—and are not clearly communicated and agreed.

Defining the Scope. Scope definition, cost, time, and performance criteria are intimately related. If they are unrealistic, expectations for the project will not be met, and it will be said to fail. The strategic plan for attaining the project's objectives must also be developed in a comprehensive manner from the start. If the project objectives change, the scope definition and investment criteria must be reconsidered.

Setting Functional Strategies. The setting of a project's functional strategies must be handled with great care, and requires the determination of the design, the technology to be used, the method of its implementation, and eventual operation best suited to achieving the objectives. The design standards selected will affect the difficulty of construction and eventual operation of the plant. Technical risk in particular needs to be assessed. Technical problems can have a huge impact on the likelihood of project overrun.[13]

Managing the Design Process. No design is ever complete; technology is always improving. A key challenge is to achieve a balance between meeting the schedule and making the design that fits better. Central to modern project management is the orderly progression of the design and its technical basis through a sequence of review stages. At each stage, the level of detail is refined, with strict control of technical interfaces and changes (through *Configuration Management,* Chap. 7, and through end-of-stage reviews, Chap. 18). Changes can result in extensive rework, as people on other parts of the project may have based their assumptions on the agreed design. You should therefore aim to achieve a progressive design freeze as soon as possible. This is usually feasible in traditional engineering projects, but an early design freeze may conflict with meeting the customer's requirements (see Chap. 7), especially in organizational development, high technology, and information systems projects. In setting up projects, care should be taken to appraise technical risk, prove new technologies, and validate the project design, before freezing the design and moving into implementation. The management of the design process is described in Chap. 11.

Resources. It is no good defining what you want to achieve if you do not have the right number of good, committed people, sufficient money, adequate infrastructure, and so on. In fact, getting adequate resources, managing them well and ensuring that the context is supportive are at the heart of successful strategic management, yet are rarely addressed by the literature on strategy. I cover resources under both the project's internal organization and its external context in Chap. 6.

Attitudes. This is probably the most important force. The chances of success are substantially diminished unless

- There is a major commitment to making the project a success.
- The motivation of everyone working on the project is high.
- Attitudes are supportive and positive.

To achieve positive attitudes it is vital to develop a clear vision, by linking projects plans to business plans, and by functional and task managers being seen to cooperate to achieve the same objectives. It is particularly important that the project receive visible commitment and support from the top, without which it is probably doomed. However, while commitment is important, it must be towards viable ends. Great leaders can become great dictators. If sensible projects are to be initiated, they must not be insulated from criticism. Critique the project at the specification stage, and ensure it continues to receive frank reviews.

People Issues. Projects usually demand extraordinary effort from the people working on them, (often for modest reward, and with the prospect of working oneself out of a job). In

Chap. 4, we will discuss how significant institutional resistance can be overcome in order for the factors listed here to be achieved. This puts enormous demands on the qualities of those working on the project, from senior management through the professional teams to artisans. The initial stages of a project may require considerable leadership and championing to get started. Beware though of unchecked champions and leaders: of the hype and optimism which too often surrounds projects in their early stages. The sponsor must be responsible for providing the objective check on the feasibility. The sponsor might be considered as the person providing the business case and the resources. Evidently they ought to be convinced of the merits of the project on as objective a basis as possible. We should recognise the importance of team working, of handling the conflicts which arise on projects positively, and of good communications. Consideration should be given to formal start-up sessions at the beginning of a team's work (mixing planning with team building) (Chap. 11). The composition of the team should be looked at from the social angle as well as the technical: People play social roles on teams, and these will be required to vary as the project evolves (Chap. 4).

Planning and Control Systems. Appropriate systems must be used to plan and control all the significant functions, including scope, quality, cost, time, risk, and other elements identified as appropriate. Table 1.2 lists many of the tools and techniques used. Plans should be prepared by those technically responsible for their work, and integrated by the project support office (Chap. 16). Initial planning should be at a broad, systems level with detail only being provided where essential, and in general on a rolling-wave basis (Chap. 5). Similarly, cost estimates should be prepared by work breakdown element, detail being provided as appropriate (Chap. 8). Cost control should be in terms of physical progress, and not in terms of invoiced value (Chap. 13). Cost should be related to finance, and be assembled into forecast out-turn cost, related both to the forecast actual construction price and to the actual product sales price. All changes to the proposed project baseline, proposed as well as actual, should be monitored extremely carefully. Implementation of systems and procedures should be planned carefully so that all those working on the project understand them properly. Start-up meetings should develop the systems procedures in outline, and begin substantive planning while simultaneously "building" the project team (Chaps. 11 and 19).

Project Organization. There are three organization issues which must be considered at the earliest stages.

Management Structure. A project structure is expensive on resources (Chap. 6.) Many projects begin and end with a functional line structure, but change to a matrix during implementation. Implementing a matrix takes time, and effort must be put into developing the appropriate organizational climate. (The issues in selecting a structure are described in Chap. 6.)

Client Involvement. The issue is the extent to which the client continues to be involved, even after hiring contractors to undertake the work. They may feel they have a legal or moral responsibility to ensure it is done to a certain standard, or may just want to ensure it is for their own comfort. The dilemma is between not being involved at all, versus constantly tinkering with the design, both frustrating the contractor and adding expense. The balance will depend on the nature of the project. A solution is to schedule milestone review points and limit owner involvement to those reviews.

Use of Contractors. No organization has the skills or resources to undertake all its project work, and must therefore buy in goods and services. At an early stage of project definition it is necessary to determine the contract and procurement strategy. Indeed, financiers may not lend money without knowing who suppliers will be, so they can judge their reliability. The selection of contractors and contract strategy are beyond the scope of this book.[14]

The Project Excellence Model

Another model of project strategy is the project excellence model (Fig. 3.4). This was first developed by Eddie Westerweld,[3] but a very similar model has been developed by the International Project Management Association (IPMA), for their project excellence award. This model shows both success factors and success criteria in one model, with success factors on the left-hand side and success criteria on the right-hand side. In their project excellence award, IPMA award projects 500 points for how well they address each side of the model. I am not going to discuss this any further, since the elements of the model have been covered by much of the above discussion.

3.5 PRINCIPLES OF PROJECT MANAGEMENT

The remainder of this book focuses on five of the seven forces: definition, attitudes, people, systems, and organization. The two forces from the external context are beyond the scope of this book.[14] The book describes a process-based approach to the management of projects, as outlined in Chap. 1, first describing the project management functions, the management of scope, organization, stakeholders, quality, cost, time, and risk, and then describing the project management process through the life cycle, covering definition, implementation, control, and close-out. In order to successfully address the seven forces and avoid the pitfalls, the approach described in this book is based on five principles of good project management:

- Manage through a structured breakdown, with single point responsibility.
- Focus on results: what to achieve, not how to do it.
- Balance results through the breakdown structure.
- Organize the project by negotiating a contract with the parties involved.
- Adopt a clear and simple management reporting structure.

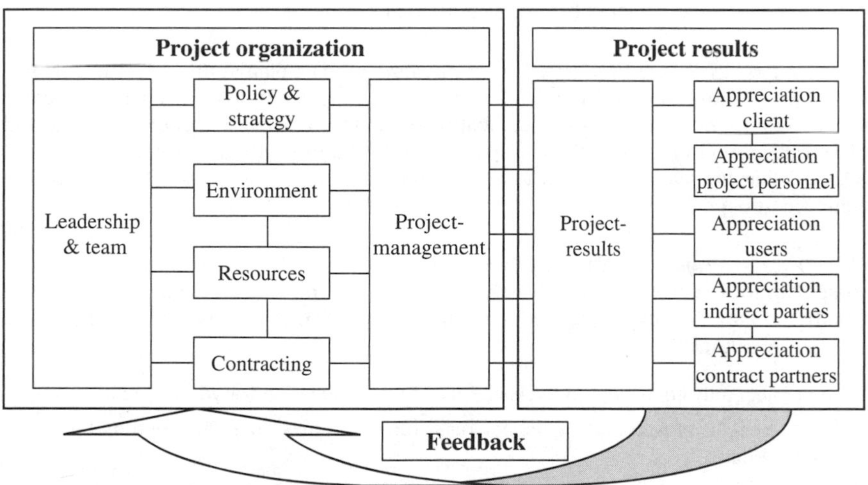

FIGURE 3.4 The project excellence model.

Structured Breakdown. Almost everything we do in life, we plan over several levels, breaking our understanding down in a structured way. Projects are no different. Using a breakdown structure lets us

- Define and control the scope
- Isolate changes
- Isolate risk

By breaking the facility down in a structured way, we can determine its essential components required to achieve our project and business objectives. We then do work because we know it is going to deliver a result we need, not because it seems like a good idea. By dividing the project up in this way we can ring-fence elements of work and help to ring-fence changes and risk, as with changes to the journey described in Sec. 3.3. The breakdown structure is the core of project management and almost all the planning and control systems are based on it. Hence the project organization is very closely linked to the breakdown structure, and it is common to identify one person or team as being responsible for the successful delivery of each element of work at a given level. A person or team is given *single-point responsibility* for each element of work.

Focus on Results. The primary breakdown structure is the *product breakdown structure (PBS)* by which we break the facility up into its components. We plan the project in terms of the results, or *deliverables*, we want to achieve rather than the work to be done. The reason for this is it makes the plan robust but flexible, and because it gives better control of the scope: The plan is robust or stable, because the definition of the expected results should be stable. If the definition of the results changes substantially, the project changes as well. Even where the configuration or specification of the results may be poorly defined (Fig. 1.12 and Sec. 7.3,) we can still plan in terms of deliverables, the precise specification of which is yet to be determined. On the other hand, if we plan in terms of the work, the plan can be constantly changing, especially if the goals or methods are poorly understood, in which case the early stages of the project will define the work to be done in the later stages. It also gives better control of the scope because we only do work which delivers results we know we need to achieve. Planning in terms of the work, it is possible to define work that seems like a good idea, but which in fact does not deliver useful results.

Balance Results through the Breakdown Structure. The plan at the strategic level can be used to ensure that proper emphasis is given to all areas of work, to balance the levels of ambition for different areas of technical work, and for changes to people, systems, and organization, and to ensure they are appropriate to the project's purpose. I suggested in Sec. 2.2 that the team's attention can focus on the technical work. A balance must be achieved through the strategic plan.

Organize the Project by Negotiating a Contract. Nobody is altruistic; nobody does something for nothing. People will only work on your project because they expect some benefit in return. The expected benefit can take several forms, positive returns or absence of negative returns:

- The project may contribute to the success of the organization for which you all work.
- Working on the project may be the person's job, and if they do not they will not get their annual bonus.
- They may like and respect you, and expect that if they contribute to your project, you will contribute to theirs.

Whatever the expected benefit, in asking for someone's contribution to your project, you must negotiate their contribution, which means

1. You must trade their inputs against expected benefits, as just discussed.
2. The agreement must be reached through open discussion.
3. The agreements must be represented through clear, simple, open, visible plans which represent the expected contribution and the promised returns.

It is not uncommon for project managers to plan their projects on their own, and then to tell the project team what they are expected to do. However, a contract is not agreed by one party telling the other party the answer; it is agreed through discussion, and trading of positions. It must be the same with the project plan. This also allows the project team to contribute their ideas, and the experts to retain their integrity by determining how they will achieve the milestones for which they are responsible. I describe group planning in Chaps. 5, 6, and 12.

Clear and Simple Reporting Structure. The plans must also be clear and simple so that the project team members can see precisely what their contribution is, and how that contributes to the objectives of the parent organization. Complex plans confuse (see the quote in Sec. 3.4); and they confuse the project team as much as they confuse the outside world. Single page reporting means

- You try to represent the project objectives and the business purpose on a single page.
- You develop a single page strategic, or milestone plan, representing the overall approach to the project through one to two dozen milestones.
- For each milestone you develop a list of activities, showing how that milestone is going to be achieved.

SUMMARY

1. There are two elements of project success:
   - *Success criteria*—How we will judge the project to be successful.
   - *Success factors*—The elements of the project we can influence to increase the chance of success.
2. Different stakeholders judge the project to be successful in different ways. It is important to achieve a balance of those different criteria, meeting the needs of the different stakeholders.
3. Criteria for judging project success include
   - The project increases the shareholder value of the parent organization.
   - The project generates a profit.
   - The project provides the desired performance improvement.
   - The new asset works as expected.
   - The new asset produces a product or provides a service that consumers want to buy.
   - The new asset is easy to operate.
   - The project is finished on time, to budget, and with the desired quality.
   - The project team had a satisfactory experience and the project met their needs.
   - The contractors made a profit.
4. Overall the project will be successful if it delivers the desired performance improvement, or better, at a time and cost that provides value for the organization.

5. In a classic piece of work, Jeffrey Pinto identified 10 success factors on projects:
   - Project mission
   - Top management support
   - Schedule and plans
   - Client consultation
   - Personnel
   - Technical tasks
   - Client acceptance
   - Monitoring and feedback
   - Communication
   - Troubleshooting
6. In setting the project up you need to consider success factors under
   - Establishing the project
   - Planning the project
   - Organizing and implementing the project
   - Controlling the project
7. There are seven forces which influence your choice of project strategy
   - Two from the context
     - PESTLE
     - Sponsorship
   - Two from the parent organization
     - Definition
     - Attitudes
   - Three internal project drivers
     - People
     - Systems
     - Organization
8. The approach to project management followed in this book is based on five principles:
   - Manage through a structured breakdown
   - Focus on results
   - Balance results
   - Organize a contract between parties involved
   - Keep it simple

REFERENCES

1. Wateridge, J.H., "IT projects: a basis for success," *International Journal of Project Management*, 13(3), 169–172, 1995.
2. Dumaine, B., "How managers can succeed through speed," *Fortune,* 1988.
3. Westerveld, E. and Gaya-Walters, D., *Het Verbeteren van uw Projectorganizatie: Het Project Excellence Model in de Praktijk*. Dementen, NL: Kluwer, 2001.
4. Turner, J. R. & Müller, R., *Choosing Appropriate Project Managers: Matching their Leadership Style to the Type of Project*, Newton Square, Pa.: Project Management Institute, 2006.
5. Hartman, F.T, *Don't Park Your Brain Outside: A Practical Guide to Improving Shareholder Value with SMART Management,* Newtown Square, Pa.: Project Management Institute, 2000.
6. Andersen, E.S., Grude, K.V., Haug, T., Katagiri, M. and Turner, J.R., *Goal Directed Project Management,* 3rd ed., London: Kogan Page/Coopers & Lybrand, 2004.
7. Pinto, J.K. and Slevin, D.P., "Critical success factors in effective project implementation," in Cleland, D.I. and King, W.R., (eds.), *Project Management Handbook,* 2d ed., New York, N.Y.:Van Nostrand Reinhold, 1988.

8. Cooke-Davies, T., "The 'real' project success factors," *International Journal of Project Management,* 20(3), 185–190, 2001.

9. Johnson, J., *My Life Is Failure: 100 Things You Should Know to Be a Successful Project Leader,* Boston, Ma.: Standish Group International, 2006.

10. Deal, T.E., and Kennedy, A.A., *Corporate Cultures: The Rites and Rituals of Corporate Life,* New York, N.Y.: Addison-Welsey, 1986.

11. Flyvbjerg, B., "From Nobel prize to project management: getting risks right," *Project Management Journal,* 37(3), 5–15, 2006.

12. Turner, J.R. and Müller, R., "Communication and cooperation on projects between the project owner as principal and the project manager as agent," *The European Management Journal,* 22(3), 327–336, 2004.

13. Morris, P.W.G. and Hough, G.H., *The Anatomy of Major Projects: A Study of the Reality of Project Management,* New York, N.Y.: Wiley, 1987.

14. Turner, J.R and Wright, D., *The Commercial Management of Projects,* Aldershot, U.K.: Gower, 2009.

CHAPTER 4
THE PEOPLE INVOLVED

In this chapter, I consider the third element of the project's context (Fig. 1.11), the people involved. The project is proposed to introduce change to deliver performance improvement, but not everybody in the organization shares the project's objectives. Although the project is intended to benefit the organization, it may not be beneficial for everybody in it; there will be winners and losers for most projects. In this chapter, I consider reactions to change in organizations, and then how to identify stakeholders and their reactions to the change, and persuade them to support the project. That includes the development of a communication plan to communicate with stakeholders. I also describe project teams, and the leadership of the project manager. Although more part of the project than the project's context, this is where people from the project's context become part of the project, and the project manager has to lead people in the context as much as he or she has to lead the project team.

4.1 REACTIONS TO CHANGE

There are fundamentally three potential levels of change within organizations measured by the way it can impact on people's lives:

Background change: All the time an element of background change is taking place: people retire or leave the organization, new people join; minor changes are made to existing products, production machinery, or computer systems. It is all part of life, and people accept it as natural. It can also be managed through the routine organization. The work is undertaken by giving a temporary assignment to somebody in the routine organization, not by creating a temporary organization to undertake it. This sort of change can lead to small levels of performance improvement within the functional organization, what in Sec. 2.4 I called habitual incremental improvement.

Normal change: Then there is the normal change that is the primary focus of this book. The organization wants to achieve a step-wise level of performance improvement that requires it to undertake a significant change, and that requires the organization to assemble a temporary organization, a project, to undertake that change. People do not view this as natural, and the emphasis is on trying to win support for the project and the change it is introducing.

Extreme, life-modifying change: Finally there is change that has a significant impact on people's lives, perhaps making significant numbers of people redundant, or requiring them to join new organizational units with significant impact on their working relationships. The required performance improvement requires significant structural changes totally changing people's lives. This sort of change has a significant emotional impact on people.

Vic Dulewicz and Malcolm Higgs,[1] quoted in my work on project leadership with Ralf Müller, described later in this chapter, propose types of change, which they describe as relatively stable, significant, and transformational, requiring three types of leadership. For the purposes of this discussion, I would group all three of these into the middle level of change above; they are more than background change and less than life modifying. This classification relates more to the complexity of the change itself, than the impact on people's lives, though relatively stable change will be on the border between background and normal change, and transformation on the border between normal and life-modifying change.

I discuss the second and third levels of change further.

Normal Change

The issue with normal change is twofold: winning people's commitment to the change and getting them to internalize it as something they think is the right thing to do; and overcoming resistance. In reality, these are two sides of the same coin, but I want to discuss each separately because they highlight different issues.

Winning Commitment. The need to win people's commitment to the change is illustrated by the following quotation:

Every new idea goes through three stages:

1. First, people think it is stupid.
2. Then they think it is dangerous.
3. Then they think they believed that all along.

For many years I couldn't find where this quotation was from, but then in quick succession I heard three authors had used variants of it: Mark Twain, Arthur Schopenhauer, and Mahatma Ghandhi. For the second stage, Mark Twain suggested they think it is against the bible, and Mahatma Gandhi had a previous stage, that first they try to ignore it. However, when searching on Google, I could only find it credited to Arthur Schopenhauer. There is a science program on British television called *Horizon*, which, when discussing a controversial new scientific idea divides the program into three parts. In the first scientists are wheeled out to say how stupid the idea is, then they are wheeled out to say that if the idea gains credence they are going to have to rewrite all the text books and change what they have been teaching for 80 years, which they don't want to do. Finally they are presented, self-righteously saying it was obvious all along (see example 4.1).

Example 4.1 is light-hearted; it makes good television. The point is when introducing a change you must understand that people need to be given time to go through these three stages. Do not expect them to go in one step from hearing about a new idea, to internalizing it and accepting it as obvious. They need to have time to understand the idea and see why it is sensible and relevant to the organization's needs. Then they will begin to realize the impact on their working lives, and they may not like that. They need to be given time to deal with their concerns, and be shown the changes will actually be in the best interest of themselves and all concerned. Then they will accept the new truth as self-evident, and internalize and accept it. You need to help people through this process and allow time for each step.

Example 4.1 Three stages for a new idea

One *Horizon* episode dealt with the idea that Europeans had settled in North America about 15,000 years ago. The evidence was that early North American tribes had used a design of flint axe that originated in France about that time. This theory of course contradicted two established theories: that North America had been inhabited by people coming across a land bridge from what is now Russia through Alaska, and that the first wave of settlement was 11,000 years ago. Further, Europeans 15,000 years ago didn't have boats and so weren't capable of crossing the Atlantic. So the first third of the program was devoted to academics rubbishing the proposed theory. Then evidence was produced showing that there were two waves of people coming through Alaska, the first during an interglacial period 18,000 years ago, and then the second at the start of the current interglacial period 11,000 years ago, so in fact North America had been inhabited for 18,000 years, longer than previously thought. DNA evidence was then produced showing that some original North American tribes, especially from the East coast, had 15,000-year-old European DNA in them. The second third of the program was devoted to native North Americans saying they didn't want to be descended from Europeans because they hated what they did to them starting in the middle of the last millennium. Anyway, how did Europeans get across the Atlantic without boats? The clue there is in the word "interglacial." 15,000 years ago the Atlantic was frozen half way down and so they walked across. Well, that was obvious all along; nobody ever believed anything different!!!

Resistance to Change. I use two quotations to illustrate resistance to change. The first is due to Charles Handy[3]:

> If you put a frog in cold water and slowly raise the temperature, it will allow itself to be boiled to death.

Handy uses this as a metaphor for people in organizations. They are surrounded by an environment that they find familiar and comforting. You come along and say, "We need to achieve performance improvement and so we need to change," and they say, "But we have done it this way for the last 5 years; it has been OK for the last 5 years, it is OK now." They cannot see that outside the organization they know and love the world is changing, that the business environment is beginning to boil. If you try to change people too quickly you will just get resistance.

When I first started working for Coopers and Lybrand as a consultant, I would go into client organizations and say, "What you are doing is wrong; you need to do it this completely different way." The reaction I would get from the directors of the client organizations would be that they were intelligent people; if their problems were so easy to spot they would have spotted them. But the problem the directors suffer from is they can't see the point where what they are doing goes from being OK to being a problem. Things change slowly and they cannot see the point where they cross that line from being good to needing to achieve performance improvement. I learnt a lot from my boss in Coopers and Lybrand. He would start by building a relationship with the client and winning the client's trust. Then he would begin to ask the client questions about how they thought they were doing. Well, they had called in the consultants because they had a sense of unease but were not quite sure why. So through a series of questions my boss would get the client to identify the problems for themselves and identify the solutions for themselves. In this way the client would have much greater acceptance and ownership of the proposed solutions.

Many Western managers hit their staff with logic, "You have to do as I say, because; because; because I say so." As long as 2500 years ago Aristotle suggested that you should start by building relationships with people. Once you have done that you can sell them your values and vision, the need for performance improvement. Once you have done that, then and only then can you persuade them with the logic of the best way of achieving the vision. The American President Ronald Reagan was very good at this; the British Prime Minister John Major always went straight in with the logic and didn't persuade people.

The second quotation I find useful is from Machiavelli[4]:

> There is nothing more difficult to arrange, nor doubtful of success, and more dangerous to carry through, than initiating changes in a state's constitution. The innovator makes enemies of all those who prospered under the old order, but receives only lukewarm support from those who would prosper under the new. Their support is lukewarm partly from fear of their adversaries, and partly because they do not trust the new order until they have tested it by experience.

Machiavelli says that trying to introduce change is dangerous. Part of the reason is, as I said above, there are winners and losers from the change. The losers are trying to stop you from being successful. You might think that is balanced by the winners, who are supporting the change. But, no; the winners are sitting on the fence. They don't want to come out strongly in favour of the change, in case it doesn't work. They don't want to make enemies of the losers in case they win. So the winners sit on the fence, waiting to see what happens. So initially you are on your own.

Extreme, Life-Modifying Change

Extreme, life-modifying change leads to much more severe emotional responses (Fig. 4.1 and Table 4.1). The emotional response shown in Fig. 4.1 was first identified in people who have been told they have a fatal illness (Example 4.2), but it is now recognized that anybody going through an extreme, life-modifying change follows a similar cycle. This can include losing or even changing your job, getting divorced, or losing a loved one. It is relevant in a change context, when the change is extreme and life modifying, where people lose their jobs, or have to make significant changes to their working environment, such as moving location or suffering significant changes in work colleagues. Example 4.3 describes one such situation. Again the change manager needs to recognize that people have to go through these stages and need to be given time to deal with each one. Don't try to rush people to acceptance. Give them time to deal with their denial and anger, but try to help them through the depression stage quickly to testing and acceptance. Also the manager's style needs to change through the cycle. There is a saying (from Shakespeare's play *Hamlet*) that "you have to be cruel to be kind." It is not kind during the early stages to give people false hope. You have to make it clear through the denial, anger, and bargaining stages that there is no

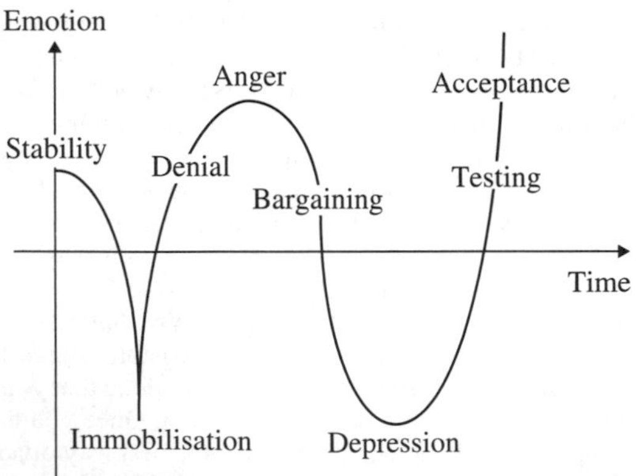

FIGURE 4.1 Response to extreme, life-modifying change.

TABLE 4.1 Response to Extreme, Life-Modifying Change

Stage	Response
Stability	Management communicate their vision, the need for change, and the consequences.
Immobilisation	People are taken by surprise. Their reaction is anxiety and confusion.
Denial	People defend themselves against the threat to their life or livelihood: "They can't mean me!!!" "Is that what we get for years of loyalty!!!" "Management is overreacting; it can't be that bad!!!"
Anger	Openly displayed anger towards management emerges. People try to take control, through their power base in the organization, trade unions, etc. Alliances are formed; efforts to divide management are made; all means to reverse the situation. Management must persistently argue the case, and not indulge in warfare.
Bargaining	People begin to try for a modified solution. All kinds of remedies will be proposed in order to try to reduce the impact of the change: "If we take a cut in salary?" "If we increase our productivity?"
Depression	Frustration, and a feeling of having lost spreads. People find it difficult to work, and organizational paralysis sets in. Management must help. They must have plans containing supporting packages, and must actively assist individuals in taking responsibility for themselves in the new situation.
Testing	The individual and the organization start working with alternative exit strategies to try to facilitate the individual's transition: "Did you say I could have 6 months pay while looking for a new job?" "Being paid through a year's MBA program would help the transition." Management helps to find realistic alternatives.
Acceptance	Individuals and the organization deal realistically with the situation. They may not like it but they accept it. Management gives recognition and support towards future plans. New stability is achieved.

alternative, you are sorry they are upset but this is the way it has to be. If you give people false hope, and then have to let them down again later (as happens in Example 4.3), the second letdown will hurt more than the first (Example 4.4). It is better to be resolute. However, during the depression, testing, and acceptance stages, you have to be supportive, helping people test out alternatives for the future. Reversing the change is not an alternative, but there may be many other options. In Example 4.4, the military began to help the town to use the facilities in the redundant army base to attract new industry to the town. At the time of a major redundancy program many organizations set up an out-placement department to help redundant employees develop their resumé and seek new work.

Example 4.2 Homer's odyssey

There is an episode of *the Simpsons* where Homer thinks he has eaten the poisonous bit of Japanese fish. He rushes to hospital, and the doctor says he will be immobilized. Homer says, "I can't move." Then the doctor says he will go into denial. Homer says, "Perhaps I didn't eat it." Then the doctor says he will be angry. Homer says, "Why me?" Together they go right through the cycle. I had been teaching this for several years when I saw this episode and was amused.

Example 4.3 Parade ground management

There is a case study in a former book of mine of the Norwegian army shutting an army base.[5] Based on a combination of the peace dividend at the end of the cold war and new technology the Norwegian government decided to halve the number of personnel in the army. This involved shutting half the army bases. The case study describes the shutting of the first base. Half the people working on the base were military personnel, half civilians. The military personnel would not be made redundant. The army was reducing numbers through natural wastage, with a very high churn rate. But the civilians were all redundant. Further, they represented about one third of the working population in the local town. Where you have a large employer like that, for every job in the base there is a job in the town: shopkeepers, doctors, taxi drivers. Two thirds of the jobs in the town are threatened. When communicating the decision to the base employees, the base commander assembled everybody, military and civilian staff. He said, "Base, attention! We are shutting the base. Military personnel, you will receive your transfer details. Civilian personnel, you are all redundant. Dismissed!"Before the meeting there was no inkling; the civilian staff were all stable. Immediately after, they were first immobilized. "Did we hear right? I think the commander said they are shutting the base." Then denial set in. "It can't be right. They must be shutting the base up the road." Then anger. "Why aren't they shutting the base up the road? Why us?" Then bargaining. There was a piece of technology on the base that accounted for about 40 percent of the operation, that only one other base in Norway had. The staff went to ask the base commander if that bit was shutting. He gave the wrong answer. He said he would check back with Oslo. This was interpreted as meaning that part of the base would stay open. When the answer came back, they were shutting half the bases in Norway, one other base in Norway has this technology, it would stay open, this one would close, the pain was greater than before as raised expectations were dashed.At about this point, as things were spiralling out of control Kristoffer Grude was bought in as a consultant, and a much more humane approach was adopted. He encouraged the army to work with local politicians and the unions to find options for the future. The base had to close but new industry could be attracted to the town.

Example 4.4 Being clear in the messages

When I worked for a company called Imperial Chemical Industries, ICI, in the early 1980s, they merged the two engineering departments in the northeast of England, and made one third of the staff redundant. There were two divisions in the northeast, Petrochemical and Agricultural Divisions, which used very similar technology. Each had its own engineering department and it was decided that this was a waste and the two would merge. About a month before the announcement was made a rumour began to circulate that this would happen, and half the staff would be made redundant. When the actual announcement was made that it was only a third, people's reaction was, "Phew, it was not as bad as we feared. That's all right then." We suspected afterwards that the directors had seeded the rumour of a half so people would be much more accepting of the actual announcement.When the directors were making the announcement at a staff meeting, they said, "Just to put in perspective what it means, look at yourself, look at the person on your left, look at the person on your right. One of you is going!!!" The communication was clear. About 5 years later a privatized company in the U.K. was making one in five members of its staff redundant, based on new technology and organizational changes. However, they thought that this number would be unacceptable, so they fudged the message and made it sound like 1 in 10. When it turned out it really was one in five, people were more angry than they otherwise would have been.

4.2 MANAGING STAKEHOLDERS

Figure 4.2 illustrates a stakeholder management process. I suggested in Sec. 3.4 that a necessary condition for project success is to agree the success criteria with all the stakeholders before you start. This process helps you do that.

Identify Interested Parties

A stakeholder can be defined as anybody who has an interest in the project, its work, outputs, outcomes, or ultimate goals. Table 4.2 contains a list of potential stakeholders. Most of these require no further discussion. The media can be very dangerous. A couple of years ago I did some work with a public sector organization which lived in fear of the tabloid press. If a project went wrong they would be ridiculed in the tabloid press. This is a competency trap (Chap. 17). They may have two ways of doing a project, an excellent way with a 90 percent chance of success and a mediocre way with a 100 percent chance of success. They would choose the mediocre way. If they chose the excellent way and were unlucky and the project went wrong, the tabloid press wouldn't say they did the project the right way and were unlucky. They would just say they had wasted public money on a failed project. They wasted public money doing projects certain but mediocre ways for fear of the tabloid press. The tabloid press aren't interested in telling the truth; they are only interested in selling newspapers.

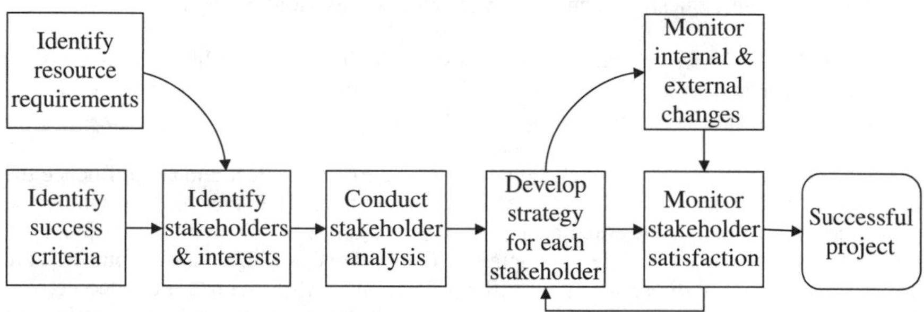

FIGURE 4.2 Stakeholder management process.

TABLE 4.2 Potential Stakeholders on a Project

Potential stakeholders
Employees (users, operators, bystanders)
Management
Shareholders
Resource providers
Customers (internal and external)
Suppliers (internal and external)
Neighbours
Government (local, national, continental)
Opinion formers (media)

Identify Success Criteria

Table 3.1 shows that different stakeholders have different perspectives on the success criteria for a project. You need to identify the different views different stakeholders potentially have on project success. I said when discussing Table 3.1, that potentially the different success criteria are incompatible, and they will be if you wait until the end of the project to try to make them consistent. You have a much better chance of balancing the various success criteria if you negotiate agreement before you start the project.

Identify the Stakeholders and their Interests

From these two steps we are in a position to identify the various stakeholders and their interest. You can begin to compile the stakeholder register (Table 4.3).

Analyse Stakeholders

You are now in a position to analyse the stakeholders. You do that by asking three questions about each stakeholder:

1. Are they for or against the project?
2. Can they influence the outcome?
3. Are they knowledgeable or ignorant about the project?

The answers to these questions can be entered into the stakeholder register.

For or Against? There are three potential answers to this question: the stakeholder is for the project, against it, or they don't care whether it is successful or not. There are some contractors who don't care, just as long as they get paid.

Influence the Outcome? The stakeholders who are for the project and can influence the outcome are the ones you like. You want to encourage them. The ones who are against the project and can influence the outcome are the ones you don't like. You either need to try to reduce their influence, or change their opinion. Those who can't influence the outcome are not so important. With those who are for the project but can't influence the outcome, you might try to find ways to make them more involved. With those who are against the project but can't influence the outcome you might try to change their opinion, or you might try to ensure they have no influence. There is one other type of stakeholder who is quite dangerous: People who can influence the outcome but are for the project for the wrong reason. They will be trying to take the project off course to achieve their own covert objectives and not the project's overt or stated objectives. The technology manager in Table 4.3 may be like that.

SWOT Analysis. A strengths, weaknesses, opportunities, and threats (SWOT) analysis (or more correctly OTSW analysis) of each stakeholder can help you answer the first two questions. You ask yourself about each stakeholder, Do they view the project as an opportunity or a threat? If they view it as an opportunity, then presumably they are for it, and if they view it as a threat they are against it. Then you ask yourself what are their strengths and weaknesses. If they have strengths then presumably they can influence the outcome and you will be trying to reinforce the strengths of your proponents, those who are for the project. If they have weaknesses they won't be able to influence the outcome and you will be trying to reinforce the weaknesses of your opponents.

TABLE 4.3 Stakeholder Register for the CRMO Rationalization Project

TriMagi Stakeholder register					
Stakeholder	Objectives	For/against	Influence	Informed	Communication strategy
Board	Expand operations Improved customer service Improved profitability	For	High	Must be	Regular briefing Explain solution and benefits
Operations managers	Improved customer service Excellent support	For	Medium	Must be	Regular briefing Explain solution and benefits
Maintenance managers	Operation that works Maintain position and influence	For	High	Must be	Seek opinions Regular consultation Confirm solution with them
Maintenance staff	Ease of operation Maintain jobs	For	Medium	Not at start	Briefings/company newspaper Consultation Explain solution
Operations staff	Support their work Minimum disruption	Ambivalent	Low	Not at start	Briefings/company newspaper Explain solution
Technology manager	Influence technical solution Develop power base	For	Medium	Must be	Regular consultation Seek opinions Explain solution
Suppliers	Make profit Ongoing business with company	For	Medium	Not at start	Open channels Regular consultation
Customers	Good service	For	Low	Not at start	Customer newsletters
Local community	Minimum disruption to environment	Ambivalent	Low	Low	Local newspaper advertisements

There is a moral issue here. You might identify while doing the analysis that a stakeholder is in favour of the project but it is a potential threat to them: Do you tell them? My view is that if the project is mainly an opportunity to them, but there is a small potential threat you should openly discuss that with them, and try to solve the problem. If you don't, then when they find out (and they will find out) they will be against, but if you have tried to solve the problem then you may be able to resolve the issue. On the other hand, if they are for the project but you realize it is only a threat to them (they are going to be made redundant but they don't know yet) then you may try to keep it from them for as long as possible. When they find out they are going to be against the project, whether you tell them now or later. That is the moral issue, whether failing to volunteer information is being deceptive.

Knowledgeable or Ignorant? You need to think about whether the stakeholder is knowledgeable or ignorant about the project; think about where they are now and where you want them to be. I believe you should aim to be the first to tell people about the project. If they hear about the project as a rumour they will be against it. Their thinking will go something like this: "I heard about the project as a rumour, therefore management must be trying to keep it secret. If management are trying to keep it secret it must be bad, and so I am against it." Once somebody has decided they are against the project based on incomplete or incorrect information, it will be very difficult to change their opinion. It is a quirk of human nature that if people form an incorrect opinion based on incomplete or incorrect information, they find it very difficult to change their opinion later (Example 4.5). You want to be the first to tell them so they have positive opinions from the start. But you mustn't tell people too early; you need to clarify your own thinking first.

Example 4.5 Discovering one's own mistakes

Psychologists have done experiments where they have shown people pictures progressively out of focus, and ask them to identify the picture. In this way they establish how far out of focus the picture has to be before the subjects will get it wrong more often than not. The experimenters then show the subjects a picture well out of focus, and ask the subject to say what it is. The experimenters then slowly bring the picture into focus and ask the subjects to say if they want to change their minds. The pictures have to be brought well into focus beyond the point where the person would normally make a correct identification before they will change an incorrect diagnosis.

This happened in an incident on a nuclear power station at Three Mile Island in the eastern United States in the 1980s. On the plant there was one faulty instrument which should have been indicating a fault, but was not working. A second alarm started and the operators made what would have been a correct diagnosis of the fault based on the information they had, the second alarm sounding but not the first, and reacted accordingly. But it was the wrong diagnosis because they had incorrect information. A third alarm started which should have told them their diagnosis was wrong, but they continued to react according to their original diagnosis. The whole station started to shout at them, "wrong diagnosis," but it was only when the emergency team came in with a fresh perspective that they discovered the true fault and saved the situation. The operators were like the frog in hot water, about to boil to death.

Develop a Stakeholder Influence Strategy

There are several different ways of categorizing stakeholders to determine the influence strategy.

Knowledge-Support. The first is based on whether they are for or against the project, and knowledgeable or ignorant about the project (Fig. 4.3).

Knowledgeable-support: These people must not be taken for granted. You should continue to work with them and keep them informed about the project to maintain their support.

Ignorant-support: These people you assume will support the project when they know about it, but they are currently ignorant. These are the people that you want to ensure that they hear about the project first from you so they end up in the knowledgeable-support box.

Ignorant-oppose: These are the difficult people, people who don't really know what the project is about, and perhaps oppose it for the wrong reason. The problem is, people these days can hold very strong beliefs when in reality they are quite ignorant (Example 4.6). Example 4.5 suggests it is very difficult, and even impossible, to change their views. Another important issue is to talk to people in layman's terms, talk to them in terms they can understand (Example 4.7). If you want people to support you, don't blind them with science. Talk to them in their language.

Knowledgeable-oppose: These people are easier to deal with; they are against the project for good reason. You can either try to find a way of changing the project to win them over, or you need to try to isolate them, and make sure they can't influence the outcome.

Example 4.6 Strongly held opinions

When I was at Henley Management College, I had somebody to talk on my courses about stakeholder management who worked for a publicly owned company called Nyrex, which has the popular task of disposing of low- to medium-level nuclear waste. He worked for the publicity department and had the job of delivering lectures around the country. He related the story that after one lecture a woman came up to berate him for about 10 minutes. She told him that Nyrex was completely evil, that burying nuclear waste was completely wrong, and that the nuclear industry should be completely shut down. After about 10 minutes she asked, by the way, what does nuclear radiation look like? Is it green slime? She was completely ignorant. She had very strong opinions, gained from reading the tabloid press, but was completely ignorant.

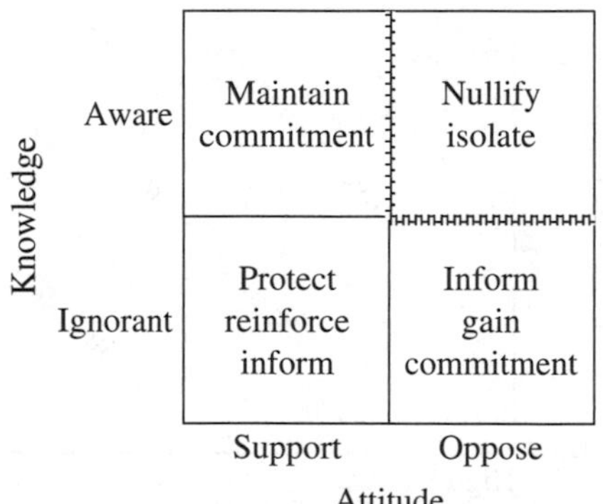

FIGURE 4.3 Stakeholder management strategy knowledge-support.

Example 4.7 Talking in terms people can understand

A British mining company was trying to develop a uranium mine in Canada, and they were holding the public enquiry. This is conducted as a quasi court case, with the mediator in the chair and lawyers questioning witnesses. One of the company's engineers was in the stand, and a lawyer representing the environmental lobby mischievously asked him what would be the radiation level of waste water leaving the site. The engineers, as engineers would, gave a precise answer, six decimal places, so many becquerels. Of course, the press were in the court and heard this. The next day the newspapers were full of the story, the waste water leaving the site would be radioactive. The locals would all die of cancer; their children would all be born with two heads. The correct answer was the radiation level would be half that of rain water—this water would be twice as pure as rain water. Even rain water has a radiation level. The company immediately tried to correct the wrong impression. But it was too late. The genie was out of the bottle; Pandora's box was open. There was no putting the furies back. People weren't listening any more (Example 4.5).

Power-Impact. Another way of categorizing the stakeholders is by their power within the parent organization, and their impact on the project. This leads to four influence strategies (Fig. 4.4).

Support-Agree. The last way of categorizing stakeholders discussed here is by how committed they are to the goals of the project and how much they agree with the way they are being achieved[6] (Fig. 4.5). The passives often represent about 40 to 60 percent of the stakeholders, and the way they feel about the project is usually influenced by the waverers. The waverers in turn are influenced by the golden triangle. You might think that the zealots are your best allies, but they just give unthinking support, which often does not help very much. The golden triangle, on the other hand, by questioning the project help to improve it.

Monitor Stakeholder Satisfaction

As the project progresses, you use the stakeholder register to monitor stakeholder satisfaction. If everything goes according to plan, then hopefully that leads to a successful project.

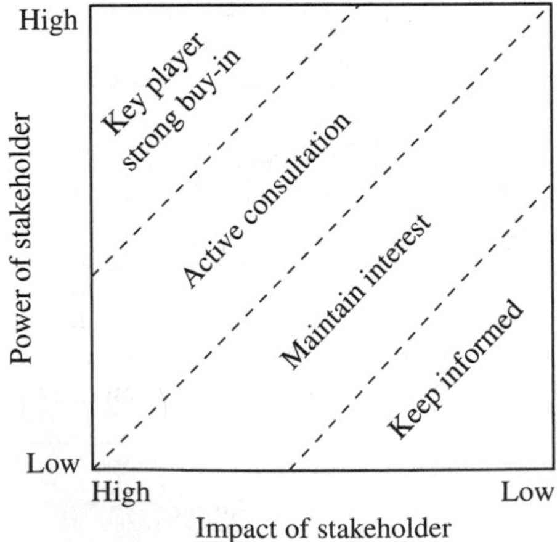

FIGURE 4.4 Power-impact matrix.

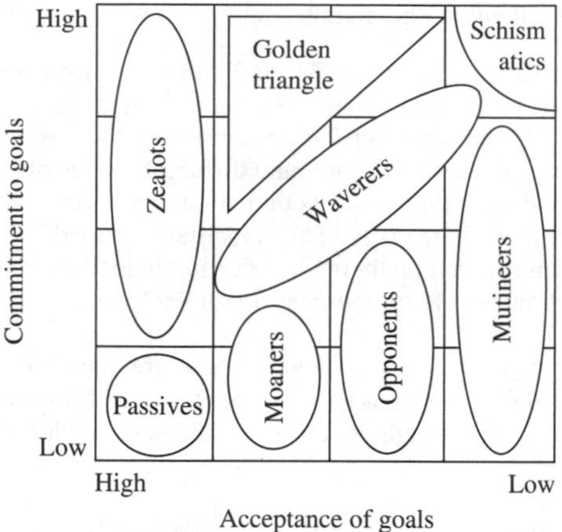

FIGURE 4.5 Leadership competencies contributing to project success.

If the stakeholders are not behaving as expected, you may need to change your influence strategy. Or something may happen which changes the stakeholders' views, and again you may need to change the influence strategy.

Table 4.3 is a stakeholder register for the CRMO Rationalization Project.

4.3 COMMUNICATING WITH STAKEHOLDERS

The next step is to develop a communication plan to communicate with stakeholders. Susan Foreman suggested that we need to market the project to the rest of the organization.[7] When developing a communication plan there are several questions you need to ask yourself.

What are the Objectives of the Communication? The objectives of the communication may include

1. To raise awareness of the project and thereby gain commitment from key stakeholders
2. To inform other business areas and promote key messages about the project, particularly the benefit to the organization, demonstrating the planned performance improvement
3. To make two-way communication to ensure a common understanding of the project and its objectives to negotiate agreement with the stakeholders
4. To maximize the benefits from the project by having everybody working for its success

Who are the Target Audience? You need to research the organization to try to understand who are the key players, and their objectives. You need to understand how the organization works and what motivates people. You should have done some of this analysis in developing the stakeholder register. In particular, you need to segment the target audience. For different recipients of your communication, there will be different objectives, and correspondingly different messages, different ways of structuring the messages, and different modes of communication (see Example 4.8).

Example 4.8 Segmenting the market

The directors of British grocery chain Tesco felt that they were not communicating well with their staff. So they hired a firm of consultants to investigate why this might be. The consultants concluded that although Tesco segmented the customers for their products very well, designing different messages and different modes of communication for different groups, they treated their staff as one amorphous mass, and sent the same message to all of them in the same way. The consultants helped Tesco segment their staff into six distinct groups based on their lifestyle, and helped them develop different messages and different modes of communication for each group.

What are the Key Messages? The messages to be communicated need to be designed to achieve the objectives. Different messages will be designed for each segment of the target audience. The communication needs identified in the stakeholder register will help in the design of the messages.

What Information will be Communicated and by Whom? The messages will indicate what information should actually be communicated. Different messages are best sent by different people. The project manager may inform stakeholders about the scope of the project, and when various things will be done. But information about the desired performance improvement and the benefit to the organization are better coming from the project sponsor, or even more senior managers at key points of the project so they can show their commitment, and the importance of the project to the organization.

When will the Information be Given? Timing can be critical. You want the sponsor and senior managers to show their commitment early on. The project manager can take responsibility for the later communication. Also, as I have said, you want key stakeholders to hear about the project from you or the sponsor first, and not as a rumour, so they gain positive views about the project from the start.

What Mechanisms will be Used? A range of possible mechanisms are available, including

- Seminars and workshops
- Press, television, and other media
- Bulletins, briefings, press releases, Web pages
- Site exhibitions
- Video and CDs

Again you will choose the media depending on the target audience, the objectives of the communication and the messages you want to convey. Different types of information are communicated using particular types of channel, and different stakeholders will be more receptive to some channels than others.

How will Feedback be Encouraged? Communication should not be one way; you should talk with people, not at people. If you want people committed to your project and the change it will introduce, they must feel involved, and feel that they have some influence over the design. (It is important that they feel they have some influence, whether or not they actually do is not so important.) But for this reason it is important that you are seen to be looking for and listening to feedback.

What will be Done with the Feedback? So it must be obvious to the stakeholders that their feedback is being used. The best way of achieving this is of course to be seen to be answering the stakeholders' questions. Not straight away as they give their feedback in a way that convinces nobody that you are actually listening; but later in a considered way, perhaps by making changes to the design of the project, but particularly by incorporating responses to the feedback in later communications. If you incorporate responses in future communications it demonstrates that you have actually heard and remembered what was said, remembered well enough and long enough to actually incorporate responses into the later communications.

4.4 PROJECT TEAMS

In forming the project team, the project manager brings together a group of people and develops amongst them a perceived sense of common identity, so that they can work together using a set of common values or norms to deliver the project's objectives. Charles Handy[8] says this concept of perceived identity is critical to team formation; without it the group of people remain a collection of random individuals. What sets project teams apart is that a group of people, who may never have worked together before, have to come together quickly and effectively in order to achieve a task which nobody has done before. The novelty, uniqueness, risk, and transience are all inherent features of projects (Chap. 1). Because the team is novel, it has no perceived identity, *ab initio*, and no set of values or norms to work to. It takes time to develop the identity and norms, which delays achievement of the team's objective. Furthermore, because the objective is novel, and carries considerable risk, it takes time to define, and, if the project is to be successful, this must be done before the team begins to function effectively.

Team Formation and Maintenance. The process of forming a team identity and a set of values takes time. Project teams typically go through five stages of formation called forming, storming, norming, performing, and mourning.[9] During these five stages, the team's motivation and effectiveness goes through a cycle in which it first decreases, before increasing to reach a plateau, and then either increasing or decreasing towards the end. The manager's role is to structure the team formation processes in such a way that this plateau is reached as quickly as possible, the effectiveness at the plateau is as high as possible, and the effectiveness is maintained right to the very end of the task.

Forming: The team comes together with a sense of anticipation and commitment. Their motivation is high at being selected for the project, their effectiveness moderate because they are unsure of each other.

Storming: As the team begins to work together, they find that they have differences about the best way of achieving the project's objectives, perhaps even differences about its overall aims. They also find that they have different approaches to working on projects. These differences may cause argument, or even conflict, in the team, which causes both the motivation and the effectiveness of the team to fall.

Norming: Hopefully some accommodation is achieved. The team members begin to reach agreement over these various issues. This will be by a process of negotiation, compromise, and finding areas of commonality. As a result of this accommodation, the team begins to develop a sense of identity, and a set of norms or values. These form a basis on which the team members can work together, and effectiveness and motivation begin to increase again towards the plateau. Although norming is important for the ultimate performance of the team, it can have a negative side effect. If the team norm too well they can become very introspective, and isolate themselves from the rest of the organization. They work very well together, but produce something the rest of the organization do not want.

Performing: Once performance reaches the plateau, the team can work together effectively for the duration of the project. The manager has a role of maintaining this plateau of performance. For instance, after the team has been together for too long, the members can begin to become complacent, and their effectiveness fall. If this happens the manager may need to change the structure or composition of the team.

Mourning: As the team reaches the end of its task, one of two things can happen. Either the effectiveness can rise, as the members make one concerted effort to complete the task, or it can fall, as the team members regret the end of the task and the breaking up of the relationships they have formed. The latter will be the case if the future is uncertain. Again, it is the manager's role to ensure that the former rather than the latter happens.

These five stages of team formation may need to be repeated at each stage of the project life cycle if there is a significant change in the composition of the project team. There are several group-working techniques which the manager can use to shorten the forming, storming, and norming stages, such as the application of the start-up processes described in Chap. 12, and in particular the use of start-up workshops.

Having formed the group, the manager's role is to ensure it continues to operate at the plateau of effectiveness. I next describe how the project manager can motivate a team of knowledge workers. But first the manager must be able to determine just how effective the team really is. On a simple level, this can be assessed by the way in which the team achieves its agreed targets, and by the way in which the individuals' and group's aspirations and motivational needs have been satisfied. The team leader and functional managers must ensure both corporate and personal objectives are met. If only the corporate goal is met, then with time there will be an erosion of morale and effectiveness followed by staff attrition. Often, however, it is only possible to measure achievement of these objectives at the end of the project, by when it is too late to take corrective action. Hence, we must also have measures by which to judge the cohesion and strength of a group during the project. Indicators of team effectiveness include

Attendance: Low absenteeism, sickness, accident rates, work interruptions, and labour turnover.

Goal clarity: Individual targets are set, understood, and achieved; the aims of the group are understood; each member of the team has a clear knowledge of the role of the group.

High outputs: Commitment to goal achievement, a search for real solutions, analytical, critical problem-solving using knowledge and skill, the search for widely tested and supported solutions.

Strong group cohesion: Openness and trust among members, sharing of ideas and knowledge, lively and constructive meetings, shared goal.

Motivating the Project Team. How does the manager motivate the members of a team of professional, knowledge workers, to build and maintain their effectiveness and commitment to the project? In the project environment, without the functional hierarchies, distinctions of title, rank, symbols of power and status do not exist, so many factors which are traditionally viewed as providing value to motivate professional staff are not available. In the project environment, managers must find new motivational factors which will be valued by their staff. There are three features of the project environment which have a significant impact on the motivation of personnel:

Matrix organization structures: Within a matrix organization, people do not have the clear indicators of title, status, and rank, as described. They also have reporting lines to two people, a short-term (project) boss, and long-term (functional) boss. Although the project manager tries to motivate the individuals towards the project goals, they often give their primary loyalty to their functional manager. It is that manager who writes their annual appraisal, and has greatest influence over long-term career development. This is exacerbated if annual performance objectives are aligned with the functional hierarchy because projects are of shorter duration than the timescale over which they are set.

Flatter organization structures: With flatter hierarchies being adopted by project organizations, individuals have less opportunity for career advancement, as there are fewer levels to occupy. They spend longer on each level before progressing, which means they have fewer opportunities to measure progress against career milestones, and are less able to judge how their contribution is viewed by the organization. Words of encouragement are not enough, because individuals can only judge their perceived value by progression, which means promotion. With decision-making processes bypassing the centre, individuals may also feel less able to influence their careers, as they no longer have direct contact with senior managers making career decisions. They rely very much on their project managers or their functional managers to act as their salesmen on career matters. This feeling of detachment can be heightened if the individual does not entirely understand the direction or strategy of the company, or how their project contributes to it. Having no direct contact with the centre through their work, they will not have the opportunity regularly to question the reasons for strategic decisions, or to suggest alternatives. This can exacerbate all the previous problems if they perceive their manager as the cause of their isolation.

The transient nature of projects: The transient nature of projects means that an individual's annual performance objectives tend to be aligned with their functional responsibilities rather than their project ones. Similarly, because projects only last a short time, they cannot satisfy an individual's long-term development needs in their own right. They can only be a stepping stone. It is the functional hierarchy which provides the focus for the individual's development, and if the individual is to be committed to projects, they must be assigned to projects which they view as fulfilling their development requirements.

So how do you motivate knowledge workers in this environment to give their commitment to projects? A traditional view of motivation is Maslow's *Hierarchy of Needs*.[10] Maslow proposed that people have five essential needs (higher levels first):

1. Achievement
2. Esteem
3. Belonging
4. Protection
5. Sustenance

People are motivated initially by lower needs. However, as they satisfy one, that reduces in importance, and they become motivated by the next higher. As their needs move up the list the lower ones lose effect. Many of the traditional views on motivation are not valid in the project environment. However, Maslow's hierarchy continues to provide a basis for motivational factors. Many people working on projects have now passed the point at which belonging is the primary need to be satisfied at work; they satisfy that through their leisure activities. They therefore look to satisfy their needs for esteem and achievement. This is especially true of knowledge workers, and leads to five factors for their effective motivation.

Purpose: Knowledge workers must believe in the importance of their work, and that it contributes to the development of the organization. This sense of purpose, and the linking of the work of a project to the mission of the parent organization, can help overcome the uncertainty of the dual reporting structures in a matrix organization.

Proactivity: As career paths become less clear and predictable, and as senior managers become remote, then people want to manage their own career development. Emphasizing the achievement of results, rather than fulfilling roles, and delegating professional integrity through results gives subordinates the opportunity to take responsibility for their own development. Furthermore, allowing people to choose their next project as a reward for good performance on the present one satisfies this need.

Profit sharing: Allowing people to share in the entrepreneurial culture will encourage them to value it. Many organizations now encourage employees to solve their own problems, and to take the initiative to satisfy the customer's requirements, and are allowing employees to share in the rewards. The growing band of freelance workers also shows that many people are taking this initiative into their own hands.

Progression: As people near the top of Maslow's hierarchy, they become conscious of the need for self-fulfillment. They therefore value the opportunity to increase their learning experiences. Each new project is an opportunity to learn new skills, and thereby increase esteem and self-achievement. However, in the flatter organization structures, people may have fewer career milestones to measure their progression. The one yardstick they still have is money (or other status symbols such as company cars). These things remain important, not as motivators in their own right, but as measures of achievement.

Professional recognition: Another measure of achievement is professional recognition. Knowledge workers do not want the anonymity of the bureaucrat, but want to accumulate "brownie points" to contribute to their esteem and achievement. I said above that in the flatter hierarchies of project-based organizations, managers at the centre may not be in direct contact with professional employees. Line managers must therefore ensure that their subordinates do receive due recognition.

Variation of the Motivational Factors with Life Cycle. The efficacy of these five motivators varies throughout the project life cycle (Table 4.4).

Definition: During this stage, the members of the project team try to determine what the project is about, so their focus on its purpose is high. They will try to determine how it can contribute to their development, and so the entrepreneurial spirit will be high. During definition, there will be some opportunity to demonstrate professional skill through problem solving.

Execution: During this stage, the focus switches from the purpose of the project to the work done. The learning opportunities, and chance of profit were set in the definition stage, and there is little chance to influence them during execution. However, through the use of responsibility charts (Sec. 6.4), people can be given responsibility for

TABLE 4.4 Variation of the Motivational Factors Throughout the Project Management Life Cycle

Factor	Definition	Execution	Close-out
Purpose	High	Low	High
Proactivity	High	Medium	High
Profit sharing	High	Low	High
Progression	High	Low	High
Professional recognition	Medium	Medium	High

achieving milestones, and so have some opportunity for demonstrating their professional skill.

Close-out: During close-out, all five factors come back into focus: The purpose becomes important again during commissioning; people deliver their results and receive their due reward, if the project has been profitable people complete their learning experience, and look forward to the next; they receive their professional recognition. During close-out, individuals can be given career counselling to help manage their careers. Individuals should be helped to define their development needs, plan how they are to be achieved, and to develop networks, internal and external to the organization, to be used in their career progression.

Virtual Teams. Virtual teams are very common in a project context.[11] The very nature of assembling a unique, novel, and transient teams means we look for the best people to do the work, and they may not always be colocated. Many people talk as if virtual teams are something new, enabled by modern technology. My own view is they have been around for centuries, modern technology just makes them more cost effective and so they are becoming more common. Indeed one definition of a *virtual team* is that it is a team where the team members are not colocated, and uses modern information and communication technology to communicate with each other. Another definition of a *virtual team* is any team where there is a boundary within the team that increases the cost of communication across that boundary. The boundary can be distance, time zone, language, culture, or professional differences (marketing versus engineers). We then see that modern information and communication technology is reducing the cost of communication across many boundaries. Virtual teams are assembled because the benefit outweighs the additional cost of communication, but with cheaper communication, we can realize the benefit more often, and hence the greater pervasiveness of virtual teams.

There are four factors that increase the effectiveness of virtual teams. Three we have met already, but they have particular significance for virtual teams. The four are

- Communication
- Trust
- Cohesion
- Goal clarity

4.5 LEADING PROJECTS

Through all of the foregoing discussion is the implied importance of the leadership of the project manager. Project managers have to lead in several directions:

- Upwards to maintain the support of the sponsor and owner
- Outwards to win the support of resource providers, professional colleagues, and the range of stakeholders listed in Table 4.2
- Downwards to lead the project team, winning the commitment to the project of people he or she may not have direct line responsibility over

Over the past century, a number of leadership theories have been proposed, and these have been interpreted into the context of projects. Most recently authors have tried to identify the competencies of leaders,[1] and I have been involved in research which attempted to identify the competencies of project managers for different types of project.[2]

Leadership Theories. One of the earliest leadership theorists was Confiucius, whose ideas formed the basis of the government of China for two and a half millennia.[12] He suggested that leaders exhibited four virtues (*de*):

1. Building relationships (*jen*)
2. Demonstrating their values (*xiao, yi*)
3. Following due process (*li*)
4. Adopting the doctrine of the mean (*zhong, rong*)

We will see that the first three of these have formed the basis of many leadership theories in the subsequent 2500 years. It is a pity that many managers have lost sight of the fourth, maintaining a balance in what they do—the goldilocks principle, neither too much not too little.

Two hundred years later, Plato and Aristotle were almost the first leadership theorists in Europe. Aristotle suggested a three-faceted approach to leadership, in Greek *pathos*, *ethos*, *logos*, or building relationships, demonstrating values, and following due process, (sound familiar, but what happened to the Goldilocks principle). I said earlier in the chapter that many western managers leap in with the logic. What differentiates a leader from a manager is the leader starts by building relationships and selling the vision, and once he or she has achieved that, then and only then says, "And this is how we have to do it."

Throughout the twentieth century, six schools of leadership developed:

The Trait School. It came to prominence in the 1930s to 1940s, and suggests that leaders exhibit certain traits they are born with. Leaders are born, not made. Kirkpatrick and Lock[13] suggested that effective leaders exhibit the following traits:

- Drive and ambition
- The desire to lead and influence others
- Honesty and integrity
- Self-confidence
- Intelligence
- Technical knowledge

The Behavioural School. It was popular in the 1940s and 1950s, and assumes effective leaders display certain behaviours or styles, which can be developed. Most theories from this school characterize leaders by how much they exhibit styles based on one or more of the following parameters:

- Concern for people or relationships (*jen, pathos*)
- Concern for production or process (*li, logos*)
- Use of authority
- Involvement of the team in decision-making (formulating decisions)
- Involvement of the team in decision-taking (choosing options)
- Flexibility versus the application of rules

Blake and Mouton's[14] is one of the best-known theories. They developed a two-dimensional grid based on concern for people and concern for production, each graded on a scale of 1 to 9 and identified five leadership styles, appropriate in different circumstances:

- Impoverished (1,1)
- Authority obedience (1,9)

- Country club (9,1)
- Compromise (5,5)
- Team leader (9,9)

Most authors from the behavioural school assume different behaviours or styles are appropriate in different circumstances, but that was formalized by the contingency school.

The Contingency School. This school from the 1960s and 1970s developed the idea that different styles are appropriate in different circumstances. They suggest you should

1. Assess the characteristics of the leader.
2. Evaluate the situation in terms of key contingency variables.
3. Seek a match between the leader and the situation.

One contingency theory that has proved popular is path-goal theory.[15] The idea is the leader must help the team find the path to their goals and help them in that process. Path-goal theory identifies four leadership styles: directive, supportive, participative, and achievement-oriented.

The Visionary School. It followed in the 1980s and 1990s, and identified two types of leaders: those who focus on relationships and communicate their values, and those who focus on process, called transformational and transactional leaders, respectively.[16]

1. Transactional leadership
   - Emphasizes contingent rewards, rewarding followers for meeting performance targets
   - Manages by exception, taking action when tasks are not going to plan
2. Transformational leadership
   - Exhibits charisma, developing a vision, engendering pride, respect, and trust
   - Provides inspiration, motivating by creating high expectations, and modelling appropriate behaviours
   - Gives consideration to the individual, paying personal attention to followers, and giving them respect and personality
   - Provides intellectual stimulation, challenging followers with new ideas and approaches

Each is appropriate in different circumstances. Following the work of Vic Dulewicz and Malcolm Higgs,[1] we can predict that transformational leadership will be required in more complex change, but in fact transactional leadership will work better with simpler change, where following due process is all that is required, and too much vision will be distracting.

The Emotional Intelligence School. It has developed through the 1990s and early part of this decade. It assumes all managers have a reasonable level of intelligence, and therefore what differentiates leaders is not their intelligence, but their emotional response to situations.[17] The school identifies 19 leadership competencies grouped into four dimensions:

1. Personal competencies
   - Self-awareness (mainly Confucius's moderation)
   - Self management (mainly Confucius's values)
2. Social competencies
   - Social awareness (mainly Confucius's values)
   - Relationship management (mainly Confucius's relationships)

TABLE 4.5 Fifteen Leadership Competencies

Group	Competency	Goal	Involving	Engaging
Intellectual (IQ)	Critical analysis & judgement	High	Medium	Medium
	Vision and imagination	High	High	Medium
	Strategic perspective	High	Medium	Medium
Managerial (MQ)	Engaging communication	Medium	Medium	High
	Managing resources	High	Medium	Low
	Empowering	Low	Medium	High
	Developing	Medium	Medium	High
	Achieving	High	Medium	Medium
Emotional (EQ)	Self-awareness	Medium	High	High
	Emotional resilience	High	High	High
	Motivation	High	High	High
	Sensitivity	Medium	Medium	High
	Influence	Medium	High	High
	Intuitiveness	Medium	Medium	High
	Conscientiousness	High	High	High

The idea is the leader needs to develop self-awareness first. Having developed that, he or she can then develop self-management and social awareness, and from those two develop relationship management. David Goleman and his coauthors identify six management styles, exhibiting different profiles of the competencies: visionary, coaching, affiliative, democratic, pacesetting, and commanding. Through a survey of 2000 managers, they identified situations in which these different styles are appropriate. The first four are best in certain situations, but all four are adequate in most situations medium to long term. They classify the last two styles as toxic. They say they work well in turn-around or recovery situations, but if applied medium to long term they can poison a situation, and demotivate subordinates.

The Competence School. This is the most recent school. Its says effective leaders exhibit certain competencies. It encompasses all the other schools because traits and behaviours are competencies, certain competency profiles are appropriate in different situations, and it can define competency profile of transformational and transactional leaders. After a review of the literature on leadership competencies, Vic Dulewicz and Malcolm Higgs[1] identified 15 which influence leadership performance (Table 4.5). They group the competencies into three types, intellectual (IQ), managerial (MQ), and emotional (EQ). They also identified three leadership styles, which they called goal oriented, involving, and engaging (Table 4.5). Through a study of 400 managers working on organizational change projects they showed goal-oriented leaders are best on low-complexity projects, involving leaders best on medium-complexity projects, and engaging leaders best on high-complexity projects.

The Six Schools and Project Management. These schools have been reflected in writings about the leadership skills of project managers:

The Trait School. Through work I did at Henley Management College, I identified seven traits of effective project leaders.

Problem solving. The purpose of every project is to solve a problem for the parent organization, or to exploit an opportunity (which also requires a problem to be solved). But also projects entailed risk, and so during every project you are highly likely to encounter problems. Project managers must be able to solve them.

Results orientation. Projects are about delivering beneficial change. But also, if you plan in terms of the results your plan is much more robust and stable than if you plan in terms of the work (Sec. 3.5). Thus project managers need to be focused on the results of their project.

Self-confidence. This is part of the emotional intelligence of project managers. They must believe in themselves and their ability to deliver.

Perspective. Project managers must keep their projects in perspective. A project manager must be like an eagle. They must be able to hover on high and see their project within the context of the parent organization. But they must have eagle-eyed sight to be able to see a small mouse on the ground, and to be able to sweep down and deal with it, but then also be able to rise again to hover above the project.

Communication. The project manager must be able to talk to everybody from the managing director down to the janitor. Sometimes the janitor knows more about project progress than anybody else. The janitor talks to everybody (see Example 4.9).

Negotiating ability. Project planning is a constant process of negotiation. As a project manager you ask people to work for you. You must convince them that it is worthwhile and beneficial to themselves to do that.

Energy and initiative. When the project gets into trouble, the project manager must be able to lift everybody else onto their back and lift them out of the hole.

Example 4.9 Talking to the janitor

When I was a post-doctoral research fellow, I had an office in one of a pair of houses. We had offices in one house while the other was being renovated. The plan was when the other was complete we would move into that house, while the one we were currently occupying was renovated. I was due to be away for a month to attend a conference. About a week before I was due to leave, the janitor, a retired miner called Frank, asked me when I was going to be away. From the 20th August to the 20th September, I said. Frank said that we were due to move into the other house on the 14th September, so it might be worthwhile for me to put my books in a tea-chest before I left. I said that was a good idea, but decided to check it out first with the administrator of the engineering department. I spoke to his secretary, but she denied any knowledge of the move. So I next asked the builders, but they said they would not be finished until late October or early November. I locked my office door, and went off to the conference. When I came back, I found that the door had been forced, and that the move had taken place on 14th September, the very day Frank had predicted. Of course, he had spoken to the University Estates people as they came to survey the work.

The Behaviour and Contingency Schools. David Frame identified four leadership styles of project managers, and showed different styles are appropriate at different stages of the life cycle.[18] I describe this more fully in Sec. 6.4.

The Visionary School. Anne Keegan and Deanne den Hartog[19] assumed that project managers need to have a transformational leadership style, and set out to show that to be the case. In the event they found a slight preference for transformational leadership, but not a strong preference. I think the reason for this is that complex projects need a transformational style, but less complex projects need a more conscientious, structured, transactional style, as suggested by Vic Dulewicz and Malcolm Higgs.[1] Too much strategic perspective can be a distraction on simpler projects. This has been borne out by the work I will describe below.

The Emotional Intelligence School. There is very little work setting project leadership within the context of the emotional intelligence school. However a contribution was made by Liz Lee-Kelly and K Leong almost by chance.[20] They were researching how the project manager's competence at managing five functions of management described in the next part contributed to project success. What they found was that there is a significant relationship between the leader's perception of project success and his or her personality and contingent experiences. Thus the inner confidence and self-belief from personal knowledge and experience are likely to play an important role in a manager's ability to deliver a project successfully. The project manager's inner self-confidence has a significant impact on their competence as a project leaders and hence on project success.

The Competence School. Ralf Müller and I investigated which of the competencies in Table 4.5 are correlated with project success.[2] We looked at different types of projects to see if different leadership styles are appropriate on different types of project. We found that emotional intelligence made a significant contribution to project success on almost all types of projects. The exceptions were mandatory projects and fixed price contracts where managerial competence was more significant. Thus, on most projects self-awareness and building relationships are more important than following due process. But I think we can understand why following due process may be more important on mandatory projects and fixed price contracts. On time and material contracts intellect was also important. We also looked at how the 15 individual competencies related to success. We found that some were positively correlated and some negatively. Some of our results are shown in Table 4.6.

You will see that communication gets mentioned the most, being important on all types of project except engineering and high complexity projects. On engineering projects, methodical working is important. On information systems projects it is self-awareness, and on organizational change motivation. It is interesting that information

TABLE 4.6 Leadership Competencies Contributing to Project Success

Project attribute	Project type	Important	Unimportant
	All projects	Conscientiousness Sensitivity Communication	Strategic perspective
Application area	Engineering	Motivation Conscientiousness Sensitivity	Vision
	Information systems	Self-awareness Communication	Vision
	Organizational change	Motivation Communication	Vision
Strategic importance	Mandatory	Developing	
	Renewal	Self-awareness Communication	
	Repositioning	Motivation Communication	
Complexity	Medium	Emotional resilience Communication	Vision
	High	Sensitivity	
Contract type	Fixed price	Sensitivity Communication	
	Time and materials	Self-awareness Communication	Empowering

systems projects show the same profile as renewal projects and organizational change projects as repositioning. What is controversial is that vision appears as unimportant so often, and may be inconsistent with what I have said earlier in this chapter. What I would say is that having a clear picture of the end state of the project, and setting clear goals is important. Goal clarity was identified as important for project team performance. What is meant by vision here is being able to picture many possible end points for the project, and in fact on most projects that is a distraction, and will reduce goal clarity. So it is important to have a vision of the one clear goal of the project, but don't get distracted by lots of alternatives. The message seems to be that small, simple engineering projects need managing, whereas larger, more complex information systems and organizational change projects need leading.

Ralf Müller and I do not expect that organizations would conduct a psychometric test on potential project managers at the start of every project. But we do suggest that they conduct a psychometric test at least once as part of the annual review process on project managers in the pool of project managers. They can then identify what shortfalls individual project managers have in their profile and work on developing appropriate competencies through the project management development program. If somebody's profile is totally inappropriate for the type of project they have to manage they can be dropped from the pool of project managers, but we did find that the career tends to be self-selecting and people don't stay working as project managers if their profile does not fit. Individuals can also look to enhance their competencies for the types of project they want to work on.

The two clear messages are emotional intelligence is important to being a project manager, and communication is important on all projects, which is where we started this chapter.

SUMMARY

1. People react to change differently, depending on the level of change within the organization.
2. You cannot expect people to immediately accept and internalize your proposals for change within the organization. You must lead them through carefully, getting them to appreciate the benefit of the proposed change, to see that it is sensible, and will not unduly affect their position within the organization, before getting them to fully accept it.
3. Recognize that extreme change can lead to significant emotional responses which must be managed carefully.
4. There is a seven-step process for stakeholder management:
   - Identify interested parties.
   - Identify possible success criteria.
   - Identify stakeholders and their interests.
   - Develop a stakeholder persuasion strategy.
   - Monitor their response.
   - Monitor the impact of the environment.
   - Make changes to the strategy if necessary.
5. To analyze stakeholders you need to answer three questions.
   - Are they for or against the project?
   - Can they influence the outcome?
   - Are they knowledgeable or ignorant about the project?
6. The stakeholder management strategy will depend on the answers to these questions.

7. When developing a communication plan for a project, answer the following questions:
   - What are the objectives of each communication?
   - Who are the target audience?
   - What are the key messages?
   - What information will be communicated by whom?
   - When will the information be given?
   - What mechanisms will be used?
   - How will feedback be encouraged?
   - What will be done with the feedback?
8. Project teams are unique, novel, and transient, like projects. There are five steps to team formation and maintenance:
   - Form
   - Storm
   - Norm
   - Perform
   - Mourn
9. Measures of team performance are
   - Attendance
   - Goal clarity
   - Outputs
   - Cohesion
10. Knowledge workers on projects are motivated by
    - A sense of purpose
    - Control of their own destiny
    - A share in the benefits of the project
    - Measures of progression
    - Professional recognition
11. There are six schools of leadership
    - The trait school
    - The behaviour school
    - The contingency school
    - The visionary school
    - The emotional intelligence school
    - The competence school
12. Emotional intelligence and communication are significant competencies for project leaders, having the greatest contribution to project success.

REFERENCES

1. Dulewicz, V. and Higgs, M., "Assessing leadership dimensions, styles and organizational context," *Journal of Managerial Psychology*, 20(2).
2. Turner, J.R. and Müller, R., *Choosing Appropriate Project Managers: Matching Their Leadership Style to the Type of Project*, Newtown Square, Pa.: Project Management Institute, 2006.
3. Handy, C.B., *The Age of Unreason*, paperback edition, London, U.K.: Business Books, 1998.
4. Machiavelli, N., *The Prince*, translated by Bull, G., London, U.K.: Penguin, 1981.
5. Grude, K.V., "The norwegian defences: a case study in downsizing and restructuring," in Turner, J.R., Grude, K.V., and Thurloway, L., (eds), *The Project Manager as Change Agent*, London, U.K.: McGraw-Hill, 1996.
6. D'Herbemont, O. and César, B., *Managing Sensitive Projects: A Lateral Approach*, London, U.K.: Macmillan Business, 1998.

7. Foreman, S.E., "Internal marketing", in Turner, J.R., Grude, K.V., and Thurloway, L., *The Project Manager as Change Agent*, London, U.K.: McGraw-Hill, 1996.
8. Handy, C.B., *Understanding Organizations*, 3rd ed., London, U.K.: Penguin, 1987.
9. Tuckman, B.W., "Development sequence in small groups," *Psychology Bulletin*, 1965.
10. Maslow, A.H., *Motivation and Personality*, New York, N.Y.: Harper & Row, 1954.
11. Lee-Kelley, L. and Turner, J.R., "Virtual teams: understanding individual motivation for collective effort," paper in preparation.
12. Collinson, D, Plant, K., and Wilkinson, R., Fifty Eastern Thinkers, New York, N.Y.: Routledge, 2000.
13. Kirkpatrick, S.A. and Locke, E.A., "Leadership traits do matter," *Academy of Management Executive*, Mar, 44–60, 1991.
14. Blake, R.R. and Mouton, S.J., *The New Managerial Grid*, Houston, Tex.: Gulf, 1978.
15. House, R.J., "A path-goal theory of leader effectiveness," *Administrative Science Quarterly*, Sep, 321–338, 1971.
16. Bass, B.M., "From transactional to transformational leadership: learning to share the vision," *Organizational Dynamics*, 18(3), 19–31, 1990.
17. Goleman, D., Boyatzis, R.E., and McKee, A., *The New Leaders*, Cambridge, U.K.: Harvard Business School Press, 2002.
18. Frame, J.D., *Managing Projects in Organizations*, 3rd ed., San Francisco, C.A.: Jossey-Bass, 2003.
19. Keegan, A.E. and Den Hartog, D.N., "Transformational leadership in a project-based environment: a comparative study of the leadership styles of project managers and line managers," *International Journal of Project Management*, 22(8), 609–618, 2004.
20. Lee-Kelley, L. and Leong, K.L., "Turner's five functions of project-based management and situational leadership in IT services projects," *International Journal of Project Management*, 21(8), 583–591, 2003.

P · A · R · T · 2

MANAGING PERFORMANCE

CHAPTER 5
MANAGING SCOPE

In this part, I describe methods, tools, and techniques for managing the six project management functions: scope, organization, quality, cost, time, and the risk that pervades them all. The aim of these six functions is to undertake the work of the project and to deliver the desired performance improvement at a time and a cost that provides value for the sponsor. That is, they are about managing the performance of the project, and the asset it produces. I start with scope. The next four chapters deal with the other five functions.

Scope management is mandatory; without scope management there is no project. It can be defined as the process of ensuring that

- An adequate, or sufficient, amount of work is done.
- Unnecessary work is not done.
- The work which is done delivers the desired performance improvement.

There are four essential steps to scope management:

1. Developing the concept through the project's objectives and product breakdown structure
2. Defining the scope of work through the work breakdown structure
3. Authorising and executing the work, and monitoring and controlling progress
4. Commissioning the facility to produce the product and obtain the desired benefit

Through the process of managing the scope the owner's requirements are converted first into the definition of the new asset required to produce the desired performance improvement, and then into a statement of the work required to construct and commission that asset, and then the work identified is brought to a successful conclusion. This is the *raison d'être* of project management, and so scope management is the principal project management function.

In this chapter, I describe the methods, tools, and techniques used to manage scope. I start by revisiting the principles of good project management introduced in Sec. 3.5 and show how these are achieved by the use of product and work breakdown. I then explain how the products and work are defined at the three fundamental levels of breakdown: How to define the facility required to achieve the owner's purpose and the broad areas of work required to construct that facility? how to break the facility into intermediate products, or milestones, in each of the areas of work? and how to specify the work, as activities required to produce the intermediate products? I end by illustrating the concepts with several case studies.

5.1 PRINCIPLES OF SCOPE MANAGEMENT

Four of the principles introduced in Sec. 3.5 relate to scope management:

- Manage through a breakdown structure.
- Focus on results.
- Balance objectives and levels of ambition.
- Keep it simple.

All four of these principles can be met by the use of a breakdown structure. I showed in Sec. 1.1 that breakdown is inherent in projects and follows from the definition, and so the principles support the inherent nature of projects. The second will be achieved if the primary breakdown is via a *product breakdown structure* (*PBS*). The third is achieved by ensuring that results are delivered in all areas of the project, and by balancing the work through the work breakdown structure (WBS). The fourth is achieved if we use single-page reporting at all levels of the project. In this section, I consider product and work breakdown.

Breakdown

Breakdown is a technique by which *the project is divided and subdivided for management and control purposes*. Rather than breaking the work of the project into a low level of detail in a single step, it is devolved through increasing levels of detail. Focusing on results means we start with a PBS. The PBS is developed by breaking the asset into intermediate or subproducts. The work required to produce each subproduct and the work required to assemble and commission the facility from the subproducts is then identified. In Sec. 1.1, I described three fundamental levels of breakdown: integrative, strategic, and detail. However, a WBS can be developed to many more levels and I have seen up to seven levels used on large engineering projects. Table 1.5 shows a typical structure, with several levels of deliverables, associated work elements, and possible relative durations for a project lasting about a year. This structure shows the project as part of a much larger program of work, required to deliver the company's 5- or 10-year objectives.

Advantages of Using a Breakdown Structure

There are several reasons for using breakdown:

Better Control. The use of a breakdown structure satisfies the first three principles of good project management in Sec. 3.5. One of the pitfalls in planning is to develop the work definition at a single, detailed level. Developing the definition in a structured way ensures better results. Further, by defining work through its deliverables ensures that, as the project progresses, only work necessary to produce the facility is done, not work which was envisaged some months previously but is no longer required. Hence, the plan also becomes more stable. The work required can change in changing circumstances, but only certain results build towards the required end objective. This is clearly the case in research and development projects, where the process of doing the project defines the work to be done. However, it can also be true of engineering, construction, information technology, and organizational development projects. For example, the construction of an aeroplane and a submarine involves similar activities:

- The fabrication of metal into a cylindrical pressure vessel
- Internal outfitting to support life in a hostile environment
- The fitting of propulsion equipment

On a detail level the work appears the same. However, one set of intermediate products leads to an airbus, and another to a submarine. The high levels of definition can also be used to balance areas of work. By developing the definition at a detail level only, there is a risk that we give undue emphasis to one area only. This may be technical work over cultural work (Sec. 2.2), or it may be our own area of expertise at the expense of another. On Heysham 2 Nuclear Power Station in the United Kingdom, the computer systems required to operate the plant were not given sufficient emphasis in the plan, swamped by the engineering work, and would have delayed commissioning several months, if it were not for another technical problem. A small amount of work could have kept a multibillion pound investment lying idle.[1]

Coherent Delegation. The parcelling of work in a breakdown structure is natural, because it is aimed at achieving a product. Responsibility can be assigned to individual parties for each product. In fact, they can be left to identify the actual work required, and in this way experts retain their integrity, while being set measurable targets. Sometimes this can be the only way to control progress on a research project, as the work itself is unknown, only the intermediate results can be measured. If work is defined at a detail level and amalgamated into packages, then they may not actually be natural packages of work, and the project manager can appear to be telling people more technically skilled than themselves how to do the work.

Levels of Estimating and Control. The lowest level of work breakdown appropriate for estimating and control depends on

- The size, type, and duration of the project
- The purpose for which the estimates will be used
- The current stage in the project management life cycle
- The requirement for effective control

I find on projects of a year's duration that activities of 2-weeks duration are the lowest appropriate level for planning and control. There is a law of diminishing returns which makes it inefficient to plan and estimate at lower levels, except in areas of high risk.

Lowest level of work breakdown: The activity level is the lowest level for central planning, estimating, and control. However, individuals may plan their own work at the task level. The lowest level does depend on the size of the project. On a 4-week overhaul of ammonia plants, the lowest level of planning was activities of 2 to 4 hours. On the other hand, I worked briefly on a project of 7-year duration, on which people were planning steps of 4-hour duration 6 months in advance. The plans were meaningless.

Lowest level of estimating: Because of inherent uncertainties, there is only a certain level of accuracy you can expect. It is pointless to plan in greater detail. The people on the 7-year project thought planning at lower levels improved the overall accuracy. Unfortunately, this is not the case. Probability theory tells us that the percentage error of the part as a ratio of the percentage error of the whole is inversely proportional to the square root of the size:

$$\pm e\% / \pm E\% = \sqrt{(S/s)}$$

We might expect to finish a year long project, S = 52 weeks, to within a month, E = ±10 percent. Therefore on an activity of 2-weeks duration, we need to be accurate to within 1 week, i.e., = ±50 percent. On a task of 2-day duration, we need to be accurate to within 2 days, e = ±100 percent. The accuracy on smaller steps is even more meaningless.

Planning in greater detail also requires more effort in estimating. The formula above implies that doubling the accuracy of the estimate requires four times as much planning effort, and this has been measured in the process plant industry.[2] Therefore, at early stages of the project, you want very coarse estimates, obtained by planning at high levels of breakdown, with lower levels developed only as the project is shown to be viable at the high levels. You also reach a point at about E = ±5 percent accuracy, at which it costs more to estimate than the value of the data you are getting. This sets a limit on the lowest worthwhile level of breakdown for estimating purposes. I return to this concept in Chap. 8.

Lowest level of control: Similar arguments apply to the level at which the project is controlled: Controlling at a lower level can mean more time is spent in control than doing work; controlling at a higher level means slippages can get out of hand before they are recognized. The appropriate size of activity for control is the same as the frequency of control meetings. If meetings are once a fortnight, activities should, on average, be a fortnight long. Then, at each review an activity is either not started, finished, or half finished: three simple states. If activities are very much shorter, then it will be difficult to determine what is critical for completion. If they are very much longer, then the percentage completion will be reported as the elapsed time divided by the original duration while that is less than one, and 99 percent while it is greater until the activity is actually finished.

Containment of Risk. I qualified remarks above by saying it did not apply in areas of high risk. In fact, there is no need to take the WBS down to the same level. The lowest level of WBS may vary according to the level of risk: In areas of low risk you may stop as high as the work package level; in areas of high risk you may continue to a very low level of WBS, depending on

- The uncertainty introduced by the risk
- The need to contain the risk

5.2 PROJECT DEFINITION

Project definition initiates the project and therefore relates the work of the project to the sponsor's business objectives. To achieve this, it is necessary to identify the sponsor's requirements, including the facility expected to satisfy them, and then to identify the broad areas of work required to construct the facility. The benefits map (Sec. 2.3) has already initiated this process; project definition converts them into a form the project team can work with. This requires the following three things to be defined:

- The purpose of the project
- The overall scope of work
- The outputs from the project

The Purpose. This is a statement of the business need to be achieved. As we have seen, it may be a problem to be solved, an opportunity to be exploited, a benefit to be obtained, or the elimination of an inefficiency, and will be derived from the strategic objectives of the parent organization and the desired performance improvement (Chap. 2). The statement of the purpose should be clear and precise, and contain both quantitative and qualitative measures. Once the project is underway, it will become the mission of those involved in the project, both as project team members and as resource providers. It can be a powerful motivating force if it is seen to be worthwhile and beneficial to the business, and can help to build cooperation. Of course, it can be a powerful demotivator if it is seen to conflict with individuals' self-interest (Example 5.1).

Example 5.1 Project objectives and personal objectives in conflict

I was involved with a project where the user representative on the team stood to be made redundant if the project was successful. He had been appointed by his general manager, because the project was likely to make a large proportion of his department redundant, reducing his empire. The project was not successful; and in fact came to an abrupt halt when we held a project definition workshop (Chap. 11). It was impossible to maintain the pretence. However, 2 years later it was overtaken by a larger project which merged several subsidiary companies into a larger unit. The general manager lost his job.

The Scope. This is an initial, high-level description of the way in which the purpose will be satisfied. If the purpose is viewed as a problem to be solved, the scope will identify possible solutions, and the one selected for further work; these comprise the fourth, fifth, and sixth steps in Fig. 1.6 and Table 1.4. The statement of scope includes three things:

1. The work within the remit of the project, required to solve the problem and achieve the benefits
2. The work which falls outside the remit of the project
3. Interfaces with other projects in the program

The inclusions will later be made redundant by the initial stages of work breakdown. However, it is important to include them in the statement of project definition. They are a key step in the problem-solving process, which indicate the thought processes of the people drawing up the project definition. The exclusions can arise either because the work is not required to achieve the benefits, (although it would be nice to have), or because it is being handled elsewhere. The sponsor does not have a limitless pot of gold, and so a boundary must be set on the work to be done. Sometimes the potential benefit must be reduced to match the available funds. Also, when a project is taking place as part of a larger program, it may share work with other projects. It can then be more efficient to have one project handle all the joint work. This is especially true when projects create a need for redeployment or redundancy. One project may then delegate the work to the other. For whatever reason, the exclusions must be clearly stated, so that they are understood by people joining the project later, and so that interfaces with other projects are identified and managed (Chap. 16). These exclusions will include the definition of interfaces with other projects in the program.

The Outputs. These are quantitative and qualitative measures by which completion of the project will be judged. They identify the facility to be produced by the project. If the facility is an engineering construction (factory, dam, or chemical plant, say), then the outputs may be something like:When the facility has been constructed, the supporting establishment is in place, the facility has been commissioned, and is operating to a certain percentage of capacity.

A similar statement can apply to a computer system, management development program, or organizational change. You will notice the statement implies the facility has been shown to be able to achieve some of the benefits. People are usually quite happy with this for a factory, less so for a computer system, or organizational change process. In the latter cases, the project is over once the system is commissioned, and the project team have no responsibility for ensuring that it works properly!!! It is not always possible to set the project's benefits as the objectives, as they may not be achieved until some time after the end of the project, and the facility has been commissioned. However, it is important that the outputs are likely to deliver the benefits, and the project team addreses the question of how they are to be attained. Further the outputs should

- Address all the work within the scope of the project
- Not address work outside the scope of the project
- Begin to set parameters for managing quality, cost, and time

You will see now why it is important to record the scope of the project.

Initiating Work Breakdown

The statement of the outputs completes the project definition. It is now possible to begin the process of work breakdown by defining areas of work. Each area of work delivers one of the project's objectives, linking the integrative level, level 1, to the strategic level, level 2. The areas of work may form subprojects, as in Table 1.5. In Chap. 12, I describe the *project definition report*. The statement of purpose, scope, and objectives appears in an early section, and sets the scene for the project. The areas of work appear in the section on work breakdown. It is important that the areas of work cover all the objectives, but no more.

Case Study

Table 5.1 gives the project definition for the CRMO Project from TriMagi introduced in Example 2.6. It gives statement of purpose, scope, outputs, and areas of work. The definition of the project contains a statement of the expected time scale: 5 months to the commissioning of the first offices and 9 months to completion of the project. At this stage, these are targets. People familiar with the technology should be able to say whether they are realistic, but the precise timescale would only be determined as the project plan is developed to lower levels. However, I am a great believer in being *goal directed*, aiming to achieve this target and scheduling the work appropriately, rather than allowing theoretical mathematics in the form of a network to impose a longer duration. Often tight timescales can be achieved with management effort. Similarly, there is already enough information for experts to begin to develop initial estimates of capital cost and revenue for the project.

5.3 PLANNING AT THE STRATEGIC LEVEL: MILESTONE PLANS

Having defined the project, we are in a position to develop the work breakdown structure to the second level, the strategic level. I now describe the requirements for planning at this level, and then introduce a tool, the *milestone plan*,[3] which satisfies these requirements.

TABLE 5.1 Project Definition for TriMagi's CRMO Rationalization Project

<table>
<tr><th colspan="2">TriMagi
Project definition</th></tr>
<tr><td>Purpose</td><td>The purpose of the project is to rationalize the CRMO organization:
1. To improve customer service so that:
– All customers calling the receipt offices obtain a free line
– All calls are answered within 10 seconds
– The maximum time from call receipt to arrival of an engineer on site is 2 hours
2. To improve productivity and flexibility so that:
– The costs are justified through productivity improvements
– The call receipt offices can be made part of a unified "enquiry desk"
but there are no redundancies so that all productivity improvements are achieved trough natural wastage, redeployment, or growth</td></tr>
<tr><td>Scope</td><td>The work of the project includes:
1. Changing from the existing structure of 18 area offices to 3 call receipt offices, 2 diagnostic offices, and 4 field offices
2. Investigating which of two new CRMO networking technologies is appropriate for the new structure, and to implement that chosen
3. Refurbishing the nine new offices to current standards
4. Training and redeploying staff to meet needs of operation of new CRMOs
5. Installing hardware to connect the CRMOs to the Customer Information System, and to implement a statistical package to analyse fault data
It is expected that the first call receipt and diagnostic offices will be available in 5 months time and the project will be complete in 9 months. The work of the project excludes the retrenchment of any staff who are surplus to requirements within the CRMO structure; they will be passed to central personnel for redeployment on other expansion projects; with the implementation of the new Customer Information System, the call receipt offices may within the next 2 years be incorporated into unified "enquiry desks" dealing with all customer contacts. However, it will not be the project team's responsibility to achieve that integration.</td></tr>
<tr><td>Outputs</td><td>The outputs of the CRMO Rationalization Project are:
1. When the CRMO facilities have been installed in nine offices, (three call receipt offices, two diagnostic offices, and four field offices), within 9 months
2. When appropriate networking technology have been selected and implemented, together with statistical MIS to achieve the required customer service levels
3. When appropriate operating systems have been designed and implemented, together with procedures to achieve the required customer service levels and productivity improvements
4. When staff have been trained and redeployed to fill new positions, and vacate old positions
5. With the objective that the first offices should be operational within 5 months and the work complete within 9 months.</td></tr>
</table>

(*Continued*)

TABLE 5.1 Project Definition for TriMagi's CRMO Rationalization Project (*Continued*)

TriMagi Project definition	
Areas of work	To achieve the project's objectives, the following areas of work are required: A *Accommodation*: Refurbish new offices, install hardware and furniture. (There is only one floor area available in the region large enough to take the first call receipt and fault diagnosis offices. The remaining eight offices must be housed in existing CRMO space). T *Technology*: Decide on networking technology to be used, implement statistical MIS, and implement networking technology in new offices. O *Organization*: Communicate all changes to the staff involved, define the operation of the new CRMOs, train and redeploy staff to fill new positions. P *Project*: Plan the project, organize the resources, and obtain financial approval.

Requirements for Planning at the Strategic Level

At the second level of breakdown, the manager sets the strategy for the project. The plan at this level:

- Shows how the intermediate products, or deliverables, build towards the final outputs
- Sets a stable framework, fixed goal-posts, for the team, and so provides a common vision
- Controls devolution of the management of the scope to other parties

I described above how similar activities are involved in the manufacture of an airbus or submarine, yet one set of intermediate products delivers an aircraft, another a submarine. It is at the second level of WBS that we set the strategy, showing how the intermediate products build towards the facility to be delivered by this project. Because only one set of intermediate products delivers the required final objective of this project, the plan at this level can be made stable. This can be a powerful motivating tool, giving the project team a common vision.

To build a common vision, the plan should be represented on one page. It then presents a clear picture of the strategy for the project. It is through this single page, *the milestone plan,* that the project manager communicates the overall strategy of the project upwards to the project sponsor, and downwards to the project team. This was the fifth principle of good project management introduced in Sec. 3.5. It is also at this level that focusing on the deliverables can help delegate work to subproject teams. A team accepts responsibility for the delivery of an intermediate product, and plans its own work to deliver that milestone independently of other project members. They know that they must achieve their milestone by a certain date to enable the project to proceed, but they are able to work without interference. We have seen how this can allow professional people to retain their integrity when working for a project manager from a different discipline.

Milestone Planning

It is common, when developing the plan at the second level to define the packages of work first, and then define the deliverable which results from each work package. However, for the reasons above, I suggest that you define the deliverables, or milestones first, in the form of a milestone plan.[3] The packages of work which result in each milestone are derived later. The milestone plan is a strategic plan, or framework, for a project, defined in terms of intermediate

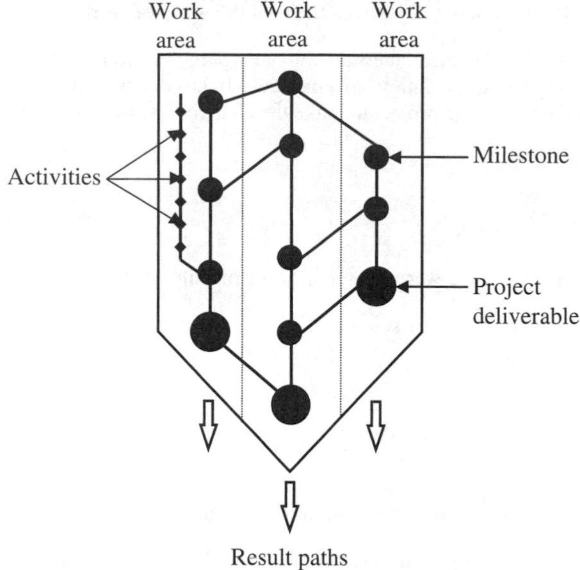

FIGURE 5.1 The milestone plan.

products, or results, to be achieved. It shows the logical sequence of the states a project must pass through to achieve the final objectives, describing what is to be achieved at each state, not how the state is to be achieved. Figure 5.1 illustrates the milestone plan, with the circles representing the milestones, and the lines joining them representing the logical dependency between them. Hence, the milestone plan represents a logical network for the project.

We return to networks in Chap. 9 where two types are described: *precedence* and *activity-on-arrow networks*. In precedence networks, work is represented by the nodes of the network. These are joined by arrows representing the logical dependency of the work. In an activity-on-arrow network, work is represented by the arrows. The nodes are events in time, and the logic is represented by the way the arrows join at the nodes. The milestone plan is a precedence network. The circles in Fig. 5.1 represent packages of work, defined by the results they deliver. The arrows show how one package follows another, and are known as *end-to-end dependencies*: The end of one package, milestone, is dependent on the end of the previous one. They say nothing about the start of the work: One package can start before the previous one has finished. This allows greater flexibility in scheduling the work.

Areas of Work

In Fig. 5.1, the milestones are grouped into vertical columns representing the areas of work. One of the principles in Sec. 3.5 was to balance the changes. I suggested that the WBS should be used to ensure that equal emphasis is given to work in different areas. The areas of work give visual representation to this. By inspecting the areas of work you can ask yourself one of two questions, as illustrated in Example 5.2:

- Have all the areas of work been covered, or has something been left out? In particular, have the cultural changes been addressed?
- Has equal emphasis been given to all areas of work?

Example 5.2 Balancing objectives through the areas of work

I did some work with a research establishment where they were installing a larger computer to store the empirical data from a particularly large experiment they were conducting. I helped them plan the project to make the change. The plan had three areas of work:

1. Hardware and software
2. The database
3. The establishment

Down the first path, there were a large number of milestones:

- Hardware and software selected
- Hardware installed
- Operating system loaded
- Database software loaded
- System tested

There were a similar number of milestones in the third path:

- Computer room ready to receive machine
- Furniture obtained
- Operating procedures written
- Operators recruited
- Operators trained

There were only two milestones in the central path:

- Data transferred
- System commissioned

Without prompting from me, the two people working with me on the development plan said, "Hold on! The purpose of this project is not to obtain new hardware and software, and not to create a new establishment. It is because the data has got too large for the old machine. We ought to be giving greater emphasis to the database." They, therefore, inserted two more milestones in the centre column. One dealt with data cleanse, that means, removing incorrect, incomplete, or redundant data. The other dealt with restructuring the database to meet future, rather than historical, requirements. (These two milestones may have made the rest redundant!!!)

Features of the Milestone Plan

A good milestone plan should satisfy several requirements.

Be Understandable to Everyone. The milestone plan is a tool to build cooperation and commitment to a common vision. It must therefore be understood by all those involved in the project. This requires the milestone descriptions to be written in clear English, not in technical jargon, only understandable to a few. Writing the plan in technical jargon shows how important you are, protects the work for yourself, and builds demarcations to make sure others aren't involved in the project. It is not good for the involvement, commitment, and cooperation of others.

Provide Quantitative and Qualitative Control. The plan is a tool for control, and so the milestone descriptions must be precise, so you can determine when they have been achieved. Technical milestones can be given a quantitative measure, such as "when the new machine is operating at design capacity." (Even that will have a quality measure built in.) Other milestones must be given a qualitative description, with some measure of quality written in. For example, it is not adequate to say "when a report is written." Two lines on the back of an envelope satisfy that. The report must

- Meet certain requirements
- Satisfy a steering committee
- Allow a decision to be made

A milestone such as "when the design is finished" is neither measurable nor achievable. You can't know the design really is finished, so you can't achieve it. A milestone such as "when the team accept the design as a basis for the next stage of the project" can be measured and achieved. Focusing on the decision provides better qualitative control.

Focus on Decisions. Milestones represent intermediate deliverables en route to the final objective. Often the interesting deliverable is not the production of a design or report. That is not the purpose of the work. It is the taking of a decision, based on the design or report, to allow more work to proceed. That is the required deliverable, and is controllable. The responsibility chart (Chap. 6) defines who is to take the decision.

Show the Logical Sequence. The milestone plan is a logical plan. It contains a network, which shows the strategy for building through the intermediate products to achieve the final objective.

Give a Single-Page Overview. The objective is to produce a plan on a single page which clearly communicates the project strategy. This is achieved if the number of milestones and areas of work is limited. The ideal number of milestones is somewhere between one and two dozen. With fewer the plan does not give a useful structure, and with more it becomes confusing. Similarly, I suggest three or four areas of work. Thus the number of milestones determines the size of the work packages, rather than the size of work packages determining the number of milestones. On small projects this will be the only level of planning. On large projects it will be the first of several.

Representing the Milestone Plan

In Fig. 5.1, the milestone plan is drawn down the page, whereas it is common to draw a network across the page (Chap. 9). However, I like to represent the milestone plan as a process flow diagram for the project, with three columns (Fig. 5.2):

1. The central one is for drawing the network.
2. The right-hand one is for writing the description of the milestones (which in themselves describe the packages of work).
3. The left-hand column is for the dates, once the work has been scheduled (Chap. 9).

The right-hand column gives adequate room to write a full description of the milestone, whereas if you draw the network across the page, you have to write small to fit the description into the box or onto the arrow. It may seem heretical to draw the network down the page, but it does allow the network and a full description of the work to be portrayed on a single page. It also represents the milestone plan as a process flow diagram for the project, emphasizing the process nature. Figure 5.2 is a milestone plan for the CRMO Rationalization Project.

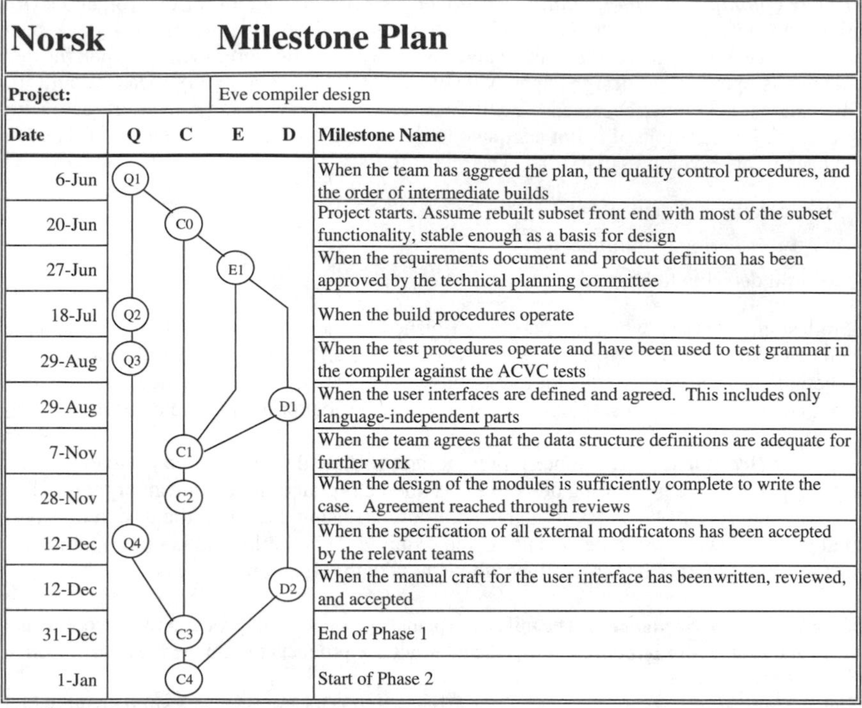

Norsk Milestone Plan

Project:	Eve compiler design				
Date	**Q**	**C**	**E**	**D**	**Milestone Name**
6-Jun	Q1				When the team has aggreed the plan, the quality control procedures, and the order of intermediate builds
20-Jun		C0			Project starts. Assume rebuilt subset front end with most of the subset functionality, stable enough as a basis for design
27-Jun			E1		When the requirements document and prodcut definition has been approved by the technical planning committee
18-Jul	Q2				When the build procedures operate
29-Aug	Q3				When the test procedures operate and have been used to test grammar in the compiler against the ACVC tests
29-Aug				D1	When the user interfaces are defined and agreed. This includes only language-independent parts
7-Nov		C1			When the team agrees that the data structure definitions are adequate for further work
28-Nov		C2			When the design of the modules is sufficiently complete to write the case. Agreement reached through reviews
12-Dec	Q4				When the specification of all external modificatons has been accepted by the relevant teams
12-Dec				D2	When the manual craft for the user interface has beenwritten, reviewed, and accepted
31-Dec		C3			End of Phase 1
1-Jan		C4			Start of Phase 2

FIGURE 5.2 Milestone plan for the CRMO Rationalization Project.

Developing the Milestone Plan

Ideally, the milestone plan should be developed in a project start-up workshop (Chap. 12), with selected managers and project personnel present. Developing the plan in a group session builds greater commitment than the project manager developing it on their own and trying to impose it on the team. However, to be effective the workshop should not have more than about six people present. The process I recommend for developing the plan has six steps:

1. Start by agreeing the final milestone, the end of the project. The project definition and benefits map should help this. If you have completed Hartman's three questions (Sec. 3.1.1), you will already have done this step.
2. Generate ideas for milestones. Brainstorm them on to flip-charts.
3. Review the milestones. Some will be part of another milestone. Some will be activities, but will generate ideas for new milestones. As you rationalize the list record your decisions, especially where you have decided that a milestone is part of a larger one.
4. Write the milestones on Post-It notes and stick them in the areas of work, in the order they occur. In this step, you may actually review the definition of the areas of work (Example 5.3).

5. Draw the logical dependencies, starting with the final objective and working back. This may make you to review the definition of milestones, add new milestones, merge milestones, or change the definition of the areas of work (Example 5.3).
6. Make a final drawing of the plan.

Example 5.3 Reviewing the areas of work

I was working with a project team who were developing a computer system. They started with four areas of work

- The hardware and software in the computer centre
- The computer network linking the computer centre to offices
- Furnishing of the offices
- Management procedures and people development

When they came to put the milestones into the areas of work, they found that there were several that didn't fit into those four areas. They included things like

- Agree the success criteria.
- Obtain approval for the budget.
- Mobilize the team.
- Measure achievement of the success criteria.

These were for them important project management milestones. So they created a fifth column to contain them. But when they came to draw the project network, they found that the computer network and office furnishing were so intertwined they combined those two areas of work and went back to four.

Work Breakdown Structure

The milestone plan, as shown in Fig. 5.2, is a communication tool to communicate the project strategy to the parties involved. It represents both the work and its logical relationship. However, we should not lose sight of the fact that we are developing level 2 of the WBS. Figure 5.3 shows the WBS tree, for the CRMO Rationalization Project. This can be represented as a simple list (Table 5.2).

5.4 PLANNING AT LOWER LEVELS

The milestone plan can be supported by plans at lower levels. These will include activity plans, work package scope statements, and subsidiary milestone plan.

Activity Plans

These detail the work packages which lead to the milestones, and describe the work at the next level of work breakdown. Following the principle of single-page reporting, the number of activities making up a work package should be limited to 15. I find six to ten a useful number. Figure 5.4 is an activity plan for Milestone P1 in the CRMO Rationalization Project.

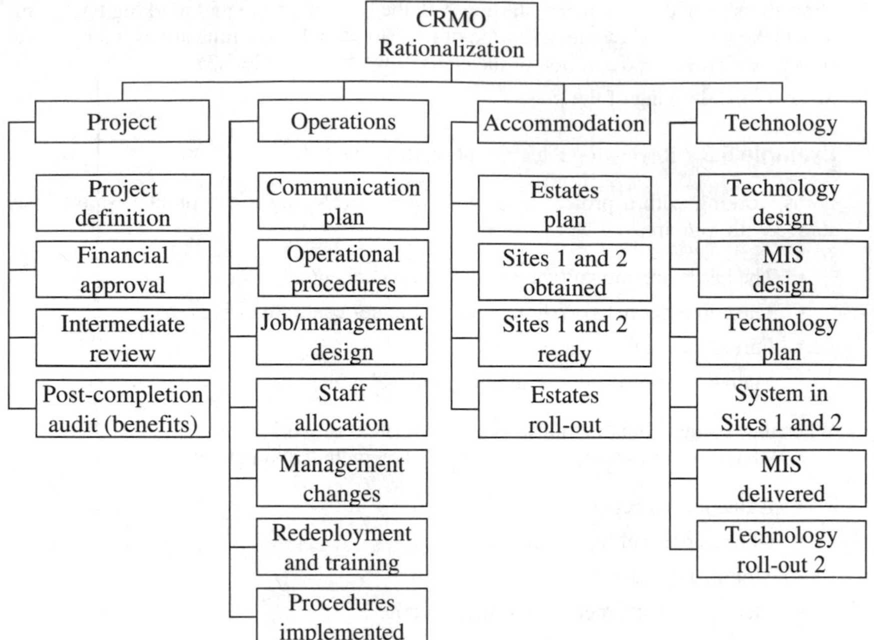

FIGURE 5.3 WBS for the CRMO Rationalization Project.

TABLE 5.2 WBS for the CRMO Rationalization Project

TriMagi Milestone List	
Accommodation	A1: Estates plan A2: Sites 1 and 2 obtained A3: Sites 1 and 2 ready A4: Estates roll-out
Technology	T1: Technology design T2: MIS design T3: Technology plan T4: System in sites 1 and 2 T5: MIS delivered T6: Technology roll out
Organization	O1: Communications plan O2: Operational procedures O3: Job/management design O4: Staff allocation O5: Management changes O6: Redeployment and training O7: Procedures implemented
Project	P1: Project definition P2: Financial approval P3: Intermediate review P4: Post-completion audit

TriMagi Activity Plan

Project:	Rationalization of the Customer Repair and Maintenance Organization				
Milestone:	D3: Personnel information pages designed				
Manager:	Rodney Turner	**Period:**	Week ending	**Target end:**	5-Mar

X	eXecutes the work
D	takes **D**ecisions solely/ultimate
d	takes **d**ecisions jointly
P	manages **P**rogress
T	on-the-job **T**raining
I	must be **I**nformed
C	must be **C**onsulted
A	may **A**dvise

			5-Feb		12-Feb		19-Feb		26-Feb		5-Mar		12-Mar				Duration	
No	**Activity Name**	1	2	3	4	5	6	7	8	9	10	11	12	13	14	15	d	**End Date**
1	Produce project proposal																	05-Feb
2	Hold definition workshop																	09-Feb
3	Define required benefits																	12-Feb
4	Draft definition report																	17-Feb
5	Hold launch workshop																	19-Feb
6	Finalize milestone plan																	24-Feb
7	Finalize responsibility chart																	24-Feb
8	Finalize risk assessment																	26-Feb
9	Finalize time estimates																	26-Feb
10	Finalize cost estimates																	26-Feb
11	Finalize revenue estimates																	26-Feb
12	Assess project viability																	26-Feb
13	Finalize definition report																	5-Mar
14	Mobilize team																	5-Mar

FIGURE 5.4 Activity plan for milestone P1 in the CRMO Rationalization Project.

Some people try to derive a full definition of the activities before any work is done. People who misuse networking systems, creating activity definition without the supporting WBS, are forced into this. However, most modern approaches to project management recommend what is called a *rolling-wave* approach to activity planning. This is core to the PRINCE2 methodology, for instance.[4] Fully detailed activity plans are only derived and maintained for those work packages which are current, or about to start. The detailing of later work packages is left until necessary, so that as much current information as possible is used to derive the activities. Some computer-based networking packages will support this approach by allowing the nesting of networks. There are several reasons for this approach:

1. You wait until you know you are likely to do the work before expending effort on detail planning. I spoke above of increasing the accuracy of the estimates during subsequent stages of the life cycle by spending increasing time on planning and design. To prepare estimates at project initiation stage you should estimate at the work-package level, and not prepare the activity definition. Some people find this uncomfortable, but I have worked in organizations which have prepared quite detailed designs and estimates for projects, only to find the project uneconomic.
2. You prepare detail activity plans when you have maximum information. If you prepare a detail plan for a yearlong project at the start, the only thing you can guarantee is you are wrong. You would have left out things which should be included, and included things which should be left out. It is better to prepare the detail activity definition when you have gathered information about the best way to achieve the milestone. This is especially true on development projects, where work in the early stages will determine work

in the latter stages. You will know what the later milestones are, if you are to reach your final objective, but you will not know how they are to be achieved. There is no point in trying to guess, because it serves no purpose and wastes time.

3. You can delegate the definition of activities to reach a milestone to the teams who will be undertaking the work.

Work Package Scope Statements

Although the detail activity planning is done on a rolling-wave basis, it is necessary to prepare some definition of the scope of each work package at an earlier stage. There are several reasons for this:

1. It is necessary to prepare some form of estimate of work content and duration for early, high-level estimating and scheduling. This should be based on some substance, even if it is only an approximate statement of the most likely outcome.
2. Work packages may include activities with a long lead time. These must be recognized and started in time.
3. While preparing the milestone plan you may decide that one proposed milestone is actually an activity in another milestone. This must be recorded.

These requirements can be satisfied by preparing work-package scope statements. These will be akin to the definition of scope and areas of work for the project as a whole, but on a smaller scale. The milestone name, remember, defines the purpose and objectives of the work package. The work-package scope statement can also include a measure of completion for configuration management purposes (Chap. 7). Table 5.3

TABLE 5.3 Work-Package Scope Statement for Milestone P1 for TriMagi's CRMO Rationalization Project

TriMagi Work-Package Scope Statement	
Milestone	P1: When the project plans have been prepared and resources assigned to the project.
Scope	The work package requires the preparation of high-level plans and estimates to be prepared, to enable resource budgets to be prepared, and their availability agreed.
Possible work	Identify key managers. Hold launch workshop. Finalize milestone plan and project responsibility chart. Estimate resource requirements and durations. Schedule resource requirements. Discuss requirements with managers. Plan and agree resource availability.
Measure of completion	Project plans approved by the steering committee. Resource managers sign agreements to resource availability.

contains a sample work-package scope statement for Milestone P1 in the CRMO Rationalization Project.

Subsidiary Milestone Plan

Sometimes there is a milestone which requires a particularly large amount of work. You may want to define intermediate milestones as control points through that work, but there may be no natural milestone to use on the level of the milestones plan. It is not sufficient to define milestones such as "when the work is 25 percent complete," because that is not measurable. In these circumstances, it may be worthwhile to derive a subsidiary milestone plan for that package of work. In effect, the work package is treated as a miniproject. Figure 5.5 is the milestone plan for developing a compiler for a computer language. Milestone C1 is of the type described, requiring 5 months of work to achieve it. However, there are no natural milestones on the level of this plan to define control points through the work. The team therefore derived a subsidiary plan (Fig. 5.6) for that milestone alone.

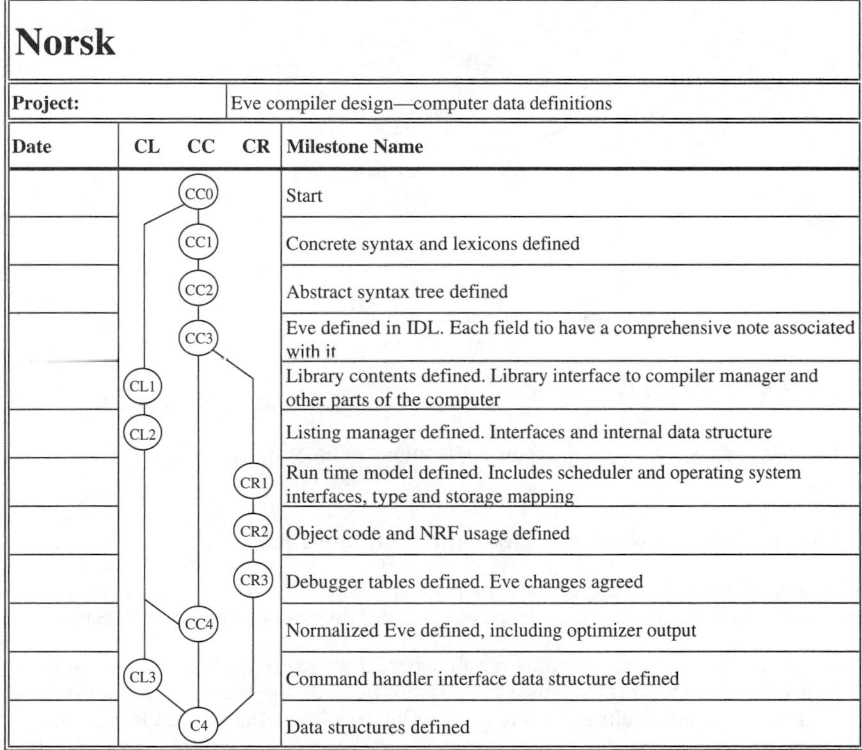

Norsk

Project:	Eve compiler design—computer data definitions			
Date	**CL**	**CC**	**CR**	**Milestone Name**
		CC0		Start
		CC1		Concrete syntax and lexicons defined
		CC2		Abstract syntax tree defined
		CC3		Eve defined in IDL. Each field tio have a comprehensive note associated with it
	CL1			Library contents defined. Library interface to compiler manager and other parts of the computer
	CL2			Listing manager defined. Interfaces and internal data structure
			CR1	Run time model defined. Includes scheduler and operating system interfaces, type and storage mapping
			CR2	Object code and NRF usage defined
			CR3	Debugger tables defined. Eve changes agreed
		CC4		Normalized Eve defined, including optimizer output
	CL3			Command handler interface data structure defined
		C4		Data structures defined

FIGURE 5.5 Milestone plan for developing a compiler language.

TriMagi Milestone Plan

Project:	Rationalization of the Customer Repair and Maintenance Organization
Project Sponsor:	Steve Kenny
Project Manager:	Rodney Turner

Date	P	O	A	T	Milestone Name	Short Name	End Date
5-Mar	P1				When the project definition is complete including benefits map, milestone plan, and responsibility chart	Project definition	
30-Apr				T1	When the technical solution including appropriate networking and switching technology has been designed	Technology design	
22-Mar		O1			When a plan for communicating the changes to the CRM Orgaization has been agreed	Communicaton plan	
15-May		O2			When the operational procedures in the CRM Offices has been agreed	Operational procedures	
31-May		O3			When the job design and management design is complete and agreed	Job and management design	
31-May				T2	When the functional specification for the supportiong management information system (MIS) has been agreed	MIS funcational spec	
15-Jun		O4			When the allocation of staff to the new offices, and recruitment and redeployment requirements have been designed and agreed	Staff allocation	
15-Jun				T3	When the technical roll-out stratgey has been defined and agreed	Technical roll-out plan	
15-Jun			A1		When the estates roll-out stratgey has been designed and agreed	Estates roll-out plan	
30-Jun	P2				When the budget for implementation has been determined and provisional fiananacial authority obtained	Financial approval	
15-Jul			A2		When sites 1 and 2 are available	Sites 1 and 2 available	
15-Jul		O5			When the management changes for sites 1 and 2 are in place (first call receipt and first diagnostic offices)	Management changes	
31-Aug				T4	When the system is ready for service in sites 1 and 2	Systems in sites 1 and 2	
31-Aug		O6			When a minumum number of staff have been recruited and redeployed and their training is complete	Redeployment and training	
15-Sep			A3		When sites 1 and 2 are ready for occupation	Sites 1 and 2 ready	
15-Sep				T5	When the MIS system has been delivered	MIS delivered	
30-Sep		O7			When sites 1 and 2 are operational and procedures implemented	Procedures implemented	
30-Nov	P3				When a successful intermediate review has been conducted and roll-out plans revised and agreed	Intermediate review	
31-Mar			A4		When the last site is operational and procedures fully implemented	Roll-out implemented	
30-Sep	P4				When it has been shown, through a post-implementation audit that all benefit criteria have been met	Post-completion audit	

FIGURE 5.6 Subsidiary milestone plan for milestone C1.

5.5 APPLICATIONS

I close this chapter by describing some applications of milestone planning:

Different Stages of the Project Management Life Cycle

Milestone plans can be prepared for work at all stages of the project life cycle, not just the execution stage. The management emphasis changes throughout each of these stages:

1. At the early stages, the emphasis is on encouraging creativity. The milestone descriptions should enable this by allowing maximum flexibility in the way the milestones are achieved, and the results delivered, while still providing a framework for control.
2. At the later stages, the emphasis will be on completing the work. Money is being spent, and so the benefits must be obtained as quickly as possible. Therefore the milestone names will be more prescriptive, providing more rigid control.

Large Multidisciplinary Projects

I have worked on several large multidisciplinary projects which for management purposes we divided into several subprojects almost independent of each other, and each the responsibility of a separate discipline. The project team derived a milestone plan for each subproject, and each discipline was then able to work virtually independently of the other, corresponding only at key milestones. I have applied this approach to construction projects, development projects, and IT projects.

North Sea Oil Field Development. This development consisted of two phases each of £3 billion. In the first phase, the project used well-known, mainframe-based project planning software, and planned at a fairly low level of detail. Management reports were 150 pages of computer output, and the management team had no visible control. In the second phase, it was recommended that they adopt a work breakdown structure. The development was divided into several contracts, and each contract into several stages, such as

- Feasibility
- Design
- Procurement
- Construction
- Linkup
- Commissioning

A milestone plan was prepared for each contract stage. The management team monitored against the milestone plans. The project teams supported these with lower level plans.

Regional Health Authority, Regional Distribution. The Health Authority was changing from distributing supplies through each of the 15 districts, to regionally coordinated distribution. The project was divided into 22 subprojects, each with its own milestone plan, and each the responsibility of a separate discipline. There were a few easily monitored links between each plan. The projects were:

- Construction of the regional warehouse
- Creation of the warehouse establishment
- Implementation of computer systems
- Recruitment, redeployment, and training
- Switching from district buying to regional buying
- Switching from district revenue to regional revenue
- District implementation (15 districts)
- Commissioning the warehouse

Each discipline met once every 2 weeks to monitor progress against their plan. The team leaders then met every 6 weeks to monitor progress of the project overall, by comparing progress on each plan.

Computerization of the Norwegian Securities Service. This program consisted of four subprojects:

- Design and implementation of the computer system
- Creation of a company to operate it

- Registration of dealers and holders of stock
- Legal basis

An overall milestone plan was developed for the program as a whole. Subsidiary milestone plans were also prepared for the first two subprojects. This project involved one million people, and yet was managed to a successful conclusion using manual planning methods only by taking this structured approach. At one point the Norwegian government tried to delay passing the enabling legislation by 12 months. Using the top-level plan, the project team was able to demonstrate to the minister that would delay the project by 12 months, and effectively kill it. The argument won the day and the bill was passed.

Customer Service System in a Regional Supply Company of a Large Public Utility. Implementation of the customer service system (CSS) required several projects:

- Implementation of hardware and software
- Transfer of data
- Networking of buildings
- Estates refurbishment
- Writing operating procedures
- Training
- Commissioning

Again, an overall milestone plan was developed, supported by milestone plans for each subproject.

Summary. All of these projects involve a mixture of

- Construction or building work
- Information systems work
- Organizational change
- Recruitment, redevelopment, and training

They were all PSO Projects. They each also had a duration of about 2 to 3 years, and each was finished on time and to cost, while just using simple planning methods.

SUMMARY

1. The purpose of scope management is to ensure
   - Adequate work is done
   - Unnecessary work is not done
   - The project's purpose is achieved
2. Work breakdown is a process by which the work of the project is subdivided for management and control purposes.
3. The project is defined at the strategic level, through
   - *The purpose*: The problem to be solved, or the opportunity to be exploited, and the benefit to be obtained

- *The scope*: The solutions to the problem, and covering the inclusions, (work within the remit of the project), and the exclusions (work outside the remit, because it is deemed unnecessary, or because it is shared with other projects)
- *The outputs*: The facility to be measured, quantitative and qualitative measures of when the project is complete

4. At the strategic level, the milestone plan
- Shows how the intermediate products, or deliverables, build towards the final objectives of the project
- Sets a stable framework and fixed goal-posts for the project team, and thereby provides a common vision
- Controls devolution of the management of the scope

5. A good milestone plan
- Is understandable to everyone
- Is controllable
- Focuses on necessary decisions
- Is logical
- Gives an overview to build cooperation and commitment of all the parties involved

6. There are seven steps in milestone planning:
- Agree the final milestone
- Brain-storm milestones
- Review the list
- Experiment with result paths (areas of work)
- Draw the logical dependencies
- Make the final plan

7. Plans at lower levels of work breakdown include
- Subsidiary milestone plans
- Work-package scope statements
- Activity plans developed on a rolling-wave basis

REFERENCES

1. Morris, P.W.G. and Hough, G.H., *The Anatomy of Major Projects: The Reality of Project Management*, New York, N.Y.: Wiley, 1987.
2. Gerrard, A.M., *Guide to Capital Cost Estimating*, Rugby, U.K.: Institution of Chemical Engineers, 2000.
3. Andersen, E.S., Grude, K.V., Haug, T., Katagiri, M., and Turner, J.R., Goal Directed Project Management, 3rd ed., London, U.K.: Kogan Page/Coopers & Lybrand, 2004.
4. Office of Government Commerce, *Managing Successful Projects with PRINCE2*, 4th ed., London, U.K.: The Stationery Office, 2005.

CHAPTER 6
MANAGING PROJECT ORGANIZATION

I now turn to the second function of project management, managing project organization. This is also a mandatory function: Without an organization there are no resources to undertake the project. Through the organization the manager defines the type and level of resource input required to achieve the project's objective. Once the organization has been defined, the project team can determine how much the project will cost and how long it will take, thus providing a baseline for managing quality, cost, and time. The definition of scope and organization together make a contract between the project and the parent organization, that is, between the contractor and owner in Fig. 6.1. It is through this contract that project managers negotiate their authority. The purpose of project organization is *to marshal adequate resources (human, material, and financial) of an appropriate type to undertake the work of the project, so as to deliver its objectives successfully*. The use of the word "adequate" implies that the resources should be of sufficient number, but only just sufficient: Too few, and the organization will be ineffective and the project will flounder; too many, and the organization will be inefficient. This chapter focuses primarily on human resources.

In the next section, I recall the principles of managing the project organization, and the processes of negotiating a contract between project and business. I then describe two levels of project organization. The first I call the external organization,[1] which is the relationship between the project and the parent organization. I describe types of external organization available, including a range of line, matrix, and versatile approaches. The second is the relationship between the project team members, which I call the internal organization. I then introduce the *responsibility chart* as the primary tool for defining the project organization and negotiating the contract, and show that this satisfies the principle of single-page reporting. In order to agree the contract, the responsibility chart requires the manager to identify both the type of resource input and the level of effort, the *work content*. I describe how to incorporate estimates of work content and close the chapter by explaining the use of equipment and drawing registers to manage nonhuman resources.

6.1 PRINCIPLES

Three of the five principles of good project management, introduced in Sec. 3.5, relate to managing the project organization:

- Negotiate a contract between parties involved.
- Assign roles and responsibilities at all levels of work breakdown.
- Adopt a clear and simple reporting structure.

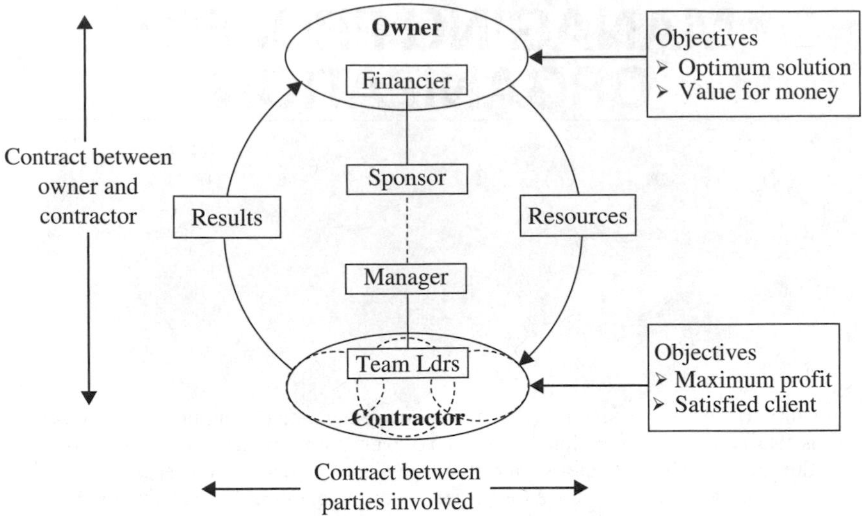

FIGURE 6.1 The owner-contractor model.

Negotiating a Contract

I focus on two elements of this: the nature of the contracts on two levels and negotiating them.

Organizing the Contract. Figure 6.1 is an enhancement of Fig. 1.2, the owner-contractor model. It illustrates that the project needs to be organized on two levels, and that we need to agree a contract at both levels (Table 6.1):

1. *Strategic level*: The first level defines the relationship between the project and the parent organization. It shows that the project will deliver beneficial change to the parent organization, and in order for it to be able to do that, the parent organization makes available resources, in the form of people, money, and materials. The contract agrees

TABLE 6.1 Two Contracts

Level	Planning and controlling scope	Organizing and implementing	Contract
Strategic or management External organization	The milestone plan defines what the project will deliver to the parent organization, and checks that it does	The responsibility chart defines the resources the parent organization will make available and when	Between the project and parent organization
Tactical or operational Internal organization	The activity plan defines what the project team will do to deliver the milestone they are responsible for	The responsibility chart defines how they will share the work between them	Between the project team members

the beneficial change and how the project will deliver that, represented by the milestone plan, and the resources that will be made available, and when, represented by the responsibility chart (Sec. 6.4). I call this the external organization,[1] the relationship between the project and its context.

2. *Tactical level*: The second level defines the relationship between the project team members, and how they will work together to do the work and deliver the results they are responsible for. The responsibility chart, used at the activity level, can also be used to represent this contract. I call this the internal organization,[1] the relationship between the people that are part of the temporary organization.

Negotiating. I said at the start of Chap. 3 that a necessary condition of project success is to agree the success criteria with all the stakeholders before you start. The project manager cannot impose these contracts on the parent organization or the project team members. They must be agreed by a process of discussion and negotiation. For me, the whole process of project planning is one of negotiation: Negotiating people's input and involvement to your project. People are not altruistic. They will only contribute to a project if they can see some benefit to themselves. You must work with people to help them understand what their involvement entails and to see the benefit. The benefit might be

- They can see the benefit of the project to the parent organization and want to work for a successful company.
- It is their job to work on projects, and their annual bonus may depend on it, so you want to convince them that your organization offers the best opportunities.
- If they help you, you will help them.

The project manager negotiates the contract by building a clear mission or vision for the project and cascading that mission down to objectives at each level of the Organization Breakdown Structure (OBS). Cooperation can then be gained by building a commitment to the objectives. The negotiation should go something like this

1. Do you believe that the purpose of the project is worthwhile?
2. Do you believe that to achieve that purpose we need to achieve the identified end and intermediate objectives?
3. Do you believe that it is the responsibility of your group to deliver some or all of those objectives?

If the answer to the first question is "no," the project manager needs to find some way of making the project of value to the people concerned. If the answer to the second question is "no," then you can involve the group of people in the planning process to gain their views. If the answer to the third question is "no," then you can gain their opinion on whose responsibility it might be. If you cannot gain agreement on the second and third question, then you must doubt the group's answer to the first, and work further on making the project beneficial to them.

Defining Roles and Responsibilities

The contract is defined by defining roles and responsibilities of the parties involved for the work elements at each level of breakdown. Many project management systems focus on just one role: Who is to do the work? There are several roles and responsibilities on a project. Table 6.2 lists some.

TABLE 6.2 Roles and Responsibilities

Responsibility	Role
For work	Who is to undertake the project's tasks
For management	Who is to take decisions
	Who is to manage progress
	Who is to guide and coach new resources
For communication	Who must provide information and opinions
	Who may provide information and knowledge
	Who must be informed of outcomes

Keep it Simple

Below, I introduce the responsibility chart as a single-page document to define resources and their input. It defines the contract at all levels of breakdown (Table 6.1), and is the document against which it is negotiated and agreed. The responsibility chart can be used to build cooperation and to ensure the novel organization of a project is brought into operation quickly and effectively. However, before describing the responsibility chart more fully, I describe the types of organization which can be used for managing a project.

6.2 THE EXTERNAL ORGANIZATION

There are several issues in choosing the external organization.

Types of Project Organization

Figure 6.2 illustrates a range of potential project organization, from line to matrix and back to line. The original work on which this model is based suggested five types of

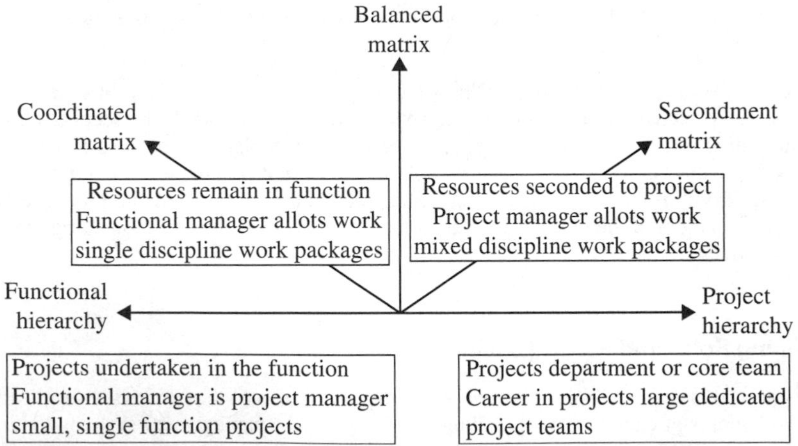

FIGURE 6.2 A range of project organization structures.

project organization.[2] However, I think that the middle one, the balanced matrix, is unstable, and so only suggest four. All four are appropriate in the right circumstances, and so you need to choose what is appropriate for your project. In fact you don't need to use just one organization type for a given project. You can mix and match, choosing a different organization type work package by work package. (I don't think that you will mix organization types within a work package.) It is sensible to choose work packages so that each work package requires a single organization type. The four organization types are as follows. First we consider the two types of line organization.

Functional Line. If the project is small enough, it can be organized wholly within the functional line organization. This only works if the project uses resources from just one group or department within the functional line organization and the resources can be drawn just from that one group. The manager of the group can then assign people from within his or her department to a project wholly within his or her department. If a limited number of resources are needed from another department, then the departmental manager can negotiate with the other departmental manager. It is his or her responsibility, not the project manager's, and it will rely on the personal relationship between the two departmental managers.

Project Line. Going next to the other extreme we look at the project line. If the project is big enough, or if the parent organization is a project-based organization, such as a construction company or software house doing nothing but projects for external clients, the parent company may create a project function within the company for doing projects. People will work permanently for the project function, and projects will be assigned to the project function for delivery.

For the vast majority of projects they are not small enough to fit just within one function and not big enough that all the project team members can work within the project hierarchy, and some form of matrix structure is necessary. Under the matrix structure, people from the line organization are given project responsibilities for the duration of their involvement in the project. However, I firmly believe that people should be receiving instructions from just one manager, either the project manager or line manager, and that is the fundamental difference between the two matrix structures I suggest.

Secondment Matrix. The project team member is seconded onto the project for the duration of his or her involvement in the project. While working on the project, he or she receives instruction from the project manager about what work he or she will do day by day. The project team member may only be working on the project for a limited period, for the duration of the work package only, and may only be working part-time, 3 days a week say, but while working on the project, he or she receives instruction from the project manager. This form of working is necessary if the work package involves the input of more than one type of resource. You cannot have several functional managers trying to coordinate the work of several different resources; you must have just one project manager.

Coordinated Matrix. If the work package involves the input of just one resource type, then it can be assigned to the functional manager to resource, and he or she can be made responsible for delivering the milestone by the due date. The resource manager may have work packages from several projects to assign people to, as well as ongoing functional duties, and can balance priorities between those different demands to deliver the project milestones within the requirements of the different projects. This only works if the work package involves the input of one function. It might work if it involves the input of one person from another function and there is a good working relationship between the two functional managers.

Balanced Matrix. Gobeli and Larson[2] also suggested the balanced matrix. Here the project manager and functional manager share responsibility, and the team member receives instruction from both. I don't think this will work; people can only have one boss.

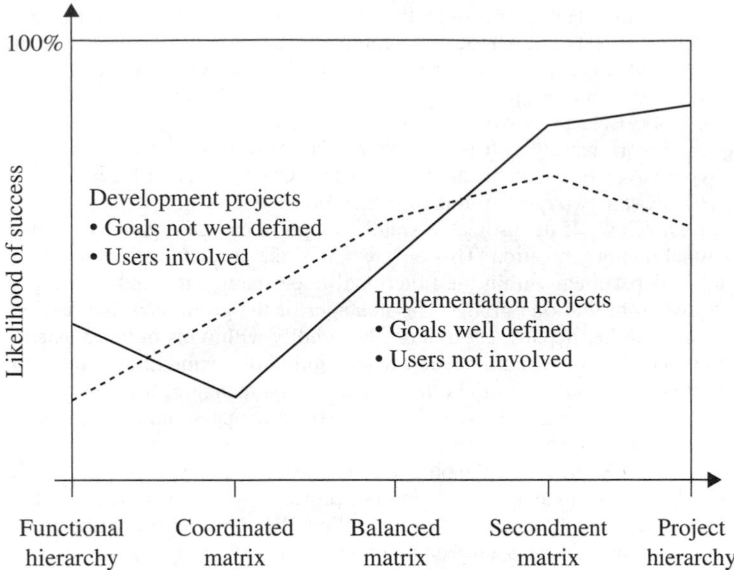

FIGURE 6.3 Success of the different organization types as measured by Gobeli and Larson.[2]

The project team members will try to play the two managers off against each other, and the more charismatic one will win, or the line manager will win because they control the annual bonus.

Gobeli and Larson[2] measured the success of about 2000 projects and matched it to the project organization type being used (Fig. 6.3). They divided the results between development projects and implementation projects. For the latter there is a clear preference for the project line. The design is finished, changes are expensive and should not be made, and so the project team should just focus on the task and get on with it without interference from users. But in the development phases the involvement of users is usually essential, so there is a preference for the secondment matrix. I don't think the low success rates for the functional line and coordinated matrix are due to faults in that approach. They work well if used properly. I assume it was because they were being used inappropriately.

As I said, you can mix and match organization types of projects (Example 6.1). Table 6.3 describes the project organization type to be adopted by TriMagi for their CRMO Rationalization Project. This shows that not only can you mix and match on a project, people from one department may be assigned to the project in different ways.

Example 6.1 Mixing project organization types

NASA uses a project hierarchy for coordinating the design, development, assembly, and launch of their satellites. Engineers responsible for designing and developing the technology to make the satellite work are assigned to the project on a secondment matrix basis. Scientists designing and developing the experiments to make the measurements work either on a secondment matrix basis or a coordinated matrix basis.

TABLE 6.3 Project Organization Type for TriMagi's CRMO Rationalization Project

TriMagi Background
Contractors doing building work will work on a project hierarchy basis.
Engineers and information systems professional responsible for the development of technology will be seconded on to the project.
Users and people from the human resources department writing job descriptions for workers in the new structure will be seconded on to the project.
People from the human resources department doing redeployment and training will work on a coordinated matrix basis.

Integrated versus Isolated Resources

Another issue with the secondment matrix and project hierarchy is whether people are moved physically to work on the project, or do they remain at their normal desk. I call these isolated or integrated resources. There are advantages and disadvantages of each approach.

Isolated. The advantages are that the project team member can work without distraction and on secret work. The disadvantages are that users seconded to the team losc contact with normal operations, users not seconded mistrust the project, operational managers are reluctant to release their best people, and it is inflexible (Example 6.2).

Integrated. The advantages and disadvantages above are reversed. To be successful this requires the manager to give his subordinate space to work on project tasks, the individual to focus on the task at hand without distraction, and the environment not to impose on the individual while working on project tasks.

Intermediate positions are possible, giving advantages of both models. Individuals seconded part-time to a project can be given a quiet room, close to their normal place of work, to use while working on the project.

Example 6.2 Isolated project teams

A public utility adopted this approach for the design and development of their integrated customer database system. People were seconded from the districts into a central design team. The development process took 2 years, at the end of which the design was 2 years out of date. Furthermore, many users seconded to the development team were given temporary promotions. When they returned to operations they expected their promotions to be made substantive, but were often of less use to their districts than before they left as their experience was now also 2 years out of date. However, the alternative, the integrated team, is extremely unlikely to have delivered the design in anything like 2 years, so the isolated approach was the only option.

When I described this story to a group of Russian managers on a course at Henley Management College, they said the people should have taken greater responsibility for managing their own careers. When I described it to a group of managers from the company concerned, one of them said he had been a member of the task force, and he had tried to manage his career, but still his earlier boss did not want to know. He changed departments.

The Versatile Project Organization

The forgoing discussion described how to overlay a project organization onto an existing functional, hierarchical, line management structure. Up until the early 1990s, this represented the vast majority of organizations. It may still represent a simple majority, but many organizations are now project-based, and some have adopted flexible, versatile approaches.[3,4,5] Indeed, Roland Gareis[5] argues matrix organization structures will not work, because of the inherent conflict of loyalties of project team members as described above.

In the pure project-based organization, the firm does away with the functional hierarchy, and people belong to project teams only. This was a popular approach in the late 1980s. However, Anne Keegan and I showed that it can cause the parent organization to lose cohesiveness.[4] You also need functions to be responsible for knowledge management. Without functions the organization can forget how to do its business. Reza Peymai and I suggested the adoption of a versatile organization[3] (Fig. 6.4). The versatile organization assumes the parent organization is operating in the top half of Fig. 1.10. Most people belong either to process teams or project teams. However, in the background, supporting the teams, and providing people to the teams, is the functional organization. Both types of team are the primary medium by which work is done and products delivered to customers. Process teams do fundamentally routine work, project teams do fundamentally novel work (and some teams do work that is somewhere in between). However, the size and composition of both types of team is constantly changing. The project teams are unique, novel, and transient. But even process teams need to change as customers' requirements, though fundamentally repetitive, can still vary. The idea of some quality nerds that an organization's procedures are unchanging is quite absurd (Chap. 7). Different customers have different requirements and hence processes and procedures need to constantly adapt, and indeed since quality is about continuous improvement, they must be constantly enhanced. As the size and composition of teams change, people move between them, or between them and the functional organization. In the versatile organization, process teams effectively operate as a coordinated matrix and project teams as a secondment matrix. However, in the versatile organization people have only one boss; they either belong to a team, in which case they take their instructions from the team leader, or they are in the functional organization, in

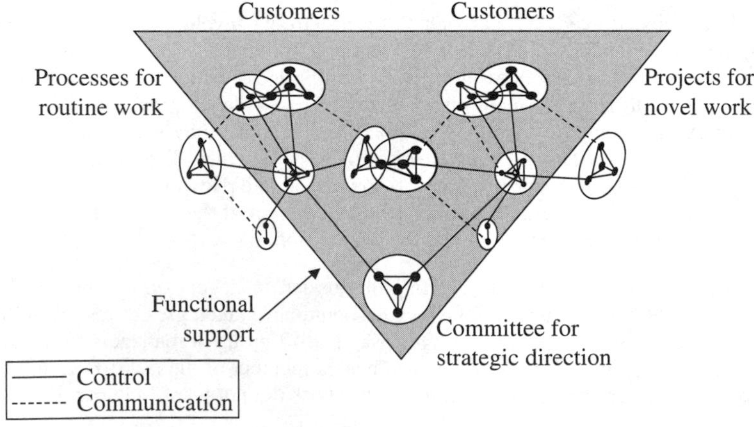

FIGURE 6.4 The versatile project organization.

which case they take their instructions from the functional manager. The advantages of the versatile approach are

- The size and composition of the teams can be changed in response to changing customer requirements, enabling the organization to reengineer its capabilities quickly.
- It provides a process focus; the organization's procedures are written to describe how it processes products to satisfy customers' requirements, not how functions perform, enabling responsiveness to changing customer requirements.
- People have one boss, avoiding the problems of split loyalties.
- It retains the functional organization, avoiding the problems outlined below which arise with the pure project-based approach.

It is essential to retain the functional organization, because

- It provides a career structure. Transient teams cannot provide a career, just learning experiences as part of a person's development.[6]
- It retains the knowledge of the organization.[7]
- It provides a resting place between projects. The chance of one project starting as another ends is slight, and so between projects people can spend time capturing their knowledge.
- It can share people between projects when they are only partly utilised.
- It develops new systems and procedures. Systems and procedures are an overhead, and since each project manager will try to minimise the cost of his or her project will not develop new ones.

Without a functional organization structure, with time the organization loses its knowledge and culture, and withers and dies.

6.3 THE INTERNAL ORGANIZATION

The responsibility chart described in the next section is used to define the internal organization. What I wish to focus on here is the use of different team types and leadership styles at different stages of the project. In a now classic work, David Frame defined four leadership styles for use on projects (Table 6.4), based on three parameters.[8]

1. How much the manager involves the team in formulating decisions (the first five steps of Fig. 1.6)?
2. How much the manager involves the team in selecting the option for implementation (the sixth step in Fig. 1.6)?
3. The manager's flexibility.

TABLE 6.4 Four Leadership Styles

Leadership style	Decision making	Decision taking	Flexibility
Laissez-faire	High	High	High
Democratic	High	Low	High
Autocratic	Low	Low	High
Bureaucratic	Low	Low	Low

TABLE 6.5 Leadership Styles and Team Types through the Project Life Cycle

Stage	Team members	Team type	Leadership style
Feasibility	Experts	Egoless	Laissez-faire
Design	Professionals	Matrix	Democratic
Execution	Implementers	Line	Autocratic
Close	Task force	Surgical	Bureaucratic

Frame also showed that different styles are appropriate at different stages of the life cycle and correspondingly different team types are appropriate (Table 6.5).

Feasibility: During feasibility, the team will be composed of experts, most of whom will be more expert than the project manager in their particular field. On some feasibility studies, for instance in the preparation of a proposal for a larger contract, some experts may actually be more senior managers than the manager of the study. The study manager's expertise is in the management of the study, and so his or her role is to guide the study. On the preparation of a proposal the team may comprise a design manager, contracts manager, and potential project manager for the implementation phase, all more senior than the bid manager from the commercial department. So the manager cannot direct or instruct the team. The team will be fully involved in formulating decisions and choosing the options for implementation. The manager's role is just to lead and guide the team.

Design: During design, the team comprises technical professionals. The nature of the team is a matrix. That means there are several technicians of different expertise, working simultaneously on the design of several components of the asset (perhaps relating to different areas of work). So the way the team relates to the task is a matrix. This is not matrix management; there may only be one design manager. The designers formulate the design of the asset, so need to formulate the decisions. But the design manager has to coordinate both their designs and the design of the asset, so the design manager has to choose the final options in a consistent manner. I call this democratic management. The design manager listens to what the team members suggest, and his or her final decisions are influenced by the team members' opinions, but the design manager takes the final decision and imposes it.

Execution: During execution the team may comprise a project hierarchy; there are several task forces working on different components of the asset. The team members are implementers, with people from several disciplines in one task force working on a particular component of the asset. By this stage of the project all the decisions should have been made and the design finalized. Changes now cost money and cause delay. There is no such thing as a "nice to have" any more. No changes should be made unless not making the change will cause the project to fail. The execution manager should be an autocrat; just tell the team what to do and get on with it. However, the manager still needs to be flexible to respond to risk.

Close: During close out, the size of the team reduces, to the point where there is just one task force working on the whole asset, pulling the final threads together and commissioning it. David Frame describes this as a surgical team. The commissioning manager is like a surgeon, operating on the patient (the asset) supported by several different professionals with specific roles to play. During commissioning there will be a number of checklists to go through: outstanding work, quality checks, client acceptance tests. These checks need to be worked through in a bureaucratic way, making sure there are ticks in all the right boxes.

Can one person change their style, as they move through the project? Sometimes they have to (see Example 6.3). Above, I called the four managers the study manager, design manager, execution manager, and commissioning manager, and perhaps the project manager may have deputies whose styles are compatible to the needs of the current stage of the project.

Example 6.3 Changing leadership style

An MBA student of mine devised two tests, one to determine the manager's preferred style, and another to determine what style the team would like from their manager. He looked only at how much the manager involved the team in each of decision-making and decision-taking, and so only looked at the first three styles. He gave the test to 10 managers, each of whom had two teams, to see if the manager's style was compatible with what the teams wanted. On a scale of 0 to 10, one manager scored (10,7), a very laissez-faire style. One of his teams wanted (10,10), so his style was quite compatible. But the other team wanted (0,0), totally autocratic management. Perhaps what they really wanted was (3,3) but were being driven to being more extreme in frustration at the manager's laissez-faire style.

6.4 RESPONSIBILITY CHARTS

The use of responsibility charts to define the project organization is widespread. The PMI PMBoK calls them Responsibility Assignment Matrices.[9] Typically a chart is a matrix with deliverables shown as rows and organizational units as columns. Symbols are placed in the body of the matrix to represent the involvement of each resource type in the work element required to produce the deliverable. The matrix can be used at any level of breakdown. This provides a one-to-one correspondence between the levels in the product breakdown structure (PBS) and the OBS (as one might expect). Even though the responsibility chart is a matrix, it can be used to describe any one of the four organization types, or any mixture of them or it can be used with the versatile organization. The use of a responsibility matrix does not imply a matrix organization.

Use of Symbols to Describe Types of Involvement

Different authors suggest a range of different symbols to use in the responsibility chart: numbers, letters, or geometric shapes. I find letters that suggest the nature of the role the most useful and suggest the letters shown (Table 6.6).[10] The letters are used as follows:

TABLE 6.6 Use of Symbols in the Responsibility Chart

Letter	Responsibility
X	e*X*ecutes the work
D	takes *D*ecision solely or ultimately
d	takes *d*ecision jointly or partly
P	Controls *P*rogress
T	Provides *T*uition on the job
C	must be *C*onsulted
A	available to *A*dvise
I	must be *I*nformed

Responsibility for Work. *X*: e*X*ecutes the work. This is self-explanatory.

Management Roles. *D* takes *D*ecision solely or ultimately and *d* takes *d*ecision jointly or partly. There are various modes of decision-making (Table 6.7). An example of D2 might be the selection of a financial management system. The financial manager agrees it meets the company's financial requirements. The IT manager agrees it meets the company's systems strategy. If they fail to agree the decision is referred to the financial director, their joint boss. In decision D3, there can be a fine line between being consulted, C (as shown in decision D4), and truly closing options, d. This may be the case with the trade union representatives with no authority but significant disruptive power. You have to use the symbols to represent the way you want to manage your project.

P controls *P*rogress. This is the person responsible for ensuring that the work is planned, organized, implemented, and controlled. The project manager is ultimately responsible, but uses the symbol to delegate responsibility at lower levels of the WBS.

T provides *T*uition on the job. This recognizes that the people doing the work may not have sufficient skill, so they are coached on the job. As their skill grows the *T* may change to *P*.

Communication Channels. *C* must be *C*onsulted. These people must be consulted in the course of the work. They have information or opinions which the project must take account of. However, they do not have decision-taking responsibility: Their opinions can be ignored.

A is available to *A*dvise. These people may have information or opinions which the project team may want to use, but cannot know until they reach that part of the project. In effect the symbol represents "may be consulted."

I must be *I*nformed. These people must be provided with information about the outcome on one part of the project to enable them to do work or take a decision on another part.

C, *I*, and *A* control the flow of information. If people feel they should be consulted or informed that is negotiable as part of the contract.

The symbols must be used flexibly and imaginatively. Nothing is served by being pedantic. The project team paint the picture they want to paint, and use the chart as a communication tool. For instance, in a training course is the trainer *T* and the tutee *X*, or is the trainer *X* and the tutee *I*? It doesn't matter as long as everybody understands.

Use of the Responsibility Chart

The responsibility chart can be used at all levels of the breakdown: the integrative, strategic, and tactical. I have already suggested how the charts at the strategic and tactical levels

TABLE 6.7 Four Modes of Decision-Taking

	Person			
Decision Mode	A	B	C	Description
D1	D			A takes the decision alone.
D2		D	d	B and C share the decision. If they agree the decision stands. If not it is referred up the usual management channels.
D3	D	D	d	B and C close options and recommend. A has the ultimate authority.
D4	D	d	C	C's opinion must be sought, but can be ignored. B closes options and A has the ultimate authority.

help define the contract at the organizational and team levels and so define the external and internal organization, respectively (Table 6.1). But it can also be used at the integrative level to define project procedures and integrate the project into higher levels of planning. I discuss each in turn.

Project Level: Procedural Responsibility Chart. At this level, the chart can be used to define procedures, principles, or policies for managing the project. For example, that may be

- Procedures for monitoring and control
- Change control procedures
- Quality control procedures
- Configuration management procedures

Figure 6.5 is a procedure for monitoring and control. At this level, the chart might also be used to show how the project integrates into a program or portfolio of projects.

Strategic or Milestone Level: Project Responsibility Chart. I call the chart at this level the project responsibility chart. It is used to define roles and responsibilities for achieving each milestone. The resources at this level of breakdown tend to be departments, management responsibilities, and external companies (contractors or consultants). Figure 6.6 is a chart for the CRMO Rationalization Project. Figure 6.6 also includes a time schedule. I think in most projects, the schedule can be drawn in manually; you don't need fancy mathematics. You cannot know how long the work will take until you know who will do it, so it is only after filling in the responsibility chart that you can draw the schedule. I discuss scheduling further in Chap. 9.

Tactical Level: Activity Schedule. At this level, the chart defines roles and responsibilities of named people to do work to achieve a milestone. Because activity schedules are prepared on a rolling-wave basis during execution, the people involved can be named. They are unlikely to change on the timescale of a work package, and if they do the work should be replanned. Furthermore, because the activities are now more certain, more effort can be put into ensuring that the chart is correct. Figure 6.7 is an activity schedule for Milestone P1 in the CRMO Rationalization Project.

Developing the Responsibility Chart

I described in Sec. 5.3 how the milestone plan is best developed through group working, specifically at a project launch or definition workshop (Chap. 12). The same applies to the responsibility chart. You can draw a blank form on to a whiteboard or flip chart. Or you can project an excel spreadsheet on to a whiteboard using a data projector. The chart can then be filled in with everyone engaged. One person entering the symbols directly on to a paper form or into an excel spreadsheet that no one else can see can isolate members of the group, with the result that they may not accept the end product. But using the open approach, as the chart is completed line by line there can be a huge amount of discussion about whose responsibility each item is. But when a person allows a symbol to remain under his or her name, they internalize the result, and accept that as their responsibility; or if a symbol is under somebody else's name he or she accepts he or she is not involved. Estimates and schedules can be entered on the projected form in the same way.

This is one of several forms where half the benefit comes from the process of filling the form. However, it is worth copying the responsibility chart down. This is a very effective

TriMagi Procedural Responsibility Chart

Project:	Procedure for monitoring and control
Project Sponsor:	Steve Kenny
Project Manager:	Rodney Turner

Period: Day **Target end:**

X	eXecutes the work
D	takes Decisions solely/ultimately
d	takes decisions jointly
P	manages Progress
T	on-the-job Training
I	must be Informed
C	must be Consulted
A	may Advise

No	Milestone Name	Project manager	Team leaders	Project members	Project support office	Steering committee	Project sponsor	Friday week 0	Monday week 1	Tuesday week 1	Wednesday week 1	Thursday week 1	Friday week 1	Monday week 2	Tuesday week 2	Wednesday week 2	Thursday week 2	Friday week 2	Monday week 3	Tuesday week 3			Duration
								1	2	3	4	5	6	7	8	9	10	11	12	13	14	15	d
	Develop milestone plan	PX	X		I	d	D																
	Create high-level network	PX			X																		
	Develop new activity schedules	DP	X	X	I																		
	Update network	P	C	C	X																		
	Issue work-to lists			I	X																		
	Do work	P	PX	X																			
	Return turnaround documents		P	X	I																		
	Activity review meeting	I	PX	X																			
	Identify variances (activities)	I	PX	X																			
	Plan recovery	DP	X	X	I																		
	Issue activity progress reports	PI	X																				
	Review progress against milestone	PX			X	I																	
	Milestone progress meeting	PX			X	X																	
	Identify variances (milestones)	PX			X	DX																	
	Plan recovery	PX	X		X	DX																	
	Issue milestone progress report	PX			X	C	I																
	Approve progress						D																

FIGURE 6.5 Procedural responsibility chart for monitoring and control.

TriMagi Project Responsibility Chart

Project:	Rationalization of the Customer Repair and Maintenance Organization
Project Sponsor:	Steve Kenny
Project Manager:	Rodney Turner

X	eXecutes the work
D	takes **D**ecisions solely/ultimately
d	takes **d**ecisions jointly
P	manages **P**rogress
T	on-the-job **T**raining
I	must be **I**nformed
C	must be **C**onsulted
A	may **A**dvise

No	Milestone Name	Regional board	Operations director	CRMO managers	CRMO team leader	CRMO staff	Project manager	Project support office	Estates manager	Estates department	Network manager	Networks department	IS department	Operators	Personnel	Suppliers	
P1	Project defintion	D	D	dX	dX	I	PX	X	X	I	X	I	C	C	C		
T1	Technology design					C					PX	X	X	X		A	
O1	Communicaton plan	I	D	d	PX												
O2	Operational procedures	I	D	d	PX	X											
O3	Job and management design	I	D	d	PX	C									TX		
T2	MIS functional spec	I	D	d	dX						PX		X				
O4	Staff allocation	I	D	d	PX	C									TX		
T3	Technical roll-out plan	D	d	C			C	X	C	I	PX	X	X	I		C	
A1	Estates roll-out plan	D	d	C	X	I	C	X	PX	X	C	I	ISD	I	I	C	
P2	Financial approval	D	d	I			PX	X	C		C		C	A	A	C	
A2	Sites 1 and 2 available		I	I	I				PX	X	I						
O5	Management changes	I	DX	X	PX	I											
T4	Systems in sites 1 and 2		I	I	I					X	PX	X	X	I		X	
O6	Redeployment and training			D	PX										TX		
A3	Sites 1 and 2 ready		I	I	X	X	F			X		X	X	I		X	
T5	MIS delivered	I	D	I	X						PX		X			X	
O7	Procedures implemented	D	D	PX	X							A	A	I	X		
P3	Intermediate review	D	d	C	C		PX	X	A		A		A	A	A		
A4	Roll-out implemented	I	D	dX	dX	X	PX	X	I	X	I	X	X	I	X	X	
P4	Post-completion audit	D	d	C	C		PX	X						C			

Project Schedule

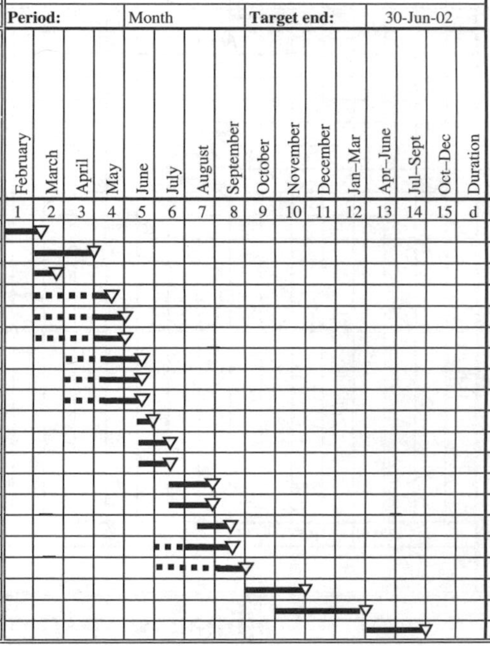

Period:	Month	Target end:	30-Jun-02

No	February	March	April	May	June	July	August	September	October	November	December	Jan–Mar	Apr–June	Jul–Sept	Oct–Dec	Duration	End Date
	1	2	3	4	5	6	7	8	9	10	11	12	13	14	15	d	
P1																	5-Mar
T1																	30-Apr
O1																	22-Mar
O2																	15-May
O3																	31-May
T2																	31-May
O4																	15-Jun
T3																	15-Jun
A1																	15-Jun
P2																	30-Jun
A2																	15-Jul
O5																	15-Jul
T4																	31-Aug
O6																	31-Aug
A3																	15-Sep
T5																	15-Sep
O7																	30-Sep
P3																	30-Nov
A4																	31-Mar
P4																	30-Sep

FIGURE 6.6 Project responsibility chart for CRMO Rationalization Project.

TriMagi Activity Plan

Activity Schedule

Project:	Rationalization of the Customer Repair and Maintenance Organization
Milestone:	D3: Personnel information pages designed
Manager:	Rodney Turner

Period: Week ending **Target end:** 5-Mar

X	e**X**ecutes the work
D	takes **D**ecisions solely/ultimately
d	takes **d**ecisions jointly
P	manages **P**rogress
T	on-the-job **T**raining
I	must be **I**nformed
C	must be **C**onsulted
A	may **A**dvise

No	Activity Name	Regional board	Operations director	CRMO managers	CRMO team leader	Project manager	Project support office	Estates manager	Estates department	Network manager	Networks department	IS department	Operators	Personnel	End Date
1	Produce project proposal	C	D	d	dX	PX	A	A		A		A	A	A	5-Feb
2	Hold definition workshop		DX		X	PX	X								9-Feb
3	Define required benefits	C	D	d	dX	PX									12-Feb
4	Draft definition report	C	D	d	dX	PX	X	I		I		I	I	I	17-Feb
5	Hold launch workshop	C		X	dX	PX	X	X		X		X			19-Feb
6	Finalize milestone plan		D	d	D	PX	X	C		C		C	A	C	24-Feb
7	Finalize responsibility chart		D	d	D	PX	X	C		C		C	A	A	24-Feb
8	Finalize risk assessment		D	d	dX	PX	X	C		C		C	C	C	26-Feb
9	Finalize time estimates				A	P	X	A		A		A		A	26-Feb
10	Finalize cost estimates				A	P	X	A		A		A		A	26-Feb
11	Finalize revenue estimates		A	A	A	PX									26-Feb
12	Assess project viability		D	d	d	PX									26-Feb
13	Finalize definition report	D	d	d	d	PX	X	C		C		C		C	5-Mar
14	Mobilize team		D	d	dX	PX	X	X	I	X	I	IX		I	5-Mar

Schedule columns: 1 2 (5-Feb) 3 4 (12-Feb) 5 6 (19-Feb) 7 8 (26-Feb) 9 10 (25-March) 11 12 (12-March) 13 14 15 d (Duration)

FIGURE 6.7 Activity schedule for milestone P1.

tool for communicating who should do what and when. The milestone plan is very good for planning the project process and strategy, and as we shall see later for tracking process. But it is not very effective for assigning work to people. The responsibility chart is best for that. Also it represents the contract between the project and the organization, so if at some later time a manager starts reneging on his or her agreement, the chart can be used to remind them of his or her previous commitment.

Incorporating Work Content

In negotiating the contract between project and business, it is necessary to include estimates of the resource requirements. Functional managers cannot commit to releasing resources without knowing what the requirement is. Two of the eight roles and responsibilities primarily consume resource:

X: e*X*ecutes the work

C: must be *C*onsulted

Estimates for these need to be included in the responsibility chart and agreed with functional managers. I describe further in Chap. 8 how to estimate the work-content and represent it using the responsibility chart. In Chap. 9, I describe how this can be used to estimate the duration of each package of work, and represent that using the chart.

SUMMARY

1. The purpose of project organization is
   - To marshal adequate and appropriate resources
   - To undertake the work of the project
   - To successfully deliver its objectives
2. The principle tools and techniques of organization management are
   - The contract between the parties involved
   - Organization breakdown structure, matching work breakdown
   - Responsibility charts
3. The project needs to be organized and the contract needs to be agreed on two levels:
   - The strategic level: agreeing the relationship between the project and the parent organization, giving the external project organization
   - The tactical level: agreeing how the project team will work together, giving the internal project organization
4. There are four types of external project organization
   - Functional line
   - Coordinated matrix
   - Secondment matrix
   - Project hierarchy
5. The versatile organization provides a flexible approach to creating project organizations.
6. Different leadership styles are appropriate at different stages of the project
   - Laissez-faire during feasibility
   - Democratic during design
   - Autocratic during execution
   - Bureaucratic during close out

7. Eight types of role or responsibility are suggested for use in the responsibility chart:
   - *X* e*X*ecutes the work
   - *D* takes *D*ecision solely or ultimately
   - *d* takes *d*ecision jointly or partly
   - *P* controls *P*rogress
   - *T* provides *T*uition on the job
   - *C* must be *C*onsulted
   - *A* is available to *A*dvise
   - *I* must be *I*nformed
8. The contract requires recording of estimates of work content, so that resource providers can commit to release of their people.
9. Drawings, materials, plant, and equipment are managed using registers, lists against the activities in which they are required.

REFERENCES

1. Andersen, E.S., "Managing project organization," in Turner, J.R. (ed.), *The Gower Handbook of Project Management*, Aldershot, U.K.: Gower, 2007.
2. Gobeli, D. and Larson, E., "Relative effectiveness of different project structures," *Project Management Journal*, 18, 81–85, 1989.
3. Turner, J.R. and Peymai, R., "Organizing for change: a versatile approach," in Turner, J.R., Grude, K.V., and Thurloway, L., (eds), *The Project Manager as Change Agent*, London, U.K.: McGraw-Hill, 1996.
4. Turner, J.R. and Keegan, A.E., "The versatile project-based organization: governance and operational control," *European Management Journal*, 17(3), 296–309, 1999
5. Gareis, R., Mainz, Happy Projects!, Vienna, AT: Mainz, 2005.
6. Turner, J.R., Huemann, M., and Keegan, A.E., *Human Resource Management in the Project-Oriented Organization*, Newtown Square, Pa.: Project Management Institute, 2008.
7. Turner, J.R. and Keegan, A.E., "Managing technology: innovation, learning and maturity," in P.W.G. Morris and J.K. Pinto (eds.), *The Handbook of Managing Projects*, New York, N.Y.: Wiley, 2004.
8. Frame, J.D., *Managing Projects in Organizations*, 3rd ed., San Francisco, C.A.: Jossey-Bass, 2003.
9. Project Management Institute, *A Guide to the Project Management Body of Knowledge*, 3rd ed., Newtown Square, Pa.: Project Management Institute, 2004.
10. Andersen, E.S., Grude, K.V., Haug, T., Katagiri, M., and Turner, J.R., *Goal Directed Project Management*, 3rd ed., London, U.K.: Kogan Page/Coopers & Lybrand, 2004.

CHAPTER 7
MANAGING QUALITY

The last two chapters described two mandatory project management functions: managing scope and project organization. Let us now turn to three secondary functions or constraints: managing quality, cost, and time. Contrary to common practice, they will be addressed in that order, which is the order I believe they should be addressed during project definition. You cannot know how much it will cost nor how long it will take until you know the desired quality standards.

This chapter addresses quality. I start by considering what we understand by good quality in the context of projects. I then introduce a five-element model for achieving good quality, and describe each element of the model. I describe configuration management, which I believe is the key tool of project management for delivering the quality and functionality of the project's outputs.

7.1 QUALITY IN THE CONTEXT OF PROJECTS

It is popular to say a project is successful if it is finished on time, to cost, and to quality. We all understand how we measure cost and time, but very few people understand what they mean by good quality in the context of a project. Indeed, in spite of it being stated as one of the major three criteria of project success, surprisingly little is written about it. There are several possible definitions of good quality on a project. The project is said to be good quality if the project's output, thc new asset

- Meets the specification
- Is fit for purpose
- Meets the customer's requirements
- Satisfies the customer

Meets the Specification. The facility is produced in accordance with the written requirements laid down for it. The requirements can be specified on several levels, mapping onto levels of product breakdown structure (PBS): customer, functional, system, and detail requirements. The requirements may specify engineering or technical design standards applied within the organization. (The word specification tends to be used for something which is project specific and standards for something which applies to all projects undertaken by the organization.) The specification may also set requirements for the time and cost of the project, needed to make it viable, and also set specific parameters for the service levels required to be met by the facility. Finally, there are the various abilities of the facility: availability, reliability, maintainability, adaptability, and the like.

Is Fit for Purpose. The facility, when commissioned, produces a product which solves the problem, or exploits the opportunity intended, or better. It works for the purpose for which it was intended, and produces the desired outcome.

Meets the Customer's Requirements. The facility meets the requirements the customer had of it. Here we mean what the customer thinks they require, the thoughts they had, not the way they vocalized their thoughts as words, and not the way those words got written down as a customer requirements specification.

Satisfies the Customer. The facility and the product it produces make the customer feel satisfied. Now there is also a difference between satisfying the customer, "that's alright then," and delighting the customer, "that's wonderful." If you can delight the customer at very little extra cost, then obviously you should try to do that. However, if that is going to make your project significantly unprofitable, then clearly you should aim only to satisfy the customer. If you still cannot make a profit, you need to massage the customer's expectations to make them more realistic.

Questions

These four definitions of quality raise several questions:

Do They Mean the Same Thing? The answer to this is quite "no." I implied above the concept the customer had in his or her mind and what was written as the "customer requirements" specification are almost certainly not the same thing. Human fallibility being what it is, the chances of the customer being able to vocalize his or her actual requirements is small, and the chances of the project team writing what the customer says down, let alone capturing the customer's unvocalized concepts, is also small. A series of gaps builds up. The customer has a problem, which he or she solves in his or her mind. That is the first gap, between the real problem and the customer's imagined solution. He or she then finds a contractor and tells them his or her ideas. There are the second and third gaps. Psychologists will tell you it is impossible to perfectly vocalize your thoughts and so the second is between what the customer thinks and what he or she says. The third is between what the customer says and what the contractor hears. The contractor writes down what they hear as the specification, and that is the fourth gap, the difference between what they hear and what gets written down. Thus the chance that the specification perfectly represents what is required is small. Hopefully it is close, but it is likely not to be perfect.

What Then Is the Correct Definition of Quality? The widely accepted definition of good quality is now taken as delivering project objectives that are fit for purpose, that is, they work to achieve the desired result. It is not slavishly delivering the specification, if what is specified will not work, and it is certainly not following predefined business processes, if those processes deliver a product that will not work. (And it is most certainly not about delivering something that won't work to time and cost.)

Does This Mean We Have to Change the Specification? Yes is the simple answer. This is one of the great dilemmas of project management. There are traditional project managers who say good project management is freezing the specification on day one of the project, and then delivering it. That, in my view, is not good project management, if the end product does not deliver the desired result. On the other hand, if you change the specification frequently, you will never finish the project, and that is most definitely not good project management either. Hence, you must be willing to change the specification as you become aware that your original proposal is less than perfect, but changing it is something you must do sparingly and with great ceremony. Later in this chapter I describe *configuration management,* a technique by which the specification can be refined in a controlled manner as the project progresses to ensure that by the end of the project, it produces the desired results.

Who Is the Customer? Is it

- The sponsor or owner of the facility?
- The operators of the facility, or users of the services it provides?
- The consumers of the eventual product it produces?
- The media, or local community, or politicians?

The answer is they are all customers, and all their requirements must be satisfied. They will usually have different requirements and to satisfy them all will be a difficult juggling act. The owners must be willing to pay for it. The operators must believe it will work, and can make failure a self-fulfilling prophecy. Consumers must want to buy the product. Configuration management can be used to gain agreement from the various parties, warring factions, as the project progresses. I said in Chap. 3 that you must agree the success criteria with the stakeholders before you start, and at configuration review points throughout the project.

Do You Give the Customer What They Want or What They Need? This is another dilemma. The attitude of engineers in the 1970s was to give customers what they needed, not what they wanted; that they knew better than their customers what their requirements were. This is an arrogant attitude; it is arrogant to think you know better than your customers and it is arrogant to think you are unfailingly correct. By the late 1980s the attitude had changed. It now did not matter what trivial whim the customer had, the "customer was king," give them what they ask for. On the one hand, you give the customer what you think they need. They look at the product, say "that's not what we asked for," and refuse to use it. On the other hand, you give them what they say they want. When it does not work, you say "the customer is king," and they say "but it was your duty to advise us it would not work." The way out of this dilemma is you must use configuration management so that by the end of the project what the multiheaded customer now thinks they want, what they actually need, and what you think they need are the same thing.

What Is the Difference between Good Quality and High Quality? To consider the difference between good quality and high quality, ask yourself the question: Is Rolls-Royce a good quality motor car? Rolls-Royce is a high-quality, well-engineered car. However, if you want a car that is economical to run, easy to manoeuvre in busy city streets, and easy to park, is Rolls-Royce a good-quality car? If you want a car that can drive off the road, across farmland, and survive a collision with a kangaroo, is Rolls-Royce a good-quality car? If you want a car that represents your status as a successful manager, is Rolls-Royce a good-quality car. The answers are probably no, no, and no. It is important not to overengineer the product, but to produce something that satisfies, even delights, the customer, but is good value for money to achieve the project's goals. Often something which is overengineered will not delight the customer because it will not work.

7.2 ACHIEVING QUALITY ON PROJECTS

Figure 7.1 is a five-element model for managing quality on projects:

1. Two elements represent what we must manage the quality of: the product and the management processes.
2. Two represent how we manage their quality: through quality assurance and quality control.
3. The fifth represents the attitudes of the people involved.

Quality of the product is the ultimate goal. It is the product which satisfies all the criteria in the previous section, and which influences attitudes years after the project is finished.

Quality of the management processes is also a significant contributor to the quality of the project's product. Following well-defined, previously proven successful ways of doing things increases the chance of success. Designing new project management processes at the start of every project increases the chance of failure. We shall see below, that this means developing procedures for the organization to be used as flexible guidelines, not rigid rules.

Quality assurance is preventative medicine, steps taken to increase the likelihood of obtaining a good-quality product and management processes. It is about trying to get it *right first time*.

Quality control is curative medicine, which recognises human fallibility and takes steps to ensure that any (hopefully small) variations from standard which do occur are eliminated. This is about trying to get it *right every time*, with *zero defects*.

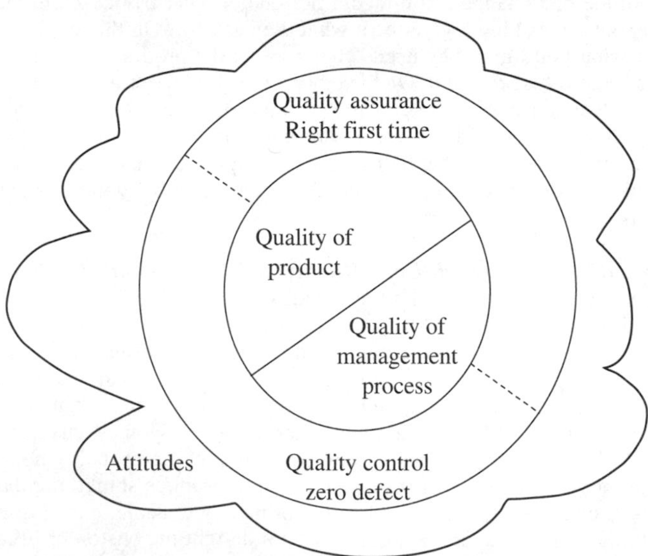

FIGURE 7.1 A five-element model for managing quality on projects.

Good attitudes are essential to successful project management. We saw this under strategy in Sec. 3.4. I used to tell Example 7.1 as a joke, but somebody in one of my courses said it happened to him. The commitment to quality must be at all levels of the organization; it cannot be delegated downwards, or pushed upwards. In the days when quality circles were popular, people implementing them had top-down teams and bottom-up teams to emphasise this point.

Example 7.1 Eliminating the culture of expecting failure

An organization ordered a batch of capacitors from a Japanese company, and specified that there should not be more than 0.5 percent faulty capacitors in the batch. The consignment arrived in a big box and a small box. They started testing the capacitors in the big box and found they were all perfect. They then tested the capacitors in the small box and found them all to be faulty. At that point they realised that the small box was 0.5 percent of the consignment!!!

Combining the two elements in each of the inner and outer circles in Fig. 7.1 leads to four steps of quality management.

Assuring the Quality of the Product

In order to assure the quality of the product it is beneficial to have

- A clear specification
- Use of defined standards
- Historical experience
- Qualified resources
- Impartial design reviews
- Change control

Clear Specification. Without a clear idea of what is to be achieved, the team has no direction. It is possible to specify both the end product of the project, and the intermediate products: milestones resulting from work packages; and deliverables of the activities at lower levels. The lower the level at which the deliverables are specified the tighter the control. However, there are risks associated with a highly detailed specification: it may be inconsistent; it may confuse rather than clarify; and the lower level products may become an end in their own right, rather than a means of delivering the new asset.

The next three are about trying to maximise the use of previous experience.

Use of Defined Standards. You can use standard designs and packages of work which, from previous experience, are known to deliver results of the required specification. One of the great differences between the project environment and routine manufacturing is that, in the latter, each day's production becomes a standard against which to improve the next day's production. In a project environment it may be some time before you repeat a process, and then the environmental conditions may be different. However, the use of standards will be beneficial in the long run.

Historical Experience. Hence, the greater the historical experience, the better will be the standards and specification. For this reason, it is not always possible to create a clear specification of R&D, high technology, and organizational development projects. However, the more historical data used the better. In the next chapter, it will be shown that there is a clear learning curve in industries with time, with it taking perhaps 50 years to build up a credible body of data.

Qualified Resources. If the people used on the project have access to that body of data, either through their own experience or training, then that makes them better able to apply standards and achieve the specification. This applies equally to professional staff (engineers, IT staff, researchers, trainers, and managers) and artisans (electricians, mechanics, and programmers). It is common in the engineering industry to put artisans through strict testing procedures before allowing them to do critical work. The use of qualified resources also applies to material and financial resources, but these can be tested against the standards.

Impartial Design Reviews. The use of auditors to check the design can help to assure that the customer's requirements are properly met. You may think that this is insulting to the design team, but there is ample evidence that human beings find it very difficult to discover their own mistakes (Example 4.5) and hence the use of auditors, sometimes called red, pink, or blue teams, to check the design can be worthwhile. However, you need to check that you do not overdo it. There are apocryphal stories about auditors outnumbering the project team, and since they are there to find fault, they find it where none exists: the design may be adequate but not perfect.

Change Control. This is vital to achieve the specification where change is necessary. It does not mean that changes are eliminated, because that can result in a product that does not meet requirements. The purpose of each change must be carefully defined, the impact on the design assessed, and the cost compared to the benefit, so only those changes that are absolutely necessary and cost-effective are adopted.

Controlling the Quality of the Product

Quality control is a process of diagnosis and cure. As the new asset is delivered it is checked against the specification to ensure that it is of the required standard, and any variances are eliminated. There are four steps in the control process (Fig. 7.2):

- Plan the work required, and do work to deliver results.
- Monitor the results achieved.
- Compare the results to the plan, to discover variances.
- Take action to eliminate variances.

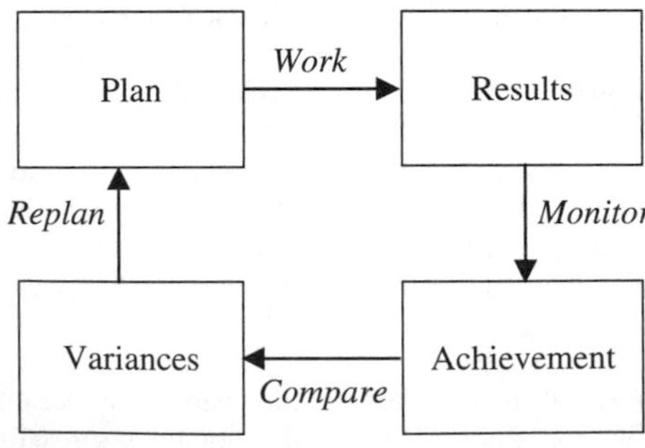

FIGURE 7.2 A four-step control cycle.

The quality plan for the project's product means understanding how every deliverable at all levels of the PBS will be judged to have been achieved. The work package scope statement (Table 5.3) had space for entering the criteria for judging achievement of the milestone. I usually say that the specification for the overall facility, (the client requirements and functional specifications), should run to several pages, that for each milestone, (in the systems specification), should be half a page, and that for the deliverable of each activity, (the detail specification), should be a couple of lines. I was challenged on this on one course. The delegates said it should be the exact opposite. The specification for each activity should be a couple of pages, and for the facility should be just six words, "every previous quality check has worked." The point they were trying to make is that if you get the detail right, there is no need to check the overall facility. What it means in practice is the specification for the new asset should run to several pages, but you check it throughout the project not right at the end.

Monitoring results and calculating variance means checking of the specification of each deliverable as it is achieved. It is important to do this from the start, from the earliest activity for the earliest milestone. It is no good waiting until the end of the project, and finding a mistake was made on the first day. Mistakes must be identified as they occur. Hence the comments from the delegates above.

Taking action from the start builds up a momentum for success, carried through the project. There is a major difference here between project and operations. In an operation where you are doing something repetitively, once the process is setup correctly, it will usually not go wrong suddenly. The process will drift. Hence you tend to monitor sparingly, using processes such as *statistical process control*. This may involve the destructive testing of, say, every 100th product. Once the process is working, the emphasis is on quality control. On a project you cannot destructively test every 100th product, you only do it once, so wrong once is wrong every time. This shifts the emphasis much more onto quality assurance and quality control at early stages of the project as described.

Assuring the Quality of the Management Process

To assure the quality of the management processes, a similar list as that for the product applies, which means having a set of defined procedures for managing projects. Procedures clearly specify how projects are to be managed by qualified resources, and are derived from standards based on historical experience. They may be derived from the company's own experience, or based on standard procedures. Many client organizations have procedures which they require their suppliers to use, and regularly audit contractors against them.

It is essential that the procedures are used, and this requires three things: They should not be bureaucratic; they must be sensible; and they must have management support. In Sec. 1.2, I suggested the procedures should describe how the organization processes product, not what the functions of the organization do (see Example 1.4). The procedures should also be flexible guidelines, not rigid rules. This means if the customer requires something different, the procedures should be changed to meet their requirements, not the requirements changed to meet the procedures. This can be achieved in a controlled way by having a procedure for changing the procedures, and by project teams regularly developing a quality plan as part of start-up. Finally, at the end of every project the procedures should be reviewed to see how well they served the project, and the organization's procedures updated if necessary. Quality is about continuous improvement, not compliance to twentieth century ways of working.

The procedures are often based on the ISO quality standards, a complete list of which are in Table 7.1, or on PRINCE2. The use of procedures manuals is described in Chap. 17.

TABLE 7.1 List of ISO, IEC, and BS Quality and Project Management Procedures

Number	Title
ISO 9000:2005	Quality management systems—Fundamentals and vocabulary
ISO 9001:2000	Quality management systems—Requirements
ISO 9004:2000	Quality management systems—Guidelines for performance improvement
ISO 10005:2005	Quality management systems—Guidelines for quality plans
ISO 10006:2003	Quality management systems—Guidelines to quality in project management
ISO10007:2003	Quality management systems—Guidelines for configuration management
ISO 10011:2002	Guidelines for auditing quality systems
PD ISO/TR 10013:2001	Guidelines for quality management system documentation
ISO 10014:2006	Quality management—Guidelines for realizing financial and economic benefits
ISO/IEC 12207:1995	Information technology—Software life-cycle processes
IEC 300:1995	Dependability management
BS6079	A guide to project management

Controlling the Quality of the Management Processes

The method of monitoring the management processes is through project audits. An audit is a detailed check of the operation of the management processes against standards of good practice, such as the organization's procedures manual or that of an external agency. (Audits are described in Chap. 17.)

The Quality Plan

At the start of the project, the manager should draw up a quality plan to define how quality will be achieved, how the company's procedures will work on this project, and how the manager intends to assure and control quality. In qualifying the procedures, it may contain new ones where items are either not covered or inadequately covered for this project in the overall procedures, and may include such things as: disputes, documentation, reporting mechanisms, customer liaison, and so on. For the quality control process, it may contain a detailed activity and resource plan. The quality plan may form a section of the project definition report (Chap. 12).

7.3 CONFIGURATION MANAGEMENT

Configuration management is a technique used to manage the refinement of the specification and work methods on development projects. The technique was first developed in the U.S. defence industry during the early 1950s to track the versions of components as they were configured in the new asset, and to control changes as they occurred. In particular, where several prototypes are being developed, configuration management tracks the design, or configuration, of each prototype. It has now become a desirable, if not essential, tool to control the functionality and quality of components in the product breakdown, and work methods in the work breakdown, to be used on software, technology, engineering, or organizational change projects.

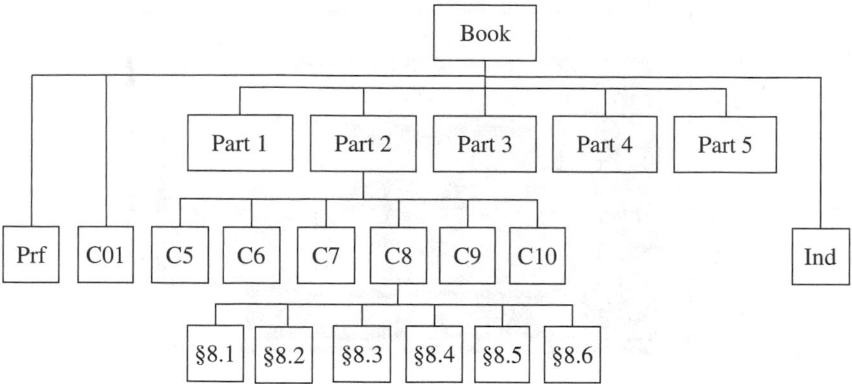

FIGURE 7.3 The configuration of a book.

So what is configuration management and how can the control of configuration be of use in a development project? Configuration management controls the specification of the product breakdown structure; it expresses the facility to be delivered by a project, as a configuration of component parts. The configuration can take various forms: a car, space shuttle, design, plan, software system, training program, or organizational structure. Each component may then be regarded as a configuration in its own right, made up of other components. This process, of course, develops the bill of materials, or product breakdown, of the project. Figure 7.3 illustrates the concept using a book as the configuration. The components are chap. 8, the subcomponents sections, and so on.

Configuration management is not a radical discovery that revolutionises the way the facility is developed and maintained. It is a set of good working practices for coping with uncertainty and change and gaining commitment of the project's participants as the design evolves. Many projects use elements of configuration management, especially in the application of change control. However, to be effective, it must be a systematic, consistent approach to managing change on complex projects. From the outset, structures must be put in place to support it. These include specified individuals with responsibility for configuration management, and procedures supported by senior management. It also involves all project participants. There may be one or more project review boards, with responsibility for approving the specification of the facility, and to approve changes to the specification. Depending on the size and complexity of the project, there may be a group of people dedicated to the function of configuration management.

Basic Approach

Figure 7.4 illustrates the basic approach to configuration management. In line with the goals and methods matrix (Fig. 1.12), we accept that there may, at the outset of the project, be some uncertainty about the specification of the project's deliverables, and some uncertainty about the methods of delivering them. Rather than trying to pretend that this uncertainty does not exist, that these things are precisely prescribed, it is better to accept the lack of clarity, and manage the refinement of our understanding.

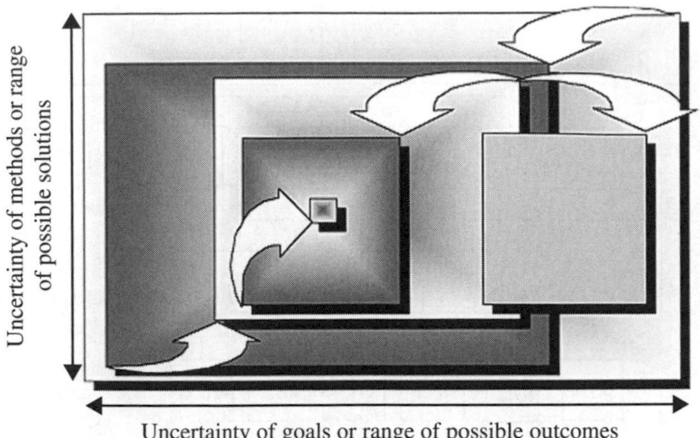

FIGURE 7.4 Configuration management.

So at the start of the project we write the specification of the deliverables, and the work methods as best as we are able, and then agree that specification with the multiheaded client: sponsors; owners; operators; users; marketing representing the consumers. We agree that the eventual solution lies somewhere within the large rectangle, but we do not know where. We then start work on the project, and refine our understanding. At a predetermined review meeting, we sit down with the multiheaded client, and agree the current status. We repeat the process, and hopefully get agreement as we home in on the eventual solution. Perhaps at a review meeting one or more of the participants disagree with the current status. Then one of two things has happened, either the previous specification was not correct, or the work to go from the previous position to the current was wrong. In the former case we need, through change control, to change the specification. Hopefully, if the problem is found early enough, the change can be made at little or no extra cost. If the change is made very late in the day, it can be inordinately expensive (see Example 7.2). If the latter is the case, we need to go back and repeat the work. Both of these are an anathema to traditional project managers: changing the specification, or doing extra work at additional cost and time. However, at the end of the day, you have to ask yourself whether it is better to finish according to arbitrary time and cost targets, or produce something that works. On some projects, like the Olympic Games, the time is imposed by external constraints and so must be achieved. But on many projects it is better to take a bit longer and pay a bit more to deliver something that works.

Example 7.2 Discovering problems early

I was working with a business school which was rewriting modules on its distance learning MBA. One module, Managing People, was being written by a sister company. The college gave the sister company a specification for the module. At that point, all contact between the two organizations stopped for the time being. The sister company

- Developed a contents page
- Identified potential authors
- Got them to write a précis
- Approved the précis
- Got the authors to write the chapters

- Reviewed the chapters
- Asked the authors to revise them
- Produced the prototype course

At that point the sister organization went back to the college and gave them the prototype. The college took one look and said it was wrong. What had happened is the sister company had interpreted Managing People as Human Resource Management. The college wanted that and Organizational Behaviour. Instead of being about 50:50 it was more like 80:20. But it was too late. It would be far too expensive to go back and start again. It could have been avoided by holding a configuration review at the end of each step in the above list. At the first configuration review the college would have seen that the balance was wrong.

Implementing Configuration Management

Implementing configuration management requires the definition of tasks to be performed and procedures to be adopted. The tasks must be allocated, which requires the organization to be established, responsibilities assigned, and appropriate resources (people, money, equipment, and accommodation) deployed. The appropriate procedures depend on the specific project, its size and complexity, but typically configuration management comprises four processes.

Configuration Identification. Configuration identification is the process of breaking a system into its component parts, or *configuration items,* each of which can be individually documented and placed under change control. Ideally, each configuration item will have maximum cohesion; that is, it would not be useful to subdivide it further for the purpose of documenting it or controlling changes to it. Also, the configuration items will have minimal coupling; that is, it would not be useful to merge two or more items to form a single item (see Example 7.3).

Example 7.3 Configuration identification

When I was writing this book, my list of configuration items was the list of section headings, as recorded in the Table of Contents (Fig. 7.3). However, I must admit that the sections did not conform precisely to the principles of cohesion and decoupling. In this chapter, the definition of the section headings was quite stable. In others, Chap. 6 for instance, the definition changed as I wrote the chapter. The chapter was perhaps therefore the configuration item. Some chapters were not configuration items on their own. The chapters in Part 4 were reconfigured as I wrote the book.

In its simplest form, configuration identification involves locating all the configuration items required to deliver the new asset so that nothing is overlooked, and then establishing the information to keep track of those items throughout the life of the project. Most systems can be broken down using a hierarchical PBS. When the system has been broken down to its lowest level, the resultant configuration items form the project inventory, or bill of material. All deliveries and revisions are tracked and controlled against two forms of configuration item recording; the planned set; and the produced/approved set.

The identification of the sets of items should cover the entire development cycle for both the facility and the supporting documentation. The definition and recording provided will support the activities of configuration control and status accounting. A complete list of all configuration items will be derived from the design specification. The configuration is complete when all items have been delivered. If extra configuration items are delivered, or some are not delivered, then this will only be acceptable if the design specification, and therefore the list of items, has been amended accordingly.

Configuration Reviews. Configuration reviews are conducted at significant points through the project. They can be conducted at the conclusion milestones or at the transition from one stage of the life cycle to the next (Sec. 18.4). The current status of the configuration is checked against the specification, and it and the specification are agreed with the multiheaded client. These reviews are automatically built into the PRINCE2 process[1] as end-of-stage reviews, and they are required by ISO 10006.

Configuration reviews conducted at the end of stages control the movement of the new asset through the life cycle. At the end of the initiation stage, the first configuration review audits that the specifications are

- *Up-to-date*: They accurately reflect the concept of the product.
- *Complete:* All the configuration management documentation that should exist at this point in the life cycle actually does exist
- *Agreed*: They have the support of all the project's participants.

At the conclusion of this stage, a requirements definition is produced, as part of the project definition report (Chap. 12) and reviewed, approved, baselined, and handed over to configuration management before it moves on to the design and appraisal stage. Similarly, at the end of design, the design specifications are produced, as part of the project requirements definition or project manual, which are again reviewed, approved, baselined, and handed over. Once the configuration identification moves into execution, it evolves from documentation into actual deliverables, whether physical or abstract. These are again reviewed at the end of this stage, to draw up the list of outstanding items for finalization and close out, and yet again at the end of this last stage, before the documentation is archived as the *as-built design*. Configuration management is the central distribution point for each stage of the life cycle, but it becomes more critical during the last stage, finalization and close out, as the facility is tested and commissioned.

Configuration Control. Controlling the baselined configuration items through each stage of the life cycle is the basis of configuration management. The project depends on the baselined items and the record of any changes. Periodically during the life of an item, the baseline may need to be revised. It should be revised whenever it becomes difficult to work with the baseline documentation and authorised changes to it. All authorised changes to the documentation should be consolidated, as should that relating to any authorised repairs and emergency modifications. When the documentation has been completed, reviewed, and approved, the baseline becomes revised. All subsequent change proposals should be made to the revised baseline.

Changes may arise internally or externally. External ones come from changes to business requirements, internal ones from forgotten requirements or problems found during the project. A procedure is required to report problems with baselined configuration items. Change control is the process of proposing, reviewing, approving, and, where necessary, implementing change to the approved and maintained items within the PBS. Through the process of change control, the impact of all changes is properly assessed, prior to deciding whether to authorise the change. Impact assessment will determine the changes in scope the change will bring out, not just in the immediate area of the change, but on the whole project. Often the change can have a far-reaching impact. The consequences for organization, quality, costs and benefit, and schedule are also assessed.

Review boards may change at different stages of the life cycle. Prior to the change review, the team determines what impact changes to configuration items has on resources, and prioritise change against requirements for all projects in the organization. The impact is documented for the board. Once a change has been approved, the person responsible for the item makes the change, and passes the rebaselined documentation to configuration management. Information on revisions to the item is recorded. The revised specification for the item is passed to all interested parties, and then secured by configuration management.

For major changes, it is sometimes desirable to adopt a top-down approach in which changes to the requirements specification are agreed prior to any work being done to define consequential changes to the specification. This, in turn, is agreed prior to changes being made to the product and component specifications. Configuration management can handle this by defining the major enhancement as a separate configuration with its own baseline. When a major enhancement becomes operational, it supersedes the current system. Until then, the current operational system continues to have its own baseline changes as necessary. This can be taken one step further, where several prototypes have their separate baselined configurations operational in parallel, each subject to separate change control. When a change is made to one, it may or may not be made to some or all of the others.

Status Accounting. Status accounting is the fourth function of configuration management. It supplies information on request about baselines, configuration items, their versions and specification, change proposal, problem reports, and repairs and modifications. For example, status accounting may identify authorised repairs and modifications awaiting the completion of amended documentation. Unless documentation is amended to be consistent with the facility, it is not accepted as being valid. Status accounting also keeps track of the complexities caused by superseding (major enhancement) configurations.

Status accounting enables people on large, volatile projects to avoid using outdated versions of documents and components. This is important for contracting companies responsible for components that need to interface with each other. It is also important for people responsible for user acceptance tests. They need the most current version of the requirements specification and the agreed functional and physical characteristic of the configuration, so they can determine whether or not the specification (quality) requirements of the contract have been met. That is the facility functions as envisaged within its environment to produce the required product and benefit.

Configuration Management and the Life Cycle

A common mistake, thankfully now made less frequently, is to confuse design management and project management. As recently as 2007 a student of mine called the Capability Maturity Model (CMM)[2] a "project management methodology." It is not; it is a design management maturity model. In the early days of project management it was common to make the chief designer project manager:

- In the software industry systems analysts were called project managers.
- In civil engineering, design contractors were labelled "the engineer," and fulfilled an advisory role which included project manager, and put them into a conflict of interest with their main role as design manager.
- In the building industry the architect worked also as project manager with similar consequences.

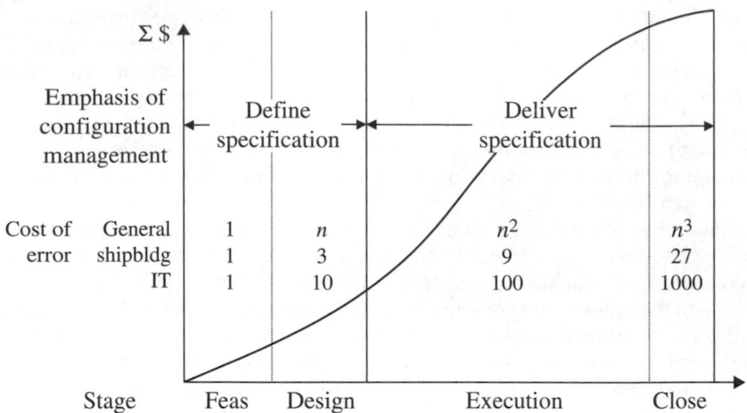

FIGURE 7.5 Configuration management and the life cycle.

Design management and project management are different, and often in conflict with each other, as the designer tries to perfect the design and the project manager tries to deliver an adequate design on time and cost. However, two techniques have as common elements life cycle and configuration management.

Figure 7.5 illustrates the evolution of configuration management through the life cycle. It shows a rule of thumb from most industries, that for every \$1 it costs you to right a mistake during feasibility, it costs you \$$n$ in design, \$$n^2$ in execution, and \$$n^3$ in close out. For the shipbuilding industry, n is said to be 3, and the ratios are 1:3:9:27, and for the software industry n is said to be 10 and the ratios are 1:10:100:1000. Hence it is a very good idea to try to agree the specification by the end of design, and move forward to execution with the design frozen (see Example 7.4). Thus the emphasis of configuration management changes as you move from design to execution. In feasibility and design, the emphasis is on gaining the commitment of the project participants to the design, and the key processes are identification, review, and change control. In execution and close out, the emphasis is on delivering the agreed design, and the key process is status accounting. That is not to say that if a show-stopper is discovered during close out, a change will not be made. But the change is made in the full knowledge of how much it will cost, and the benefit of the change must also be significant to justify it.

Example 7.4 Spending adequate time on design

A student of mine worked on a project to develop a new air traffic control system for a small country. The government wanted the project done in 18 months, so they only allowed 2 weeks for the writing of the specification. Doing it that quickly, the project team had no time to talk to any of the stakeholders, such as air traffic controllers, airport management, the airlines, pilots, and so on. They just did a desktop exercise. As a result the specification had to be changed repeatedly once the stakeholders got involved and the project took 5 years. Better to spend 2 months over writing the specification and complete the project in 18 months.

SUMMARY

1. There are four possible definitions of good quality on a project:
   - Meets the specification
   - Is fit for purpose

- Meets the customer's requirements
- Satisfies the customer

2. The four are not the same thing, and like in many areas of project management, overall optimum may not optimise any one of them. An overall compromise must be sought.
3. However, being fit for purpose is thought by many to be the primary criterion.
4. There are five elements of achieving quality of a project:
   - Quality of the product versus the management process
   - Quality assurance versus quality control
   - Good attitudes
5. Assuring the quality of the product requires
   - A clear specification
   - Use of defined standards
   - Historical experience
   - Qualified resources
   - Impartial design reviews
   - Change control
6. Controlling the quality of the product requires a clear understanding of the specification of each deliverable (at the time it is completed), and achievement of this specification must be measured, and action taken to eliminate variance.
7. Assuring the quality of the management process requires the use of procedures, which should
   - Be used as flexible guidelines, not rigid rules
   - Reflect how the product is processed not what functions the organization does
   - Be continuously improved, project by project
8. Controlling the quality of the management processes requires them to be audited.
9. Configuration management is a technique to manage the quality and functionality of the project's deliverables, and obtaining agreement of the project's participants. It has four steps:
   - Configuration identification
   - Configuration review
   - Configuration control
   - Status accounting
10. Quality is free, but not in the lifetime of a single project.

REFERENCES

1. Office of Government Commerce, *Managing Successful Projects with PRINCE2,* 4th ed., London: The Stationery Office, 2005.
2. Paulk, M.C., Curtis, B., and Chrissis, M.B., *Capability Maturity Models for Software*, Pittsburg, Pa.: Carnegie Mellon University, 1991.

CHAPTER 8
MANAGING COST

We now consider the fourth project objective, managing cost, by which the project manager ensures the project's product is financially viable and worthwhile. The next section considers the purpose of estimating costs, and shows how this leads to several types of estimate of increasing accuracy prepared at successive stages of the project life cycle. Later sections explain how the estimate is structured through the cost control cube, and describe several methods of preparing the estimate. Finally, we shall discuss how costs are controlled by comparing actual expenditure against the value of work done, and show how (S-curves) can provide a pictorial representation of this.

8.1 ESTIMATING COSTS

Over the next three sections I describe how to estimate costs. In this section, I explain the purpose of estimating and different types of estimate. In Sec. 8.2, I describe different types of cost and how to structure the estimate. Then finally, I introduce techniques for estimating in Sec. 8.3.

The Purpose of Estimating

There are several reasons why we estimate costs. Some of them are discussed in following sections.

As a Basis for Control. The estimate is prepared as a measure against which to control expenditure on the project. This measure is known as the *baseline*. The classic control cycle has four steps (Sec. 7.2):

1. Estimate future performance.
2. Monitor actual performance.
3. Calculate the difference, called the variance.
4. Take action according to the size of the variance (Sec. 13.6).

For this purpose the estimate may need to be quite detailed, prepared at a low level of breakdown.

Assess Project Viability. Before getting to a position where you need to prepare a control estimate, you need to determine whether the project is worth undertaking. You therefore prepare an estimate of the costs to compare with the estimate of returns. (Methods of assessing project viability are beyond the scope of this book.[1]) The appraisal estimate goes through various stages of increasing accuracy, at the end of each concept, feasibility, and design.

Obtain Funding. After approval has been obtained, the project must be financed. Funding will be awarded on the basis of the appraisal estimate prepared at design. (Obtaining finance is also beyond the scope of this book.[1])

Manage Cash Flow. Once funding has been obtained, and work starts, the project must be managed so that work takes place and consumes cash no faster than the rate agreed with the financiers (bankers). There are apocryphal stories about zealous project managers finishing their projects early and underspent (Fig. 8.1), and wanting a pat on the back. However, the company has gone into liquidation because the bankers called in the overdraft halfway through the project.

Allocate Resources. Human resources are a special form of project funding. The business plans their allocation in advance against the cash-flow estimate. They will be assigned to the project week by week against the control estimate.

Estimate Durations. The duration of a work element is calculated by comparing the estimate of work content to resource availability, and so the cost estimates form an input to time estimating. Time estimating, which is described in the Chap. 9, is performed for similar reasons to cost estimating and so similar types of estimate are required.

Prepare Tenders. Contracting firms tendering for bespoke contracts need to prepare estimates for the tender.

Types of Estimate

The same estimate cannot satisfy all six purposes. Five types of estimate, of varying accuracy, are required (Table 8.1). Table 8.2 summarises an idea first introduced in Sec. 5.1: you obtain increasing accuracy of estimate by estimating at lower and lower levels of

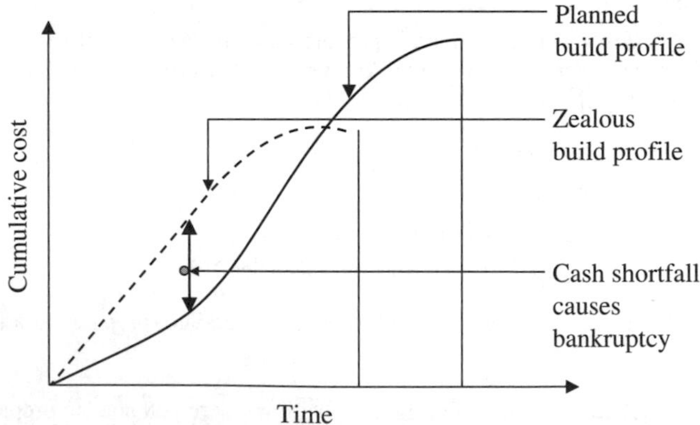

FIGURE 8.1 Different build profiles for a project.

TABLE 8.1 Types of Estimate—Purpose and Accuracy

Type of estimate	Range of accuracy	When prepared	Purpose
Proposal	±50%	Concept	Appraise viability to start feasibility
Budget	±20%	Feasibility	Appraise viability to start systems design
Sanction	±10%	Design	Appraise viability to approve project, obtain funding, schedule cash flow, and allocate resources
Control	±5%	Execution	Measure progress, assign resources
Tender		Design	Prepare tender

TABLE 8.2 Types of Estimate—Level at Which Prepared in the Work Breakdown Structure

Type of estimate	Lowest level of estimate in WBS	Accuracy of estimate of			
		Project	Work area	Work pack	Activity
Proposal	Work area	±50%	±100%	–	–
Budget	Work package	±20%	±40%	±100%	
Sanction	Scope statements	±10%	±20%	±50%	±150%
Control	Activities	+5%	±10%	±25%	±75%
Tender	Tasks	±2%	±5%	±10%	±30%
Assumed number per project		1	4	25	200

breakdown. If the estimates are truly mean values, errors cancel out. Table 8.2 implies that to obtain an estimate to the correct accuracy at the project level, you need only estimate to the order of magnitude at the currently lowest level of the breakdown. However, a consistent error will reinforce, for instance, if all activities are underestimated by 20 percent, the project will be underestimated by 20 percent. Table 8.2 can be taken to lower levels of breakdown for larger projects. On one large engineering project worth several hundred million pounds, I prepared a breakdown which had approximately 100 areas of work and a ratio of 1:10 for each subsequent level of work breakdown, down to the task level. On the same project, estimators were estimating costs accurate to the nearest pound at all levels of work breakdown structure (WBS), and yet including contingencies of several hundreds of thousands of pounds at the work-package level. This is clearly absurd. It is the right level of contingency, but the wrong level of accuracy. Table 8.3 shows appropriate levels of accuracy and contingency at different levels of the WBS for a project worth £100 million.

At any level of breakdown, there is no point calculating and quoting estimates to a greater degree of accuracy than the figure in the right-hand column. Any contingency added at that level of breakdown must be at least this amount as a level of contingency is already included through the accuracy to which figures are calculated.

When to Estimate Costs

It follows from Table 8.2 that preparing estimates of increasing accuracy requires increasing effort as you estimate at lower levels of breakdown. Table 8.2 implies that to double the accuracy at the project level requires you to estimate at a level of breakdown with four

TABLE 8.3 Levels of Estimating in a Large Engineering Project

Level of breakdown	Number in project *N*	Average cost *C*	Accuracy as ratio	Accuracy as value
Project	1	£100,000,000	±1%	±£1,000,000
Work area	100	£1,000,000	±10%	±£100,000
Work package	1000	£100,000	±30%	±£30,000
Activity	10,000	£10,000	±100%	±£10,000
Task	100,000	£1,000	±300%	±£3,000

times as many work elements, requiring four times the effort. This has been measured in the engineering industry[2] (Table 8.4). When plotted (Fig. 8.2), this is a learning curve, with greater effort giving greater accuracy, but with diminishing returns. In addition, there is a point, at 5 percent accuracy with effort 5 percent of project cost, where the effort does not justify the return. This has three consequences:

1. It is not worthwhile producing an estimate more accurate than the control estimate, because it costs more to produce than the value of the data. This is a consequence of the uniqueness of projects. In routine production, costs may be estimated to a low level of detail, because the saving is made many times over. On projects, the saving is made once only. It is not worthwhile producing plans in great detail, because the effort is not rewarded. It is better to put management effort into eliminating risk (Chap. 10), not quantifying it. The problem arises for contracting companies that when tendering must prepare estimates which will allow them to make a profit (Example 8.1).
2. The way to improve accuracy of estimates is not to put more effort into estimating, but to improve the estimating data effectively to move the curve in Fig. 8.2 to the left using historical data. (On engineering projects, 100 years of effort has gone into gathering data.[2] The information systems (IS) industry has only 40 years of experience, and didn't really start gathering estimating data until about 20 years ago.)
3. The estimate at one level should not be prepared before the estimate at the previous level. Each estimate is therefore prepared at a given stage of the life cycle (Table 8.4). Effectively, the comparison of costs and returns at the end of one stage of the life cycle justifies the commitment of resources to planning, designing, and estimating at the next stage. If the project is not viable at these high levels of estimate, work should not proceed to the next stage (Example 8.2).

TABLE 8.4 Level of Effort and Stage of Production of Project Estimates

Type of estimate	Accuracy	Level of effort as % project cost	Stage of production
Proposal	±30–±50%	0.02–0.1%	Concept
Budget	±20–±35%	0.1–0.3%	Feasibility
Sanction	±10–±25%	0.4–0.8%	Design
Control	±5–±15%	1–3%	Execution
Tender	±2–±5%	5–10%	Tender preparation

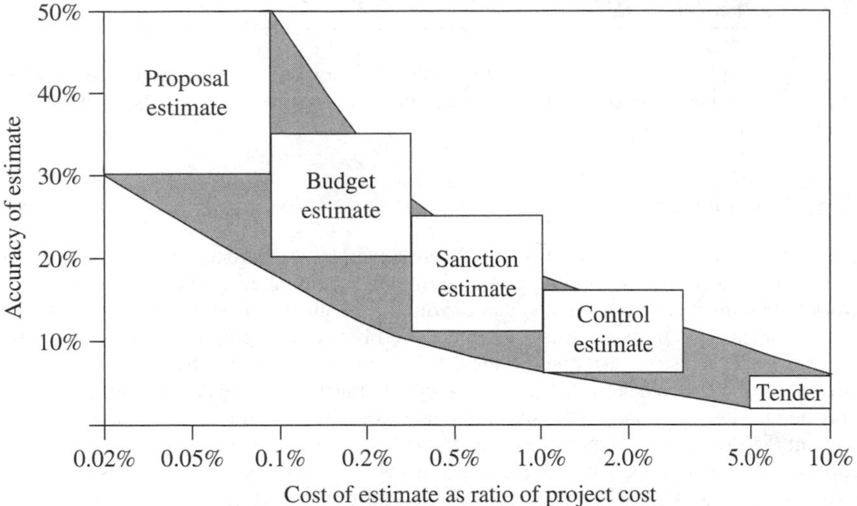

FIGURE 8.2 Accuracy of estimate versus cost of estimate (a learning curve following an inverse square law).

Example 8.1 Recovering the cost of estimating on contracts

I facilitated a bid management workshop run by a major IS vendor. They spent 3 percent of contract value preparing estimates, were successful at winning one contract in five, and had traditionally made profit margins in excess of 50 percent. The contract they won paid the estimating costs of the four they did not, but the net margin was still in excess of 35 percent. However, margins were being squeezed, and they were now lucky if they made a gross margin of 15 percent. That means they had to increase the number of contracts won, reduce the estimating costs, or make a loss. A bid manager from one of the major engineering contracting firms in the petrochemical industry spoke at the workshop. He said they had reduced the bidding costs to 0.75 percent of contract cost. They were winning one contract in five, but needed to make a margin of only 4 percent on that contract to cover the bidding costs on the five. The way they reduced bidding costs was to have a department of bid managers, who were the centre of expertise for tendering. That department could make maximum use of historical data. Effectively

- They accelerated the learning curve.
- They reduced the unique elements of projects, and so turned the bidding process into a repetitive operation.
- They achieved quality through using historical data (Chap. 7)

Example 8.2 Tailoring the estimate to the current stage of the project

I worked in a company where the IS Department prepared control estimates at the concept stage, only to find projects were not viable. If you expect an internal rate of return of 20 percent, you can only make that mistake three times per year until you cannot afford projects at all.

8.2 TYPES OF COSTS

In considering how to estimate costs, I describe the components of cost, how to structure the estimate using the cost control cube, and how that leads to simple methods of estimating using spread sheets.

Cost Components

Labour. This includes the cost of people employed by the parent company involved in executing project tasks, including people designing and delivering the new asset. I have worked in some manufacturing companies who do not attribute design labour to contracts. It is absorbed into company overheads and shared between all contracts. The result is the company only wins contracts with a high design element, and they have no control over design costs. Some other labour costs are included under other headings. The labour cost may be measured in monetary terms, or in hours worked. The latter is also called the work-content (Sec. 6.4), and is a measure of the total effort required, independent of the duration and number of people performing the task. Clearly, effort can be converted into monetary terms by applying known costs per hour worked for each resource.

Materials. This is the cost of materials consumed in delivering the new asset. This may be materials contained in the final product or consumables used on project tasks. On engineering projects materials include machinery, vessels, piping, structures, and instrumentation, but also include things like welding rods and concrete. On information systems projects, materials include main, peripheral hardware, and propriety software. On organizational development projects, materials may be more peripheral to the project, but include materials used on training programs, furniture for new offices, and stationery for new management procedures.

Plant and Equipment. These are materials used in delivering the facility, but which are not consumed, and so are available for reuse on subsequent projects. They may be bought or hired, but either way each project only pays a part of their price new. On engineering projects, plant and equipment includes welding machines and earthmoving machinery. On information systems projects it includes hardware used by programmers. On organizational development projects it may include equipment used in the preparation and delivery of training programs, and temporary accommodation used during office moves.

Subcontract. This includes the cost of labour and materials provided by outside contractors. Costs will be included in this heading where their control is not within the scope of the parent organization.

Management Overheads and Administration. This includes the cost of people and materials to manage the project. These costs are attributable to the project, but not specific tasks, and include the cost of the manager and team leaders, the project support office, a project management information system if required, and temporary site services. Management costs are typically 3 to 6 percent of the cost of the project.

Finance. Finance can be the most significant cost on a project,[1] being greater than any other single cost and yet is ignored by most project managers. It is an area where cost savings can be made by carefully scheduling cash flow.

Fees and Taxation. Fees may include insurance and finance; licence agreements and taxation may be regarded as a special type of fee.

Inflation. In these days of low inflation it can be reasonably ignored on projects lasting less than about 18 months. It is only really significant on large programs of work lasting several years, or for contractors bidding for fixed-price contracts.

Contingency. I have left contingency to last because I wish to devote a bit of space to it. I believe that the way many people estimate their projects they set themselves up to fail. Through the simple example in Table 8.5, I try to illustrate the need for contingency and what is variously called tolerance or project manager's reserve.

Table 8.5 shows a simple project comprising 10 identical work packages. Each work package is estimated to cost 100 units and so the raw estimate for the cost of the project is 1000 units. This estimate for the cost of the work packages is perhaps the most likely outcome. However, there is variability in the possible cost of the work packages. This is the nature of life; it is highly unlikely that they will all cost exactly 100. I suggest here it might cost as little as 70 and as much as 150. Different elements of work will have different ranges of variability, as I was suggesting above. Some will have very low variability, and will cost between 99.7 and 100.5. Some will have medium levels of variability, costing between 97 and 105. Here I have suggested this package of work will have high variability. In the tables above, I suggested that all the errors were equal above and below the estimate; the upside risk was the same as the downside risk. But usually the amount by which the outcome can be less than the estimate (the upside risk) is limited by the laws of physics, whereas the amount it can be more (the downside risk) is unlimited. So I have shown that the outcome cannot be less than 70, and is highly unlikely to be more than 150. I have also shown the chance of achieving the estimate is only 40 percent. (Cumm. prob. stands for cumulative probability and shows the chance that the work package will cost less than that value.) This is quite common where the downside risk is greater than the upside risk; even though 100 is the most likely outcome for the package of work, the chance that the outcome will be less than that is less than half. So we are expecting four of the packages of work to cost between 70 and 100, and six between 100 and 150.

Now when the project is finished we can actually determine how much it actually cost, and if we divide that by 10 we can work out the average cost of each package of work. But before the project starts we can make a guess at what that average will be by using what is known as the 1:4:1 formula. This guess is called the expected cost of each package of work and is calculated as:

$$\text{Expected cost} = \frac{\text{Minimum} + 4 \times \text{Most likely} + \text{Maximum}}{6}$$

TABLE 8.5 The Need for Contingency and Tolerance

Work pack	Minimum		Estimate	Expected			Maximum
Cost	70		100	105			150
Cumm. prob.	0%		40%	60%			99%
Project	**Theoretical minimum**	**Likely minimum**	**Raw estimate**	**Expected**	**Budget**	**Likely maximum**	**Theoretical maximum**
Work pack × 10	700	950	1000	1050	1100	1200	1500
Cumm. prob.	0%	1%	5%	60%	80%	99%	100%
			Contingency = 50		Tolerance = 50		

For the numbers in the second row of Table 8.5 this gives an expected cost of each package of work of about 105. So if we expect each package of work to cost 105, what is the expected cost of the project? 1000? No! It must be 1050. We are expecting the cost of the project to be 5 percent greater than the raw estimate. This is why many projects fail; estimators just use the raw estimate, but because the variability of the components we actually need to allow a contingency, typically 5 to 10 percent; here 5 percent.

Now, using techniques like Monte Carlo analysis (Chap. 10), we may be able to calculate the probability of achieving various outcomes for the project. I have shown that the project has a theoretical minimum cost of 700, but for all intents and purposes it is unlikely to cost less than 950. Likewise the theoretical maximum is 1500 but for all intents and purposes it is unlikely to cost more than 1200. I have suggested that chance of achieving the raw estimate may be as little as 5 percent. This is quite common; because the downside risk on the work elements is greater than the upside risk; the chance of achieving the raw estimate is very low, which is why many projects fail. I have suggested that the chance of achieving the expected cost, or target estimate, may be 60 percent; this is also quite common, we have just over a 50 percent chance of achieving the expected value.

But now put yourself in the shoes of either the project manager or sponsor seeking funding from the owner, or the owner about to award funding to a project. As a project manager are you happy seeking funding at a value that only has a 60 percent chance of being achieved, or as an owner awarding funding at a rate that has a 40 percent chance of failing. The answer is usually No! and No! When seeking or awarding funding for a project people suggest you should use a figure with an 80 percent chance of success. So we add a tolerance or project manager's reserve to the estimated estimate to achieve the project budget or appraisal estimate. The PRINCE2 process[3] calls this the tolerance, and in fact it is the project sponsor's reserve; the project manager is not allowed to spend it without the project sponsor's approval.

Thus we see there is a need for several estimates on a project:

- The raw estimate, sometimes called the stretch target, with typically a 5 percent chance of being achieved; this is what the project team are given to work to.
- The expected outcome, or target estimate, with typically a 60 percent chance of being achieved; this is what the project manager is working towards.
- The contingency, the difference between the raw and expected estimates.
- The budget or appraisal estimate, with typically an 80% chance of being achieved; this is the maximum the owner expects to have to spend.
- The tolerance is the difference between the expected and budget outcomes.
- Not mentioned up to now, the historical estimate, what this sort of project has typically cost in the past, which is hopefully greater than all the above—we are going to do better this time.

Example 8.3 gives a very simple example of this in practice. The CEO of a Norwegian state-owned company lost his job because he did not understand the difference between these different estimates (see Example 8.4).

Example 8.3 Uncertainty in estimates

As an example of uncertainty in estimates, I use my journey to Henley Management College. I live 40 miles from the College, and the most likely journey time is 55 minutes. I have done the journey in 40 minutes, and so this is the most optimistic. It once took me 135 minutes on a Friday evening, and the delay was due to heavy traffic, but you might call this insurable risk. Apart from that one extreme case, the journey can take up to 105 minutes. If I am teaching at nine o'clock in the morning, I must leave home by quarter past seven to virtually guarantee to be there on time. The journey home on Friday evenings can also take 90 minutes. Hence the most pessimistic journey time is 105 minutes. The distribution is what is called *bimodal*, there are two most likely outturns, one of 55 minutes corresponding to light traffic, and a lesser one of 90 minutes corresponding to heavy traffic. The median journey time is about 60 minutes and the average about 70 minutes. So if I go to the College every day for a week, how long do I expect to spend in the car during my 10 journeys: 400, 550, 600, 700, 900, or 1050 minutes? Well if I am unlucky and every journey corresponds with crawl hour, (bizarre that we should call it "rush hour"), then something like 900 minutes would be an appropriate estimate. But overall most of us would say something like 700 minutes. Yet standard project estimating would give 550 minutes, which we can see is a gross underestimate. Applying the 1:4:1 formula gives 610 minutes for 10 journeys. The cynic's version of this is the 1:4:3 formula, (the pessimistic value is multiplied by three and all divided by eight), and this gives 720 minutes. The latter is more accurate because the distribution is bimodal. Risk management is trying where possible to time my journey not to correspond with crawl hour, and so eliminate the upper tail of the distribution.

Example 8.4 Different estimates

A Norwegian state-owned company was undertaking a major development in two equal phases. The first phase cost NOK10 billion (about US$2 billion). The company's design engineers said that with the experience from Phase 1 they thought they could do Phase 2 for NOK7 billion. So this is what the CEO told the Norwegian government and that was the amount of money set aside in their spending plans and published in the press. The outturn cost for Phase 2 was NOK8 billion. So was the CEO praised for saving NOK2 billion between Phase 1 and Phase 2? No, he was sacked for the major embarrassment coming from a NOK 1 billion overspend. What the CEO should have done is set a stretch target of NOK7 billion for the team to work to, a target of NOK8 billion for the project manager to control progress against, and a budget of NOK9 billion for the government to plan to spend, against an historical figure of NOK10 billion.

This discussion and Example 8.4 suggest we have different estimates for different stakeholders. The team are working to the raw estimate or stretch target. In fact they are given the 100 for their work package. The project manager is controlling progress against the target, and the owner sets aside the budget or appraisal estimate. Some people say this is lying to the project team; I should tell them there are 105 to spend. But if you tell them there are 105 to spend, 1050 becomes the stretch target with a 5 percent chance of success. You tell the team they have 100 to spend but there is contingency if they need it. It is also perfectly acceptable to share all three figures with the owner, but tell them to set aside the budget with an 80 percent chance of success. If the owner won't accept that, then they are being foolish.

Structuring the Estimate—the Cost Control Cube

The above cost components constitute a third breakdown structure, the cost breakdown structure (CBS). The CBS is usually simpler than the other two, although one more level of breakdown can be derived under most headings. The three structures, WBS, OBS, and CBS, together form the *cost control cube* (Fig. 8.3), developed by the United States Defence Department, in the 1950s, as the basis of their cost and schedule control systems criteria (C/SCSC) methodology for controlling project costs, but which has now largely been subsumed into earned value analysis (EVA) described briefly below.

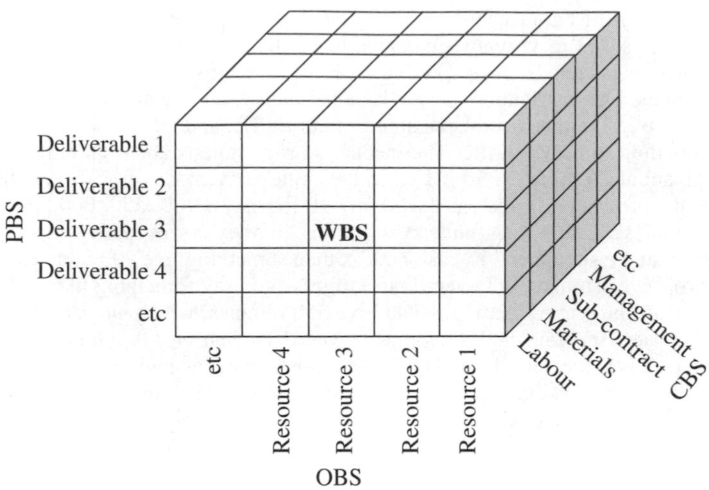

FIGURE 8.3 The cost control cube.

The front of the cost control cube is the responsibility chart (Sec. 6.4) at that level of breakdown. So for each cell, an element of work in the WBS, we can work out what will be spent by that resource to make their input to the delivery of the product indicated by the row at that level. Wherever there is an *X*, *C*, and possibly a *T* in the chart, there will be a direct labour cost. Where there is a *P* there will be a management overhead. The other symbols, *D*, *d*, *I*, and *A* don't really consume time. Then other cost types may be associated with each cell, although in reality they tend to be associated with the whole row. This lends itself to developing very simple estimates using spreadsheets. In accordance with Table 8.2, at the end of the concept stage you develop an estimate for each work area summing to the whole project. At the end of feasibility you develop and estimate against the milestone plan. Figure 8.4 shows the estimate for the CRMO Rationalization Project prepared against the project responsibility chart in Fig. 6.5. Then at the end of design you prepare an estimate at the activity level. If you adopt rolling-wave planning (Sec. 5.4) the estimates at then end of the design stage will be based on the work package scope statements (Table 5.1), and the control estimate will be based on the detailed activity definitions (Figs. 5.5 and 6.6). Figure 8.5 is the estimate against the activity plans for the CRMO Rationalization Project shown in those two figures. Note that in Figs. 8.4 and 8.5, I have estimated internal labour in terms of the number of days, and external costs, materials, and the external consultants in terms of money spent. This company is not charging internal labour to the project, but it is important to estimate how much time will be spent so it can be controlled. That gives two columns of the project estimate, one for work content for internal labour and another for money spent externally. Figure 8.4 will be prepared at the end of feasibility and confirmed at the end of design. The estimates equivalent to Fig. 8.5 will be prepared on a rolling-wave basis during execution, and the estimates used to check the figures in Fig. 8.4. If they are within the margin of error (±10 percent) they will be accepted. If they are outside, that problem has to be solved.

On larger projects, the same concept can be used to greater detail. In the mid-1980s, I estimated the cost of a £90 million to build a petrochemical plant. The resulting estimate at the project level is shown in Fig. 8.6, and for one work area in Fig. 8.7.

TriMagi Project Responsibility Chart

Project:	Rationalization of the Customer Repair and Maintenance Organization
Project Sponsor:	Steve Kenny
Project Manager:	Rodney Turner

X	eXecutes the work
D	takes Decisions solely/ultimately
d	takes decisions jointly
P	manages Progress
T	on-the-job Training
I	must be Informed
C	must be Consulted
A	may Advise

No	Milestone Name	Operations director	CRMO managers	CRMO team leader	CRMO staff	Project manager	Project support office	Estates manager	Estates department	Network manager	Networks department	IS department	Operators	Personnel	Suppliers		Sub-contract	Materials	Plant & equipment	Duration
		$K	$K	$K	$K	$K	$K	$K	$K	$K	$K	$K	$K	$K	$K		$K	$K	$K	$K
P1	Project defintion	3.0	3.0	4.0		14.0	12.0	2.0		2.0								25.0		
T1	Technology design				10.0					5.0	20.0	5.0	20.0							
O1	Communicaton plan			6.0														10.0		
O2	Operational procedures			10.0	40.0															
O3	Job and management design			10.0	20.0									50.0						
T2	MIS functional spec			10.0						10.0		40.0								
O4	Staff allocation			10.0	10.0									40.0						
T3	Technical roll-out plan			5.0		5.0	10.0	5.0		5.0	5.0	5.0					5.0			
A1	Estates roll-out plan			5.0		5.0	10.0	5.0	5.0	5.0							5.0			
P2	Financial approval			5.0		10.0	10.0	5.0		5.0		5.0					5.0	5.0		
A2	Sites 1 and 2 available							10.0	50.0									25.0		
O5	Management changes	10.0	10.0	20.0																
T4	Systems in sites 1 and 2								30.0	10.0	30.0	10.0					30.0			
O6	Redeployment and training			20.0										60.0				210.0		
A3	Sites 1 and 2 ready			10.0	10.0				10.0		10.0	10.0					20.0	180.0		
T5	MIS delivered			5.0						10.0		20.0					20.0	240.0		
O7	Procedures implemented		10.0	20.0										20.0				160.0		
P3	Intermediate review		10.0	10.0		10.0	40.0	10.0		10.0							10.0	45.0		
A4	Roll-out implemented		10.0	40.0	40.0	20.0	40.0	10.0	40.0	10.0	40.0	40.0		20.0			20.0			
P4	Postcompletion audit		20.0	40.0		40.0	40.0						20.0					900.0		
		13	73	230	130	104	162	47	135	72	105	135	40	190	0	0	115	1800	0	0

© 2008 Goal Directed Project Management Systems Ltd

Progress Report

No	Labour: Labour estimate	Labour: Actual	Labour: Estimated remaining	Labour: % Complete	Labour: Earned value	Labour: Calculated remaining	Cash: Cost total	Cash: Actual	Cash: Estimated remaining	Cash: % Complete	Cash: Earned value	Cash: Calculated remaining
	$K	$K	$K	%	$K	$K	$K	$K	$K	%	$K	$K
P1	40	40	0	1.0	40	0	25	25	0	1.0	25	0
T1	60	75	0	1.0	60	0	0		0	1.0	0	0
O1	6	5	0	1.0	6	0	10	10	0	1.0	10	0
O2	50	50	0	1.0	50	0	0	0	0	1.0	0	0
O3	80	80	0	1.0	80	0	0	0	0	1.0	0	0
T2	60	55	0	1.0	60	0	0	0	0	1.0	0	0
O4	60	65	0	1.0	60	0	0	0	0	1.0	0	0
T3	40	40	0	1.0	40	0	5	10	0	1.0	5	0
A1	35	40	0	1.0	35	0	5	5	0	1.0	5	0
P2	40	20	20	0.5	20	20	10	0	5	0.5	5	5
A2	60	20	40	0.3	18	42	25	0	25	0.3	8	18
O5	40	10	25	0.3	12	28	0		0	0.3	0	0
T4	80		80	0.0	0	80	30		30	0.0	0	30
O6	80		80	0.0	0	80	210		210	0.0	0	210
A3	50		50	0.0	0	50	200		200	0.0	0	200
T5	35		35	0.0	0	35	260		260	0.0	0	260
O7	50		50	0.0	0	50	160		160	0.0	0	160
P3	90		90	0.0	0	90	55		55	0.0	0	55
A4	320		320	0.0	0	320	20		20	0.0	0	20
P4	160		160	0.0	0	160	900		900	0.0	0	900
	1436	500	950		481	955	1915	50	1865		58	1858
Forecast Cost at Completion			1450			1455			1915			1908

FIGURE 8.4 Estimate at the milestone level for the CRMO Rationalization Project.

TriMagi Activity Plan

Activity Schedule

Project:	Rationalization of the Customer Repair and Maintenance Organization
Milestone:	D3: Personnel information pages designed
Manager:	Rodney Turner

Period: Week ending | **Target end:** 5-Mar

X	e**X**ecutes the work
D	takes **D**ecisions solely/ultimately
d	takes **d**ecisions jointly
P	manages **P**rogress
T	on-the-job **T**raining
I	must be **I**nformed
C	must be **C**onsulted
A	may **A**dvise

No	Activity Name	Regional board	Operations director	CRMO managers	CRMO team leader	Project manager	Project support office	Estates manager	Estates department	Network manager	Networks department	IS department	Operators	Personnel	Labour $K	Material $K	End Date
1	Produce project proposal				0.5	1.0	1.0								2.5		5-Feb
2	Hold definition workshop				0.5	1.0	0.5								2.0		9-Feb
3	Define required benefits		1.5		0.5	2.0									4.0		12-Feb
4	Draft definition report					2.0	3.0								5.0		17-Feb
5	Hold launch workshop		1.5	3.0	0.8	1.5	0.8	1.5		1.5					10.5	15.0	19-Feb
6	Finalize milestone plan					1.0	0.5								1.5		24-Feb
7	Finalize responsibility chart					1.0	0.5								1.5		24-Feb
8	Finalize risk assessment				0.5	1.0	0.5								2.0		26-Feb
9	Finalize time estimates						1.0								1.0		26-Feb
10	Finalize cost estimates						1.0								1.0		26-Feb
11	Finalize revenue estimates						0.5								0.5		26-Feb
12	Assess project viability					1.0									1.0		26-Feb
13	Finalize definition report				0.5	2.0	1.5								4.0		5-Mar
14	Mobilize team				0.3	0.5	0.3								1.0	10.0	5-Mar
	Contingency				0.5	1.0	1.0								2.5		
			3	3	4	15	12	2		2					40	25	

Schedule weeks 1–15 (week ending 5-Feb, 12-Feb, 19-Feb, 26-Feb, 5-Mar, 12-Mar), d = Duration

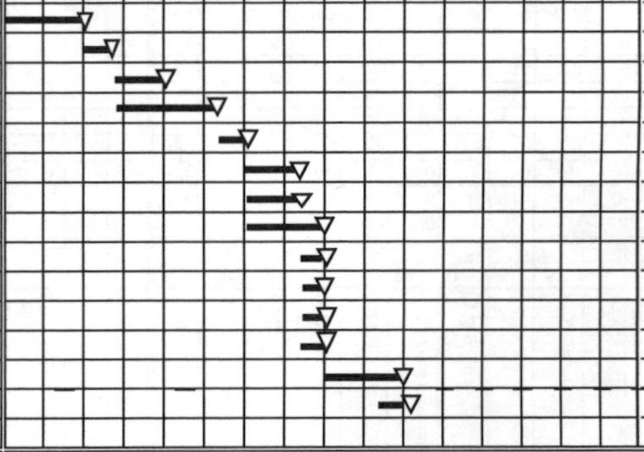

FIGURE 8.5 Estimate at the activity level for the CRMO Rationalization Project.

PROJECT ESTIMATE	NORTHERN ENERGY AND CHEMICAL INDUSTRIES PLC			2-Jan-9X
PROJECT:	Petrochemical Plant	CODE: THNS	ISSUE:	A
WORK AREA:	______________	CODE: ________	AUTHOR:	JRT
WORK PACKAGE:	______________	CODE: ________	APPRVD:	CME
ACTIVITY	______________	CODE: ________	DATE:	02-Jan-9X

	1000 tonne per day plant				SCALE	COST	1500 tonne per day plant			
	Material	Erection	Function	Plant	exponent	Factor	Material	Erection	Function	Plant
	£.000	£.000	£.000	£.000	n	15^n	£.000	£.000	£.000	£.000
Main Plant Items										
- Vessels	13.33	0.63	13.98		0.65	1.30	17.35	0.82	16.17	
- Furnace and boiler	2.89	0.14	3.03		0.70	1.33	3.84	0.18	4.02	
- Machines and drives	9.73	0.46	10.19		0.75	1.38	13.19	0.82	13.61	
- Vendor packages	6.77	0.32	7.09		0.75	1.36	9.18	0.43	9.61	
- Other	0.00	0.13	0.13		0.70	1.33	0.00	0.17	0.17	
MPI total: Materials	32.72	–	32.72				43.55	–	43.55	
MPI total: Erection	–	1.67	1.67				–	2.22	2.22	
Bulk Items										
- Piping	1.22	1.89	3.10		0.70	1.33	1.62	2.50	4.12	
- Instruments	0.64	1.10	0.14		0.60	1.28	0.82	0.13	0.94	
- Computer control system	1.56	0.88	2.44		0.70	1.33	2.07	1.17	3.24	
- Electrical	1.62	0.53	2.35		0.70	1.33	2.42	0.70	3.12	
- Structural		0.26	0.28		0.65	1.30	0.00	0.34	0.34	
- Civil		2.11	2.11		0.65	1.30	0.00	2.75	2.75	
- Painting		0.10	0.10		0.65	1.30	0.00	0.13	0.13	
- Insulation		1.50	1.50		0.65	1.30	0.00	1.85	1.95	
- Buildings		0.12	0.12		0.65	1.30	0.00	0.16	0.16	
- Plant modification		0.70	0.70		0.70	1.33	0.00	0.93	0.93	
Bulk items total	5.24	0.18	13.42				6.93	10.75	17.58	
TOTAL DIRECT COSTS				47.81						63.45
Engineering - Design			8.40		0.50	1.22			10.29	
- Software			0.53		1.20	1.63			0.86	
Construction - Management			3.22		0.65	1.30			4.19	
- Services			1.50		0.65	1.30			1.95	
Works - Start-up			6.70		0.65	1.30			8.72	
- Working capital			9.56		1.00	1.50			12.89	
Contingency			4.78						6.34	
TOTAL INDIRECT COST				34.69						45.05
CAPITAL COST OF ERECTED PLANT				82.50						108.50
Inflation			40.13						5.42	
License loos and royalties			0.41						0.54	
Insurance			0.83						1.08	
TOTAL OVERHEADS				5.36						7.05
TOTAL CAPITAL COST				87.86						115.55

FIGURE 8.6 Estimate for a petrochemical plant, plant level.

PROJECT ESTIMATE	NORTHERN ENERGY AND CHEMICAL INDUSTRIES PLC		2-Jan-9X
PROJECT: Petrochemical Plant	CODE: THNS	ISSUE: A	
WORK AREA: Synthesis	CODE: THNSS	AUTHOR: JRT	
WORK PACKAGE:	CODE:	APPRVD: CME	
ACTIVITY	CODE:	DATE: 02-Jan-9X	

	Material £.000	Erection £.000	Function £.000	Plant £.000	Parametric ratio Function % MPI	Parametric ratio Plant % MPI
Main Plant Items						
- Vessels	4.85	0.23	5.08			
- Fusion and boiler	0.00	0.00	0.00			
- Machine and drives	3.67	0.17	3.54			
- Vendor Packages	1.55	0.07	1.62			
- Other	0.00	0.00	0.00			
MPI Total Material	10.07	_	10.07		100.0%	
MPI Total Erection	_	0.47	0.47		4.7%	
Bulk Items						
- Piping			1.21		12.0%	
- Instruments			0.23		2.3%	
- Computer control system			0.62		8.1%	
- Electrical			0.61		8.0%	
- Structural			0.09		0.9%	
- CM			0.76		7.5%	
- Painting			0.03		0.3%	
- Insulation			0.50		5.0%	
- Buildings			0.05		0.6%	
- Plant modification			0.24		2.4%	
Bulk Items Total			4.74		9.47	
TOTAL DIRECT COST				15.29		1.52
Engineering - Design			1.72		17.1%	
- Software			0.00		0.9%	
Construction - Management			0.80		7.9%	
- Services			0.33		3.3%	
Works - Start-up			1.43		142%	
- Working capital			3.06		30.4%	
Contingency			1.53		15.2%	
TOTAL DIRECT COST				8.96		0.89
CAPITAL COST OF ERECTED PLANT				24.24		24.41
Inflation						
License loos and royalties						
Insurance						
TOTAL OVERHEADS				0.00		
TOTAL CAPITAL COST				24.24		

FIGURE 8.7 Estimate for a petrochemical plant, work-area level.

8.3 ESTIMATING TECHNIQUES

First I describe four basic techniques for estimating, and then show how those are realized in three different industries. I suggest for your projects you may want to use some amalgam of these.

Four Techniques

There are four basic techniques for estimating.

Detailed, Bottom Up. This is estimating from first principals. You follow the break-down in Table 8.2 to the lowest level, and estimate the detailed cost of every task and every piece of material bought. If your project is an alien (Sec. 1.1) you have no choice but to follow this approach. However it is very expensive. Figure 8.2 shows that typically the estimate accurate to ±5 percent costs 5 percent of the cost of the project; but it can cost as little as 1 percent and as much as 10 percent. If you use detailed estimating you will find it costs you 10 percent. The way to get cheaper estimates is to maximise the use of historical knowledge, which the other three methods do.

A variation of this technique, which reduces the cost somewhat, is to maintain a cost book of the cost of standard plant components. This obviously requires the collection of the historical data, and you just have to start doing that at some point. The standard costs are usually held in a computer system and so it is usually said to be computer-based.

Comparative, Top Down. The other extreme is to do comparative estimates, top down. You find similar projects you have done in the past, and extrapolate the cost, making allowances for differences in size, scope, or risk. With runners (Sec. 1.1) you may extrapolate at the project level, and with repeaters, at each stage of the life cycle. With strangers you may even be able to find familiar work packages, and extrapolate the cost of those, leaving detailed estimating for the completely unfamiliar parts.

Functional Estimating. With this technique, you identify the functions that the new asset will have, and using a wealth of historical data, estimate from past experience the cost of delivering each function. There's the rub; it requires a wealth of historical data, which you have to start collecting sometime. But you will find that in the engineering, building, and computer industries, people have been collecting that data for some time, and you can obtain ready access to it.

Parametric Estimating. You determine the cost of main components of the new asset and estimate the cost of all the standard peripheral equipment using standard ratios. Again you need to start collecting the historical data at some point, but it is published in the engineering, building, and computer industries. At later stages of the life cycle you will estimate the cost of the main components using supplier quotes, but at the earlier stages you may use functional or comparative estimating to estimate their cost.

Functional and comparative estimating cannot really deliver estimates accurate to ±5 percent and so will be used at the earlier stages of the life cycle. Parametric estimating can deliver estimates that accurate, and at a cost of 1 percent of the cost of the project.

Estimating in Different Industries

Much of the data already exists in the engineering, building, and computer industries and is readily available. Table 8.6 gives an overview of the techniques in the three industries.

TABLE 8.6 Comparison of the Estimating Methods for Three Industries

Technique	ECI	Building	Computing
Detailed	Detailed	Schedule of rates	Detailed
Computer-based	Computer-based	Bill of quantities	Computer-based
Comparative	Exponential	Empirical	Analogy
Functional—asset level	Step counting—plant level	Functional Approximate	Mathematical
Functional—lower level	Step counting—work area level	Elemental	Function point analysis
Parametric	Factorial	–	Function point analysis

Engineering Construction Industry. The engineering construction industry (ECI), or process plant industry, has a history of cost estimating going back almost 100 years, so the estimating techniques are well developed, with a wealth of historical data.[2]

Detailed estimates: These are prepared by contracting companies tendering for work, where the level of accuracy is of the same order of magnitude as the expected profit margin. At the lowest levels the costs are derived from standard cost books or from parametric data.

Computer-aided estimating: This is used to support parametric estimating and detailed estimating. They are often based on a bill of materials (BOM) or a bill of quantities (BOQ), for standard components. Possible sources of data for preparing estimates are

- Suppliers' quotations (typical, budget, detailed)
- Trade literature, technical literature, text books, and government literature
- Company historical data, standard costs, and personal records
- Computer systems

Exponential methods: These assume cost is proportional to the size of the facility, to some power. At the plant level the exponent is two-third, and so it is known as the two-thirds power law. At the equipment level the exponent is usually between 0.6 and 0.75. If you know the cost of a plant of standard size, the cost of a larger or smaller one can be derived. The law can be applied at several levels of breakdown; the lower the level, the more accurate the estimate at the plant level. Figure 8.6 extrapolates from a 1000 tonne/day plant on the left to a 1500 tonne/day plant on the right using data from Gerrard.[2]

Step-counting methods: These assume cost is function of the number of functions and plant throughput. In the engineering construction industry, standard formulae and tables have been derived from empirical data. Some of these formulae are still valid after 20 years, because of the stability of the technology. The formulae exist at several levels of breakdown, the plant level, plant area level, or main plant item (MPI) level.

Parametric methods: These assume costs are proportional to the cost of the MPI. Tables of ratios exist giving the cost of other items, such as piping, instruments, and structures, as ratios of the MPI, dependent on its value, its type, and the severity of duty. These tables exist at several levels of WBS. Figure 8.7 contains data at the plant area level, from Gerrard.[2] The techniques are so advanced in the ECI that estimates based on prices of placed order and derived at the equipment levels are sufficiently accurate for the control estimate. It is in this way that the cost of estimating is being reduced.

The estimates in Figs. 8.6 and 8.7 were in fact derived using both the computer-based and parametric methods, and taking the average. The cost of the plan derived by each

TABLE 8.7 Estimating Methods Used to Prepare Types of Estimate in the ECI

Type of estimate	Accuracy	Estimating methods
Proposal	±50%	Step-counting Exponential (plant level)
Budget	±20%	Exponential (MPI level) Parametric (plant level)
Sanction	±10%	Parametric (MPI level, vendor quotes)
Control	±5%	Parametric (MPI level, firm prices)

method was within about 1 percent of each other, though the cost associated with some of the lower level items was only to within 10 percent (illustrating the validity of Table 8.2).

Table 8.7 shows how the different techniques can be used to meet the needs of the different types of estimate in Table 8.1.

Building Industry. Methods used in the building industry include

Schedule of rates: This is not so much an estimating method, as a detailed breakdown of the cost of doing individual tasks on a building or construction site. A schedule of rates can be used for building up a detailed estimate. Or they can be used for building up costs associated with small projects, or even individual, isolated tasks, such as maintenance projects and maintenance jobs respectively. A schedule of rates will often be used on cost plus contracts.

Bill of quantities: This is equivalent to the computerized estimate. It will often be built up from a CAD drawing of the building, using standard bills of quantities for repeated elements.

Empirical estimating: Costs are extrapolated from the cost of schemes of similar size, scope, and type. Historical data is used to establish overall parameters and indicators which influence cost. These can be derived by regression analysis or curve fitting, from established data or industry standard formulae.[4]

Approximate methods: The cost is assumed to be proportional to the "lettable" floor area for a building of appropriate type, use, and quality. Tables of figures are given in Spon.[4,5] These include costs not only for the whole building but also for individual services (all related back to the area of the whole building). The cost given is proportional to floor area and can range by a factor of three for a given type of building, and so it is important to be aware of the use and quality. The user must also be aware of what services are and are not included in the costs calculated. However, the figures give estimates accurate enough for proposal estimates.

Functional methods: A coarser method of approximate estimating is to estimate in terms of the functional requirements, that is, cost per bed in a hospital, the cost per pupil in a school, or the cost per seat in an office building. These estimates have the same validity in terms of location and time as the approximate methods, and will be prepared at an earlier stage of the project than the approximate methods.

Elemental estimating: The building is broken down into major elements, and the cost estimated as a ratio of the assumed duty or floor area of that element. For instance, in a hospital the elements will be wards, theatres, radiography; in a hotel rooms, dining, kitchen, bars, and the like. The difference between this and the previous method is that the cost of each service is calculated from the size of that service, not the floor area of the whole building. This method can produce an estimate accurate enough for budget, or even sanction purposes. Once this estimate has been accepted, it can be used to generate a complete bill of quantities.

Computing Industry. The information technology (IT) industry has developed a set of estimating techniques to meet its own particular needs. There are major differences between estimating on software projects versus construction projects, for the following reasons:

- Software projects are not mechanistic, (though neither is engineering design). The activities are indeterminate and cannot be measured by simple means. Task size and complexity can be assessed by experts, but this is not normally reliable. The more complex the project, the less reliable the estimate.
- Because of the rapid change of technology, there is not a wealth of historical data. Technology changes almost faster than data can be gathered.[6]

Techniques for estimating on software projects are described by many authors,[7,8,9,10,11] and include

Bottom-up estimating: This built up from the knowledge of the design of the system. It is most effectively used to provide an estimate of the next stage of a project prepared on completion of the current stage. The technique is expensive, and has several disadvantages which mean it must almost always be used in conjunction with other techniques:

- Errors tend to compound, usually resulting in underestimation of the total cost of a system.
- It takes no account of the shortened project timescales—two people do not take half the time of one person to do a job.[12]

Top-down estimating: The estimate is made against stages of a standard life cycle and activities within the life cycle, often applying fixed percentage allocations to each stage. This approach has several advantages:

- A detailed design of the final system is not required, so the approach can be used at an early stage.
- The technique is comparatively inexpensive.
- It does not constrain the use of other techniques.

Analogy: Estimates are made by comparison to previous, similar projects. This is probably the most valid technique for many organizations, but does rely on historical records. The technique relies on the use of a consistent software development life cycle. Using the technique to extrapolate between projects of different size can also be fraught with danger, given the nonlinear relationship between size, effort, and timescale.

Mathematical models: These relate effort and time to lines of code, similar to step-counting and exponential methods. They rely on historical data, and must be tailored to an organization's needs. The models only apply to the development stage of a project. In many of the models, the equations take the following form:

$$\text{Effort} = A \times \text{size}^b$$

$$\text{Time} = C \times \text{effort}^d$$

where size is measured in thousands of lines of code, effort in months of work, and time in elapsed months. Table 8.8 contains coefficients for several models.[13] At most the exponent *b* is greater than one, giving relatively larger cost for bigger systems. Table 8.8 also shows the effort and time predicted by the different methods for a system of 40,000 lines of code, and software development costs of $8,000 per month. The figures vary wildly. Each model was developed within one organization, and therefore represents the

TABLE 8.8 Mathematical Estimating Models

Model	a	b	c	d	Effort months	Duration months	Cost $
Watson Felix, (IBM)	5.2	0.91	2.47	0.35	149	14.2	1184
Nelson, SDC	4.9	0.98	3.04	0.36	192	19.8	1456
COCOMO, organic	2.4	1.05	2.5	0.38	115	15.2	924
COCOMO, semi-d	3.0	1.12	2.5	0.35	187	15.6	1494
Frederic	2.4	1.18	–	–	186	–	1492
COCOMO, embedded	3.6	1.20	2.5	0.32	301	15.5	2410
Phister	1.0	1.28	–	–	110	–	882
Jones	1.0	1.4	–	–	175	–	1400
Halstead	0.7	1.5	–	–	177	–	1416

characteristic of that organization. This means organizations should develop their own models, and organizations should question their software development environment if their estimates are uncompetitive.

Function point analysis: The mathematical models apply only to the development stage of the project (cutting code), which typically accounts for only 50 percent of the cost. Function point analysis counts the function points, which represent the total functionality of the system.[13] Function points include

- Inputs: forms and screens
- Outputs: reports and screens
- End-user enquiries
- Logical data files
- Interfaces to other systems

Function points are converted to an estimate by:

- Comparison with previous systems: applicable to the whole life cycle
- Conversion to lines of code: applicable to the development stage only

Updating Estimates

Estimating data is only valid at a certain time, in a certain place, and in a given currency. It will often be necessary to allow for inflation, and may be necessary to convert from one country to another and one currency to another. Tables of ratios exist for these conversions.[2]

8.4 CONTROLLING COSTS: OBTAINING VALUE FOR MONEY

In describing how to control cost, I want to start with a simple method, which can be used as an extension of the spreadsheet used for estimating in Fig. 8.4. I will then show that this is compatible with the earned value method (EVM) or EVA[14] which is appropriate for larger, more complex projects.

Simple Method

In Fig. 7.2, I introduced a control cycle, and said that we need to compare what we are achieving against what we planned to achieve. In the simple, direct method of controlling cost this means comparing our forecast costs at completion against the planned cost at completion. The forecast cost at completion can be calculated quite simply as:

Forecast cost at completion = actual cost to date + estimated cost remaining

The two elements on the right-hand side of this equation can be further broken down so that:

Forecast cost at completion = actual cost of work finished + actual cost of work in progress + estimated cost to complete work in progress + estimated cost of work not yet started

Three components on the right-hand side are easily determined. We can gather information on how much we have spent to undertake the work we have done so far, both that finished and that in progress, and we can estimate the cost of the work not yet started as simply as we estimated the cost in the first place. In fact, with the work not yet started we can either use the original estimate, or update the estimate if we have some new information that suggests it might be different than what we first thought. The one element for which there is some doubt is the forecast to complete the work in progress, because that requires us to estimate how much of it we have already done, which is notoriously difficult. Techniques have been developed to do that. One is to ask people to report on their time sheet how much work they have got left to do, and hopefully they will make an honest estimate, and not just subtract what they have done from the original estimate. One of my clients assumed that work in progress was on average a third finished, and another that work in progress was on average half finished. The latter seems logically right but the former gave a better answer. However, we are only talking about a small part of the project, so the total error introduced is not great.

Table 8.9 gives a simple example of this in practice. The project consists of five packages of work, each estimated to cost 100 units, so the estimate for the project is 500. The third column shows what we expect to have spent on the mid-day of the project. It shows the amount of work we expect to have done, measured by how much we estimate it will cost. We plan to have finished A and B, done half of C, and none of D and E. The fourth column shows what we have actually spent on that day. This shows the difficulty we run in comparing the actual cost to what we have planned to spend, comparing the fourth with the third column. All we can see is that D has started early, and that A is overspent. But are A and B finished, and how much of C and D have we done for what we have spent? In the fifth column we have asked

TABLE 8.9 Simple Cost Control Example

Work pack	Estimate	Planned work	Actual	Complete	Work Rem 1	Work Rem 2
A	100	100	120	100%	0	0
B	100	100	90	100%	0	0
C	100	50	60	40%	60	90
D	100	0	20	30%	70	50
E	100	0	0	0%	100	70
Total	500	250	290		230	210
Forecast cost at completion					520	500

the team to estimate how much they have done. A and B are finished, so A is overspent and B underspent. C is running behind schedule, but worse, the work we have done has cost us 50 percent more than it should. D has started early, but the work done has cost us a third less than it should. So in the sixth column we estimate how much it will cost to finish C, D, and E, assuming the original estimate is correct. That gives us a forecast cost at completion of 520, (the 290 we have already spent and the 230 we have left to spend). However, do we believe the sixth column? We are 50 percent overspent on C, so do we expect the rest to cost the original estimate? The seventh column recalculates the estimated cost remaining. We expect the remaining 50 percent of C to cost 90. With D we are one third underspent, so at the end we expect it to have cost us 66.6. I have rounded this up to 70 and said there is 50 remaining. What information do we have about E? Here I know D and E are identical, so if D costs 70 so will E. That gives us a revised forecast cost at completion of 500 (the 290 we have already spent and the revised estimate of 210 remaining). So we are still on target for our project.

Figure 8.4 contains the control data for the CRMO Rationalization Project at a point partway through the project. For both the estimate of labour and the estimate of external cost, the second column contains the actual date, and the third column the team's estimate of how much is remaining. The fourth column contains the percentage completion. At the bottom there is the forecast completion in both instances.

I need to qualify the calculation of forecast cost at completion in three ways:

Actual Cost Is Commitment. When calculating actual cost, you need to include everything that you are committed to spend, not just what you have actually spent. For internal labour this is the same thing. The team will have completed their time sheets; you know what they have spent. But for external suppliers, they may not have submitted their invoices yet, or you may not have paid them. But once the work is done you are committed to that expenditure and so it must be included in the calculation of actual cost. I said above that actual cost is quite easily determined, but there is a proviso here. If the contractor is on a fixed-price contract, once the work is done you know the commitment (unless there has been a claim or variation). If the contractor is on a time and materials contract, you need to make sure they keep you informed of what the commitment is by reporting their actuals to you.

Contingency. Which of the three estimates in Table 8.5 do you compare the forecast cost at completion to: the raw estimate, expected cost, or appraisal or budget estimate? I suggest you compare it to the expected cost. This is the estimate the project manager is using for control. That means the estimates for work remaining should include contingency. If you have added contingency as a blanket percentage, you should continue to add it as the same percentage to the estimates of work remaining. If originally you added contingency to individual components of work, you should continue to add it to those components not started, and a proportion to those not finished. For work completed, contingency is automatically consumed through the actuals.

Earned Value. So far I have compared forecast cost at completion to the original estimate. That is quite simple, and provides a neat solution. However, many people like to compare what they have spent on the work they have done. Therefore they calculate what is known as earned value as a measure of the amount of work they have done, and compare that to the actual cost. They can then see whether they are over- or underspent on the work they have done so far. There are many ways of calculating the earned value, but following what I have done so far, the earned value can be calculated as:

Earned value = original estimate – estimated cost remaining

Thus in Table 8.9, using the first estimate of work remaining, earned value is 270 (500 – 230), and so we are 20 units overspent on the work we have done. Using the second estimate it is 290 (500 – 210) and we are exactly on plan. This is what we determined above focusing on the figures at completion. However, we can now plot out performance as we go along (Fig. 8.8) getting a highly visual picture comparing what we have spent to the work we have done.

What I have done so far is known as *forward-looking control*, I can only control the future, and so I focus on what I have left to spend; that is what I can do something about. Another way of calculating earned value is to try to add up the value of the work done so far. In fact, for the sixth column in Table 8.9 it leads to the same answer as we shall see; it is 270 units. But this is backward-looking or rearview mirror control, and using that you cannot calculate the seventh column in Table 8.9 which gives us a better picture of the project.

I am now going to describe the earned value method in greater detail, but before I do let me explain why it is called earned value. It is a measure of the amount of work done, so it is a measure of the value earned for the money spent. We control costs by measuring the value we have earned for the money we have spent; hence the title of this section.

Earned Value Analysis

Earned value analysis, or the earned value method, has now become a core, established technique of project management.[10,11] It is very powerful and recommendable (though perhaps in the simplified form described above on smaller projects). I am going to finish this chapter by giving an overview of the full technique, though as I have said, I would suggest you use the simpler technique described above on most projects, and keep the full technique only for larger more complex projects. Figure 8.8 illustrates the technique. Throughout the project we plot three things:

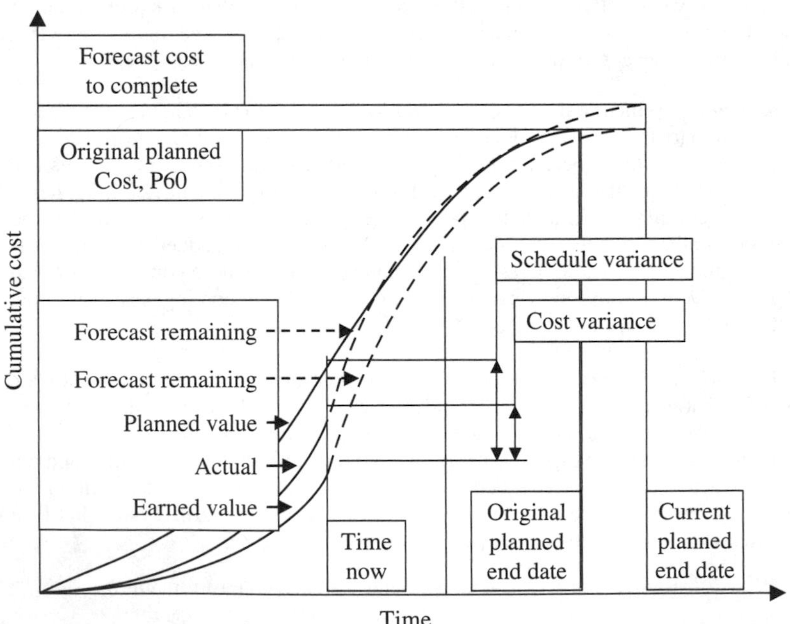

FIGURE 8.8 Earned value analysis.

Planned value (PV): Before the project starts, we estimate how much each element of work will cost, and schedule when it will be done (Chap. 9). We can therefore calculate the expected cash flow through the project. I suggested at the start of the chapter that that was one of the purposes of cost estimating. We call this the *planned value*, the value of the work we have planned to have done at each point through the project. When I was first exposed to EVA in the 1980s, this was called variously the *planned cost of work scheduled (PCWS), baselined cost of work scheduled (BCWS)*, or the *baselined cost of work planned (BCWP)*. At some point during the 1990s, somebody called it the budgeted cost of work scheduled, also BCWS, and this became incorporated into the Project Management Institute PMI PMBoK[15] and so has now become standard practice. However, I disagree with this. The project manager should control progress against the expected (baselined) cost, 1050 in Table 8.9, not the budgeted (appraisal) estimate, 1100 in Table 8.9. According to PRINCE2 the tolerance (the difference between the expected and the budget) belongs to the project board and not the project manager,[3] and so the project manager tracks progress against the expected. This is my recommendation, so the planned value should include contingency, but not tolerance, and so should be based on the expected cost, not the budget estimate or sanction value. Calling it planned value, PV simplifies the acronyms.

Actual cost (AC): On a certain day we can calculate how much money we have actually spent on the project. Remember above I said this should include all commitments, not just money actually paid. This has always been called the *actual cost of work complete (ACWC)*, but I found myself having to phrase what I said above very carefully, because it is not just the cost of work packages that are finished, but also of work in progress. Calling it the *actual cost*, or actual cost to date, AC, simplifies the acronyms again.

Earned Value (EV): The comparison of PV and AC tells us nothing. If AC is less than PV we do not know if that is because the work is underspent, or behind schedule. In fact one of the most misleading things is when the work is well behind schedule and overspent, but you are lulled into thinking it is underspent (Example 8.5). For this reason we also need to monitor how much work has already been done. We do this by calculating the value earned, that is the estimated cost of the work that has actually been done for the money spent to date. Again in the 1980s this was called the *planned cost of work complete (PCWC)*, or the *baselined cost of work complete (BCWC)*, but accepted practice is now to call it *budgeted cost of work complete*, also *BCWC*, consistent with PV.

Example 8.5 The need to monitor work complete

I audited a project that had gone disastrously wrong. An English company had bid a fixed price of US$20 million to do a job in Israel. At the start of the work they had bought US$20 million to hedge currency movements, and throughout the project the finance director was monitoring the expenditure of this money against the forecast rate of expenditure. Up until about 90 percent of the money had been spent, the rate of expenditure was running slightly behind the forecast cash flow. However, when 90 percent had been spent, the forecast was beginning to flatten off like the S-curves in Fig. 8.8, but the rate of expenditure was continuing at the same rate, and so now more money had been spent than predicted. At this point the finance director asked the project manager how much work was left to do, and the project manager said he was half finished!!! 90 percent of the money had been spent to do 50 percent of the work. The company would have become insolvent if it had not been bailed out by the parent company.

EV can be calculated in several ways. The norm is to calculate the percentage completion of all work started. EV is then the sum of percentage completion multiplied by the original estimate. This gives the answer in Col. 6 of Table 8.9, EV = 230. This calculation is done in the right-hand columns under labour and cost in Fig. 8.4. For work packages that are finished, you can add the estimate into the calculation of EV. For work packages in progress you may make a guess at percentage completion, but better is to drop to the activity level and do the EV calculation at that level for activities finished and in progress to calculate the EV of the work package. The alternative is to subtract the estimate of work yet to be done from the original estimate. If you use the original estimate of the work elements you get the same answer, but as I said above, the advantage of forward-looking control as opposed to backward-looking, rearview-mirror control is you can adjust the estimates of work yet to be done based on experience so far, to get a more realistic picture, as illustrated by Table 8.9.

The calculations really ought to be done using the raw estimates for the packages of work, 100 in Table 8.5, with the contingency held as an amount to be added at the end. But you might use the expected values for the work packages, 105 in Table 8.5.

The comparison of EV with AC tells us whether the project is over- or underspent for the work we have done. The comparison of EV with PV tells us whether on average the project is ahead or behind schedule. It is only on average; progress on the critical path (Chap. 9) tells us how we are doing on the work which will determine the duration of the project. We can calculate four further parameters to indicate overall project performance: cost variance (CV); schedule variance (SV); cost performance index (CPI); and schedule performance index (SPI).

$$CV = AC - EV$$

$$SV = PV - EV$$

$$CPI = EV/AC$$

$$SPI = EV/PV$$

If CV is positive the project is overspent, and if SV is positive the project is behind schedule. However, it is best practice to describe the two variances as favourable (negative) or unfavourable (positive). If CPI is less than one the project is overspent, and if SPI is less than one the project is late. My inclination would be to calculate the reciprocal of the two indices. I don't know why standard practice is to do it this way, but it might be because we expect most projects to be late and overspent, so it is easiest to have these numbers usually less than one—but that is too cynical for words. Using the sixth column in Table 8.9, CV = 20 unfavourable, we are 20 units overspent, and SV = 20 favourable, we are 20 units ahead of schedule. CPI = 93 percent and SPI = 108 percent.

We can also use these figures to calculate the forecast cost at completion (FCaC). We either assume the rest of the project will be done according to the estimate, in which case:

$$FCaC = Estimate + CV$$

or we assume that we continue to overspend (or underspend) at the same rate, in which case:

$$FCaC = Estimate/CPI$$

For Table 8.9, this gives 520 or 538. The first figures were also the first calculated above using the simple method; that is not coincidental. Thus for Table 8.9 we calculate three different numbers for FCaC. Using forward-looking control we calculated 500, and using backward, rearview-mirror looking control we calculated 520 or 538. I prefer forward-looking control because we can try to do something about our future. The three calculations are done for the CRMO Rationalization Project for both internal labour and external costs in Fig. 8.4.

Figure 8.8 provides a highly visual representation of how the project is performing.

SUMMARY

1. A cost estimate is prepared
   - As a basis for control
   - To assess the project's viability
   - To obtain funding
   - To allocate resources
   - To estimate durations
   - To prepare tenders for bespoke contracts
2. There are four types of estimate of increasing accuracy requiring proportionately more work to prepare:
   - Proposal estimate
   - Budget estimate
   - Sanction estimate
   - Control estimate
3. The proposal estimate is prepared at the concept stage to commit resources to the feasibility. The budget estimate is prepared during feasibility to initiate the project, and commit resources to design. The sanction estimate is prepared during design to gain funding for the project, or approval from the project sponsor. The control estimate is prepared during implementation planning.
4. There are several types of cost to be estimated, including:
   - Labour
   - Materials, plant, and equipment
   - Subcontract
   - Management, overhead and administration
   - Fees and taxation, inflation, and other contingency
5. The cost control cube, a three-dimensional matrix of the WBS × OBS × CBS provides a structure for estimating and controlling costs. The estimate is prepared by breaking the work down to an appropriate level of WBS, and then estimating the cost of each element in the cost control cube. Effectively we estimate each type of cost for each cell in the responsibility chart at that level. Spreadsheets can be used to support this process.
6. There are five techniques for estimating cost:
   - Detailed or bottom up
   - Computer supported
   - Comparative
   - Functional
   - Parametric
7. Cost is most easily controlled by forecasting cost at completion, by adding the actual work to date to the estimated cost of work remaining, and comparing that to the original estimate.
8. Cost can also be controlled by comparing the earned value, a measure of the amount of work performed to date, to the actual expenditure to date. A comparison of earned value to the originally planned cash flow helps to control elapsed time. S-curves provide a visual representation.

REFERENCES

1. Turner, J.R. and Wright, D., *The Commercial Management of Projects*, Aldershot, U.K.: Gower, 2009.
2. Gerrard, A.M. (ed.), *A Guide to Capital Cost Estimating,* Rugby: Institution Chemical Engineering, 2000.
3. Office of Government Commerce, *Managing Successful Projects with PRINCE2,* 4th ed., London: The Stationery Office, 2005.
4. Spain, B.J.D., *Spon's Budget Estimating Handbook,* London: Spon Press, 1994.
5. Mott Green & Wall *Mechanical and Electrical Services Price Book: 2002,* 2002 ed., London: Spon Press, 2002.
6. Turner, J.R., *Managing Web Projects: The Management of Large Projects and Programmes for Web-space Delivery,* Aldershot, U.K.: Gower, 2004.
7. Boehm, B.W., *Software Engineering Economics*: Prentice-Hall, 1981.
8. Boehm, B.W., Apts, C., Brown, A.W., and Chulani, S., *Software Cost Estimation with COCOMO II*: Prentice Hall, 2000.
9. DeMarco, T., *Controlling Software Projects: Management, Measurement, and Estimation,* Yourdon Monograph: Prentice-Hall, 1982.
10. Londeix, B., *Cost Estimation for Software Development,* New York: Addison-Wesley, 1987.
11. Pressman, R.S., *Software Engineering: A Practitioners Approach,* New York: McGraw-Hill, 2004.
12. Brooks, F.P., *The Mythical Man-Month,* 25th anniversary ed., New York: Addison-Wesley, 1999.
13. Albrecht, A., "Software function, source lines of code, and development effort prediction," *IEEE Transactions on Software Engineering,* November 1983.
14. Fleming, Q.W. and Koppelman, J.M., *Earned Value Project Management,* 3rd. ed., Newtown Square, PA.: Project Management Institute, 2006.
15. PMI, *A Guide to the Project Management Body of Knowledge,* 3rd. ed., Newtown Square, PA.: Project Management Institute, 2004.

CHAPTER 9
MANAGING TIME

We now turn to the fifth function, managing time, by which the project manager coordinates the efforts of those involved, delivers the change to meet market opportunities, and so ensures revenues are derived at a time which gives a satisfactory return on investment. All three of these purposes for managing time imply it is a soft constraint on most projects. Being late reduces the benefit; it does not cause the project to fail absolutely. There are only a few projects for which there is an absolute deadline. Project Giotto, the space craft which intercepted Halley's comet in 1986 was one: there was a very small time window in which to make the rendezvous, and if missed it would not reoccur for 76 years.[1] Another is the preparation for the Olympic Games. The start date is known six years in advance, to the nearest minute, and to miss that minute would be embarrassing, not to mention play havoc with the TV schedules. Such projects are rare. Unfortunately many project managers treat time management as being synonymous with project management, and much of the project management software is written on this assumption.

In the next section, I consider the purpose of managing time, define the concepts and terminology of the time schedule, and introduce tools for communicating the schedule, including activity list and bar charts. I describe how to calculate the duration of work elements, and how to use networks to calculate the overall project duration. I then show how to adjust the schedule by balancing resource requirements and resource availability, and end by describing the use of the schedule in controlling the duration of a project.

9.1 THE TIME SCHEDULE

The time schedule is a series of dates against the work of the project, which record

- When we forecast the work will occur
- When the work actually does occur

Purpose of the Schedule

The purpose of recording these dates and times is

- To ensure the benefits are obtained at a timescale which justifies the expenditure
- To coordinate the effort of resources
- To enable the resources to be made available when required
- To predict the levels of money and resources required at different times so that priorities can be assigned between projects
- To meet a rigid end date

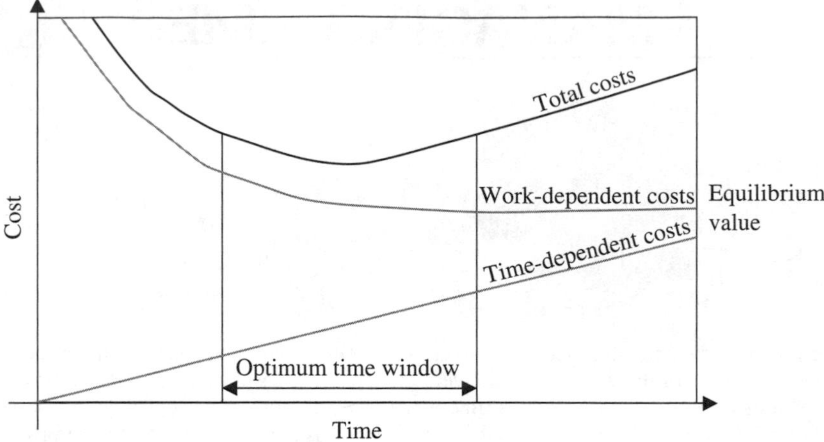

It costs you money to shorten the duration.
It costs you money to extend the duration.

FIGURE 9.1 Timing of minimum cost of a project.

The first of these is the most important. It addresses the *raison d'être* of project management, achieving the overall purpose. The second is the next most important as it enables the project to happen. The third and fourth are variations. It is the fifth item that gets most attention from project managers. They set a rigid end date, sometimes unnecessarily, and focus on this to the detriment of cost and quality. Part of the aim of managing the time is to optimize the cost and returns from the project. Figure 9.1 shows that the cost is made up of two elements:

- *A work-dependent element:* 100 days of effort is the same whether 5 people take 20 days or 2 people take 50 days.
- *A time-dependent element:* the project manager's salary for instance.

However, the work-dependent element does increase as you try to shorten the project, and people interfere with each other; 10 people take 12 days, and 20 people 8, perhaps. Adding the two gives an optimum time window for the project in which cost is minimized. Figure 9.2 shows that maximum returns may not correspond to minimum cost. The value of the asset may decay with time, because of limited market windows and hence highest profit may be made at a time earlier than minimum cost. Through the time schedule we must optimize cost and benefit.

The Schedule

On a simple level, the schedule records the planned and actual start date, finish date, and duration of each work element. We may also record whether there is any flexibility in when each element may start without delaying the completion of the project. This is called the *float*. Sophisticated schedules record up to five versions of each of the start date, finish date, duration, and float: the early, late, baseline, scheduled, and actual dates.

The Duration. This is the time required to do the work. It is common to treat a work element's duration as a fixed given. For some, it is dependent on external factors beyond the control of the team. For others, it is a variable, and can be changed by varying the number of people working on the activity, or by other means. Before work starts for each activity

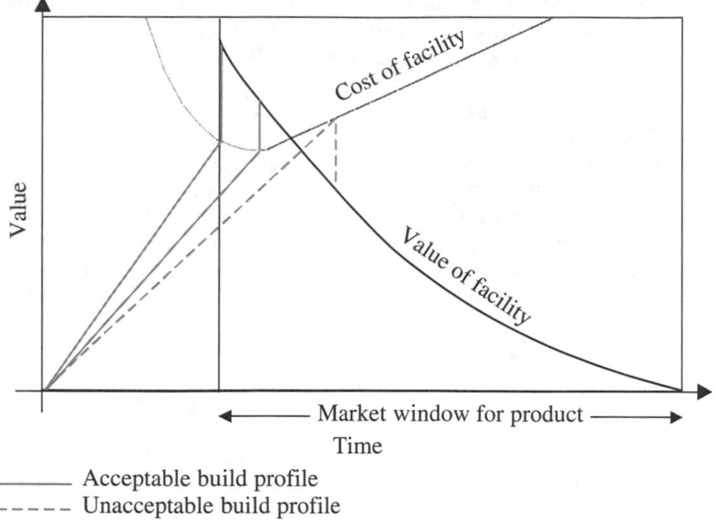

FIGURE 9.2 Timing of optimum return from a project.

we estimate its duration. Once work starts, but before it finishes we can estimate remaining duration. This may be equal to the planned duration less the time since the activity started, or we may reestimate remaining duration based on the knowledge gained from doing the work so far. Once work is complete we can record an actual duration. It is useful to record actuals because a comparison of planned and actual figures may indicate trends which may be useful in the control process.

Early and Late Dates. These are forecast from the estimated duration of all activities. The start of an activity may be dependent on other work finishing. Therefore there is an earliest date by which an activity may start. This is known as the *early start date*. The early start date plus the estimated duration is the *early finish date*, the earliest date by which the work can finish. Similarly, other work may be dependent on the activity being finished, so there is a latest date by which it can finish and not delay completion of the project. This is known as the *late finish date*, and correspondingly the late start date is this less the duration. If the late start date is different to the early start date, there is flexibility about when the element can start, the *float*:

$$\text{Float} = \text{late start date} - \text{early start date}$$

If the duration is fixed, the difference between early and late start and early and late finish is the same (and indeed this is the assumption made in most scheduling systems). However it is not too difficult to imagine situations where the duration is dependent on when the work is done and we will then get different answers if we calculate float using finish dates.

A work element with zero float is said to be critical. If a project is scheduled with minimum duration, then running through it will be a series of work elements with zero float. This series is known as the *critical path*, and the duration of the project will be equal to the sum of the durations of the work elements along the path. Work elements with large float are known as *bulk work*. They are used to smooth forecast resource usage, by filling gaps in the demands made by the critical path. There are also work elements with very small float. These are *near critical*, and should receive as much attention as the critical path. Section 9.3 describes the critical path method (CPM) networks, which are mathematical tools for calculating early and late start and finish, and float.

Planned, Baselined, and Schedule Dates. *Planned dates* are dates between the early and late dates when we choose to do the work. However, the date we planned to do a work element at the start of the project may be different to our current plan. It is important to record the original plan, because that is the measure against which we control progress. This original measure is commonly known as the *baseline date*, and the current plan as the *scheduled date*. If the baseline start is later than the early start, then the planned or baseline float will be less than the available float. Likewise, as a project progresses, if the start or finish of a work element is further delayed, then the remaining float will be less than the original float.

The Total Schedule. Hence, there are up to fifteen dates and times associated with a work element (Table 9.1). The process of scheduling the project is the assignment of values to these dates and times. First you estimate the duration and then assign start and finish dates. This is usually done by calculating the early start and late finish dates and then assigning baseline dates somewhere between these, after taking account of other factors such as resource smoothing. It is sometimes necessary to assign a finish date after the late finish and thereby delay the project. If the logic is correct it will be impossible to schedule the start before the early start.

For some projects with a well-constructed work breakdown structure (WBS), you can schedule the project manually, by nesting the schedule at lower levels within that at higher levels. To do this you need to break the project into discrete work areas and work packages, with few logical links between them and little sharing of resources. The four large multidisciplinary projects described in Sec. 5.5 were like that. In the Regional Health Authority warehouse and the Norwegian Security Centre projects, the project managers positively resisted computer systems because they felt they retained greater visibility without them. Where there are complex interdependencies and multiple shared resources, it may be necessary to use computer-aided support tools.

Communicating the Schedule

There are two accepted ways of communicating a project's schedule:

Activity Lists with Dates. This is a list of some of the work elements at a given level of the WBS, with dates listed beside them. This method of communicating the schedule gives a comprehensive check list, but is not very visible. Table 9.2 is an activity listing for a simple project to erect a statue. Although this list shows the float, I believe it should not be shown as it tends to be consumed. The responsibility chart (Figs. 6.5 and 6.6) is effectively an activity listing.

TABLE 9.1 Fifteen Time Elements of the Schedule of a Project

Early start	Duration	Early finish
Late start	Float	Late finish
Baseline start	Baseline float	Baseline finish
Schedule start	Remaining float	Schedule finish
Actual start	Remaining duration	Actual finish

Where Planned duration = planned finish – planned start
Planned float = late finish – planned finish

TABLE 9.2 Activity Listing for a Project to Erect a Statue 1

Landscape Ltd **Activity Listing**					
Project		Erect statue			
Activity		Duration	Early start	Early finish	Float
No	Name	Days	Day	Day	Days
A	Grade site	3	0	3	0
B	Cast plinth	2	3	5	0
C	Plant grass	3	3	6	1
D	Set concrete	2	5	7	0
E	Place statue	1	7	8	0

Bar Charts. The schedule can be more visibly represented using bar charts, (sometimes called Gantt charts, after Henry Gantt who pioneered their use in routine operations management). Figure 9.3(*a*) is a simple bar chart for the project in Table 9.2. This is what I think you should show to the project team to tell them when you want the work to be done. Figure 9.3(*b*) is the same bar chart with the float shown. It is also possible to show the logic in a bar chart, Figure 9.3(*c*). These second two are useful planning tools for the project manager and project planners. I do not believe in showing the project team the float, for exactly the same reason you should ask them to work on the raw estimate (Table 8.5); it tends to get consumed. You can show the team which work is critical and which is not, as shown in Fig. 9.3(*a*), but ask them to work on the planned dates, and come back and negotiate extra time for noncritical activities if they need it.

Once work has started, we can also draw a tracked bar chart, Fig. 9.3(*d*). The original schedule has now been converted into the baseline plan, the upper set of bars in each pair. The lower set is the actual dates and current schedule (actual before time now and current schedule after time now). In this way the team are given realistic dates to work to, but they can see the original schedule (baseline), and so control is maintained.

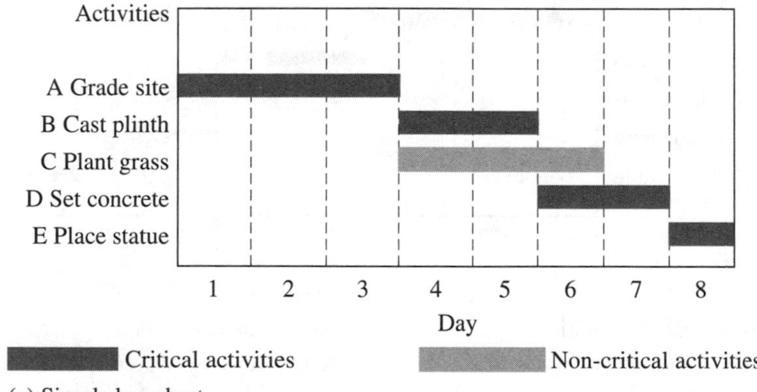

(a) Simple bar-chart

FIGURE 9.3 Bar charts for the activity listing in Table 9.2.

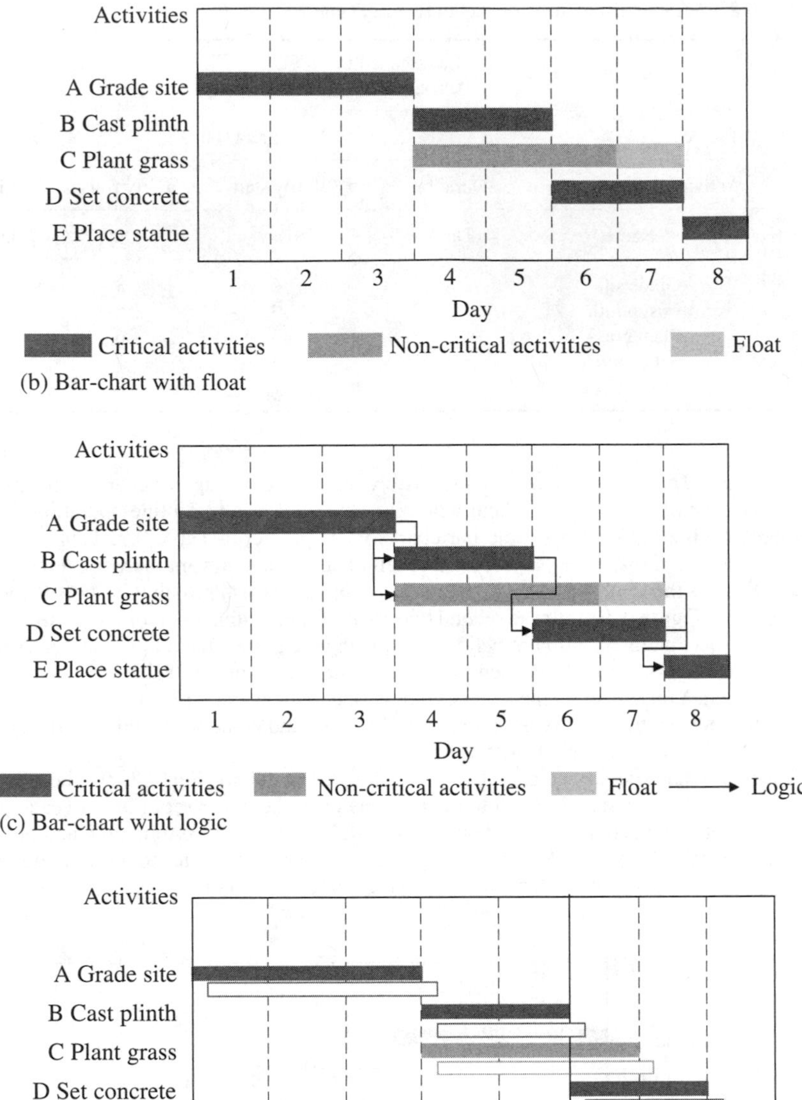

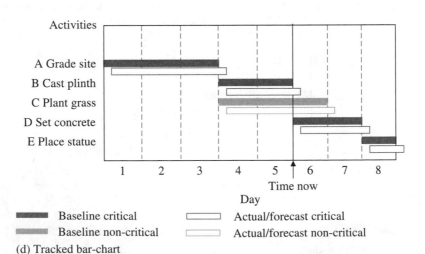

FIGURE 9.3 Bar charts for the activity listing in Table 9.2. (*Continued*)

9.2 ESTIMATING DURATION

The duration of work elements is central to scheduling, not only in relating the start and finish of a given work element, but in calculating its earliest start from the cumulative duration of the preceding activities, and the latest finish from the cumulative duration of the succeeding activities. The duration of a work element is dependent on one of three things:

1. The amount of time it physically takes to do the work involved, which in turn is dependent on the number of people available to do it.
2. The lead time, or waiting time, for the delivery of some item is independent of the number of people doing the work.
3. Some mixture of the two.

Duration Dependent on Work Content

It is often assumed the duration of a work element depends on the amount of work to do and the number of people available to do it. Nominally:

$$\text{Duration} = \frac{\text{work content(days of effort)}}{\text{Number of people available}}$$

I described the role of work content in negotiating the contract between project manager and resource providers in Chap. 6 and how to estimate it as a labour cost in Chap. 8. It is always necessary to add allowances to this raw estimate of duration, to calculate the actual duration. These allowances are to account for various factors, which include

- Time lost through nonproject activities
- Part-time working
- Interference between people doing the work
- Communication between people doing the work

Lost Time. Somebody nominally working full-time on a project is not available 5 days/week, 52 weeks/year. They lose time through holidays, sickness, training, group meetings, and the like. It is suggested that for the average project worker these consume 80 days/year; somebody assigned full-time to a project does on average 180 days of project work a year, equivalent to 70 percent availability. To allow for this 40 percent is added to the nominal duration (1.4 = 1.0/0.7). A smaller ratio will be added if the project's resource calendar allows for some lost time.

Part-Time Working. Individuals may be assigned to a project part-time. However, you must be careful not to double account. If somebody is assigned two days per week to a project, 40 percent, you must be clear whether those two days include or exclude a proportion of the lost time above before adding the 40 percent allowance.

Interference. Doubling the number of workers does not always halve the duration, because people doing work can restrict each other's access to the work face, and so reduce their effectiveness. For instance, if the task requires access to a limited space with room for just one person, adding a second person will not double the rate of working. Two will work faster than one, because they can step each other off, but only one can work at a time. Adding a third person will not increase the rate of working, and may even reduce it by distracting the other two. A third person would be most effectively used to extend the working day through a shift system.

Communication. Where more than one person work on a job, they need to communicate details of the work to each other to make progress. This is especially true of engineering design and writing software. With two people there is just one communication channel, so they may work almost twice as fast as one. With three people there are three channels, with four people six, and as the number of people grows, the channels grow as the square of the number of people. Hence, you reach a point where adding another person in fact reduces the amount of effective work (Example 9.1). The way to overcome this is to find ways of reducing the channels of communication, by using a central administrator or project support office (Chap. 16). In the office in Example 9.1, the pool was split into four pools of three secretaries. It is commonly believed that in a professional office, three is the optimum team size, balancing the additional motivation from working in a team, with the added levels of communication.

Example 9.1 Communication consumes time

In an office I worked in, there were three managers each with a secretary. As the office grew, and new managers joined, the numbers of secretaries grew, until there were about twelve working in the same pool. We reached a point where adding a new secretary seemed to make no difference to the amount of work done in the pool. If we assume a new secretary spends a quarter of an hour each day talking to each of the others, (not an unreasonable amount of time for social interaction), then each conversation consumes half an hour's work, and since he or she has twelve conversations, six hours is lost, equal to the effective working day.

Estimating Durations. Hence the estimate of duration for a work element is based on the formula above, but adjusted taking account of all the factors discussed, which may indeed dominate. This just reinforces that project management is not a mathematical exercise, but much more of a social science.

Duration Dependent on Lead Time

For some work elements the duration depends on the lead time or waiting time to obtain some item of material or information or to wait for some change to take place. This may include

- Delivery time for materials in procurement activities
- Preparation of reports
- Negotiations with clients or contractors
- Obtaining planning permission or financial approval
- Setting of concrete or watching the paint dry

Duration Dependent on Work Content and Lead Time

In some instances a work element contains lower-level activities, some of which are work dependent and some lead-time dependent (Example 9.2). The duration of the work package must be calculated from the duration of each of the activities and their logical dependence, perhaps using the networking techniques described in Sec. 9.3 in more complex cases.

Example 9.2 A work package from the CRMO Rationalisation Project containing activities of mixed type

The work-package, O5: Redeployment and Training, from the CRMO Rationalisation Project may consist of the following activities:

1. Identify training needs of staff
2. Develop training material
3. Conduct courses
4. Transfer staff to new posts

The first two of these are work dependent. The number of trainers assigned depends on the number of people requiring training and the amount of material to be developed. However, two people will not work twice as fast as one since they need to keep each other informed of progress. The duration of the third activity depends on the availability of training facilities, and the fourth on how quickly people can be assimilated into new work environments.

Estimating Sheets

The estimating sheet (Table 9.3) is a tool which can be used for estimating work content and durations. Table 9.3 shows the calculation of milestone P1 of the CRMO Rationalization Project. Example 9.3 provides a rationale.

Example 9.3 Rationale for the duration of the work package *P1: Project Definition*

The person with the most work to do is the project control officer, with 24 days of effort. The duration of the work package will be determined by his or her availability. It is assumed during project definition he or she will not take holiday. Therefore his or her availability will be greater than the average 70 percent. A figure of 80 percent is assumed. The duration is therefore 30, (24/0.8), days.

9.3 CALCULATING THE SCHEDULE WITH NETWORKS

Having estimated duration, we assign dates to work elements. I believe that on majority of projects, that can be done manually using bar charts. However, with larger more complex projects that is more difficult, and computer-aided techniques help with the calculations. The simplest of these are based on a mathematical technique called variously the critical path method (CPM), critical path analysis (CPA), or the program evaluation and review technique (PERT). The initials CPM, CPA, and PERT are used interchangeably by many people, although they do mean something slightly different. At their core is network analysis. Networks are a mathematical technique used to calculate the schedule. They are seldom useful for communicating the schedule. Bar charts or activity listings (Sec. 9.2) are best for that. Networks will only be used where the project is too complex to be scheduled manually through the WBS and so will only be used in conjunction with computer-aided systems. However, I think it is useful to know the mathematics behind the analysis.

CPM, CPA, and PERT are themselves only useful on projects of lower complexity. They are linear and deterministic; D follows C follows B follows A, with no looping back,

TABLE 9.3 Estimating Sheet with Durations Entered for the Milestone *P1: Project Definition* from the CRMO Rationalisation Project

TriMagi
Estimating Sheet

Activity			Work content			Resources, nine people					
No	Description	Duration (days)	No of steps	Effort/ step (days)	Total effort (days)	Proj Mgr	Proj Offic	CRMO TL	CRMO Mgrs	Ops Direct	Other Mgrs
						1	1	1	2	1	2
1	Produce project proposal		1	4	4	1	2	1			
2	Hold definition workshop		1	4	4	1	1	1			
3	Define required benefits		1	2	2	1		1		1	
4	Draft definition report		1	8	8	2	6				
5	Hold launch workshop	1.5	1	12	12	1.5	1.5	1.5	3	1	3
6	Finalize milestone plan		1	2	2	1	1				
7	Finalize responsibility chart		1	2	2	1	1				
8	Assess risks		1	3	3	1	1	1			
9	Prepare time estimates		20	0.1	2		2				
10	Prepare cost estimates		20	0.1	2		2				
11	Prepare revenue estimates		1	1	1		1				
12	Assess project viability		1	1	1	1					
13	Finalize definition report		1	5	5	2	3	1			
14	Mobilize team		1	3	3	0.5	0.5	0.5			
	Total effort	54 days		Subtotal	50	13	22	7	3	2	3
	Total cost, $K	40.0		Allowance, %	10	10	10	10	10	10	10
	Duration	25 days		Total effort	54	14	24	8	3	2	3
	Target start	01 Feb 0X		Unit rate, $K/day	1.0	0.5	0.5	1.0	1.5	1.0	
	Target end	05 Mar 0X		Cost, $K	14.0	12.0	4.0	3.0	3.0	3.0	

Estimating sheet with durations entered for the milestone *P1: Project Definition* from the CRMO Rationalisation Project and no branching depending on what happens at a certain step. On more complex projects still, where at a certain step several possible things can occur depending on what happens at that step, and where feedback loops can occur, more sophisticated modelling techniques are necessary,[2] but they are beyond the scope of this book.

In this section, I describe the mathematical technique of networking.

Types of Network

There are three types of network:

- Precedence networks (also called activity-on-node networks)
- Activity-on-arrow networks (sometimes called IJ networks)
- Hybrid networks

Precedence Networks. In precedence networks, work elements are represented by boxes, linked by logical dependencies, which show that one element follows another. Figure 9.4 is a simple precedence network with four activities A, B, C, and D. B and C follow A and D follows B and C. Four types of logical dependency are allowed (Fig. 9.5):

End-to-start: B cannot start until A is finished.

End-to-end: D cannot finish until C is finished.

Start to-start: D cannot start until C has started.

Start-to-end: F cannot end until E has started.

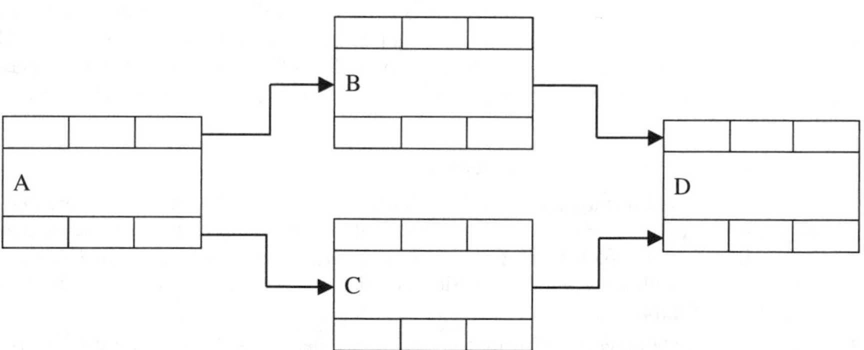

FIGURE 9.4 A simple precedence network.

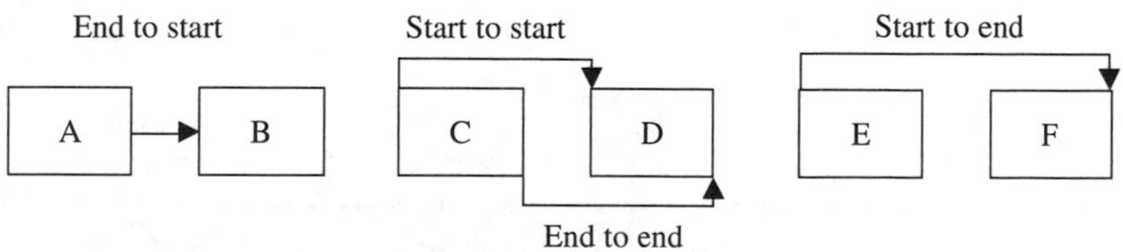

FIGURE 9.5 Four types of logical dependencies.

End-to-start dependency is usually used, (a hangover from IJ networks). End-to-end and start-to-start are the most natural and allow overlap of succeeding work elements in time. It is not uncommon to build ladders of activities like C and D. It is the use of end-to-end and start-to-start dependencies which allows fast-track or fast-build construction (Sec. 3.5). Start-to-end are only defined for mathematical completeness. I have never come across a case where it might be used. I will introduce later leads and lags on dependencies. I suggest you use only end-to-start dependencies, and use leads and lags to overlap activities. This greatly simplifies the network. The milestone plan (Sec. 5.3) is a precedence network. The circles, nodes, represent the work. The lines are end-to-end dependencies, linking the milestones.

Activity-on-Arrow Networks. These are often called IJ networks, because each activity is defined by an IJ, start/finish, number. In this type of network each work element is represented by an arrow between two nodes. The activity is known by the number of the two nodes it links. Figure 9.6 is Fig. 9.4 redrawn as an IJ network. Activity A becomes 1-2, and so on. Because activities must be uniquely defined two cannot link the same two nodes. Therefore, B and C finish in nodes 3 and 4, respectively, and these nodes are linked by a dummy activity. Because activities are linked through nodes, end-to-start logic is imposed. However, it is possible to introduce dummy activities to represent the other three logical links.

Hybrid Networks. These mix the two previous types. Work is represented by either a box (node), or a line (arrow). Furthermore, there may be boxes and lines which do not represent work, just events in time and logical dependency. A line need not join a box at its start or finish, but at any time before, during or after its duration. In advanced hybrid networks, even the distinction between nodes and lines disappears. Hybrid networks are rare, so cannot be discussed further.

Precedence versus Activity-on-Arrow Networks. You will find some people fervently committed to one or the other. The very early work on network analysis in the late 1940s was done with arrow networks, whereas precedence networks were not introduced until the mid-1950s. Therefore arrow networks tend to be the default option. However, precedence networks are often preferred by practising project managers. There are several reasons for this:

1. It is more natural to associate work with a box.
2. It is more flexible for drawing networks. All the boxes can be drawn first and the logical dependencies added later. In Sec. 5.3, I described how to develop a precedence network, milestone plan, by moving Post-Its around a flip-chart or white board. The same is not possible with an activity network since the activities are only defined by two nodes and that imposes logic.
3. It is easier to write network software for precedence networks. Most modern softwares are precedence only or both. That which is both has an algorithm to convert from precedence to IJ.

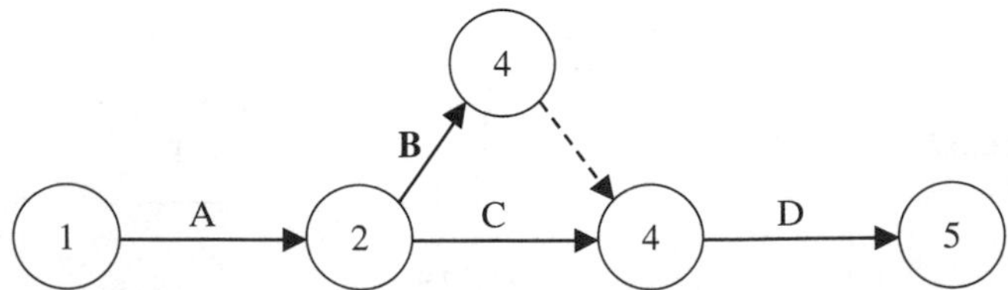

FIGURE 9.6 Activity-on-arrow network.

4. It is easier to draw a bar chart showing precedence logic with the bars representing the activity boxes and vertical lines showing the logical dependencies [Fig. 9.3(c)]. With an arrow network either more than one activity must be drawn on a line or dummies must be used to show logic, which virtually gives a precedence network. (This last statement reintroduces hybrid networks, and shows that the distinction between precedence and IJ networks really is slight.)
5. The work exists independently of the logic, and so you can draw a work breakdown structure and overlay the logic later. (People who use IJ networks have to draw the network before developing the work breakdown structure.)

Networking Technique

All networks do is to calculate the early start and finish, the late start and finish and the float of work elements in a project, given their duration and logical dependency. The reason this is useful is it allows you to explore many different options, called conducting a "what-if" analysis, assuming different durations and logical dependencies of the work elements. As I introduce networking technique, I will illustrate it by scheduling a simple project, represented by the network in Fig. 9.4. An activity listing for the network is given in Table 9.4. This is modified Table 9.2 and you will see shortly that the activity, "set concrete" has been replaced by a lag on the logical dependency from B to D.

Notation. In a precedence network, each work element is represented by a box with seven segments (Fig. 9.7). The top three segments contain the early start, duration, and early finish, respectively. The bottom three contain the late start, float, and late finish. The central one contains a description of the activity. Figure 9.8 is Fig. 9.4 with durations entered. In an arrow network the node has four segments, the identifier, the early and late time, and the float. The time is the start of the succeeding activity and the finish of the preceding activity. The duration is still associated with the activity (Fig. 9.9).

Leads and Lags. The dependencies connecting the activities in a precedence network usually have zero duration. However, they can be given positive or negative duration, and this is called *lag* or *lead,* respectively. In Table 9.4, the concrete must be left for two days to dry before erecting the statue. These two days can either be added to the duration of B (taking it to four days) or shown as a lag on the dependency. Similarly it might be possible to start planting grass on the second day after the first third of the

TABLE 9.4 Activity Listing for a Project to Erect Statue 2

Landscape Ltd **Activity Listing**				
Project		Erect statue		
Activity				
No	Name	Duration days	Preceding activity	Lead/lag days
A	Grade site	3	–	0
B	Cast plinth	2	A	–2
C	Plant grass	3	A	0
D	Place statue	1	B, C	+2, 0

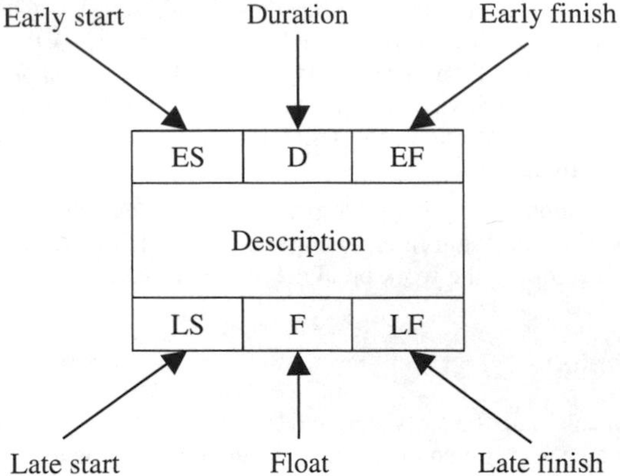

FIGURE 9.7 Activity in a precedence network.

site has been graded. This can be shown as a start-to-start dependency with a lag of 1 or a finish-to-start with a lead of −2. The latter is chosen. The leads and lags are also shown in Fig. 9.8.

Forward Pass. Early start and finish are calculated by conducting a *forward pass* through the network. The early start of the first activity is zero and the early finish is calculated by adding the duration. The early finish is transferred to subsequent activities as the early start, adding or subtracting any lead or lag, assuming a finish to start dependency. For a start-to-start dependency it is the start time which is transferred to the start, for a finish-to-finish dependency the finish time to the finish, and for a start-to-finish the start time to the finish. Where an activity has two or more preceding activities the largest number is transferred. The process is repeated throughout the network. Figure 9.10 shows the example network after a forward pass.

Back Pass. The late start and finish and float are calculated by conducting a *back pass*. The early finish of the last activity becomes its late finish. The duration is subtracted to calculate the late start. The late start is transferred back to the late finish of preceding activities. Again it is the start or finish time which is transferred to become the start or finish time

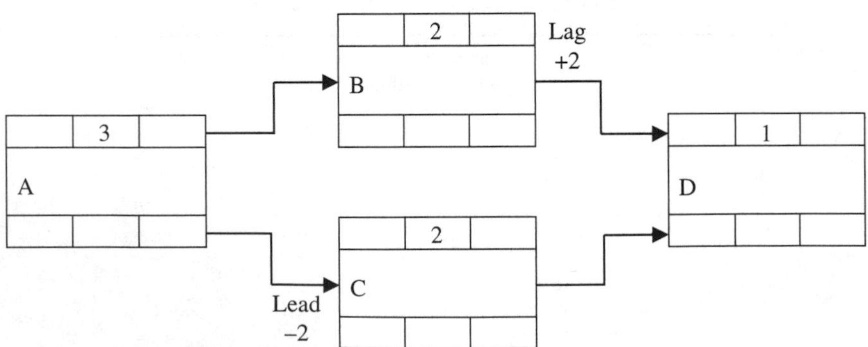

FIGURE 9.8 Precedence network: durations entered.

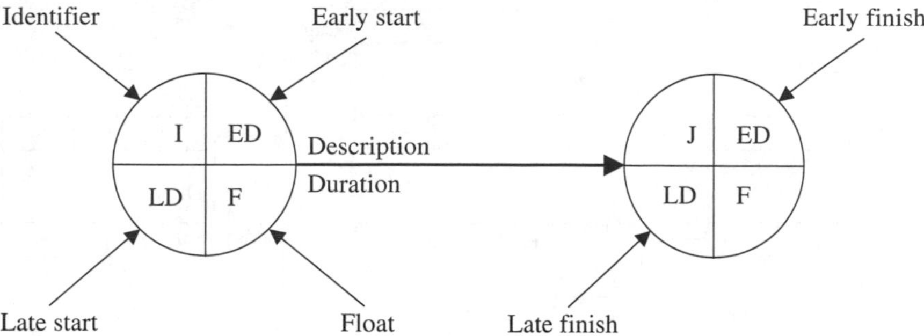

FIGURE 9.9 Activity in an IJ network.

depending on the type of dependency. Where an activity has two or more succeeding activities it is the smallest number which is transferred, after adding lags or subtracting leads. The process is repeated throughout the network. The float of each activity is calculated (Sec. 9.2). (This should be the same for both start and finish.) The float of the first and last activities should be zero. Figure 9.11 shows the network after the back pass.

Identifying the Critical Path. This is the series of activities with zero float, here A-B-D.

Arrow Networks. Figure 9.12 shows the drawn as an arrow network after forward and back pass.

Case Study Project. Figure 9.13 is the precedence network (at work-package level) for the CRMO Rationalization project.

Software Packages. Some software packages assume that if an activity has a start date of day six, (Monday say), and duration three, then it will finish on Wednesday evening, day eight. Therefore the finish is

$$\text{Finish date} = \text{start date} + \text{duration} - 1.$$

However, if there is no delay to the start of the next activity, it starts on Thursday morning, day nine, and so one is added to the finish date as it is transferred to be the start date of

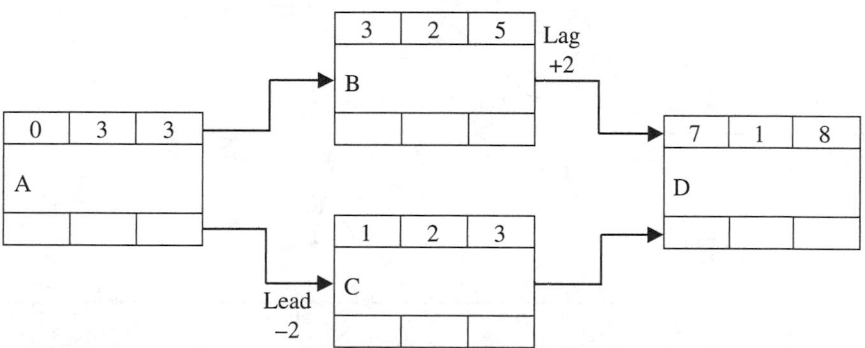

FIGURE 9.10 Network after forward pass.

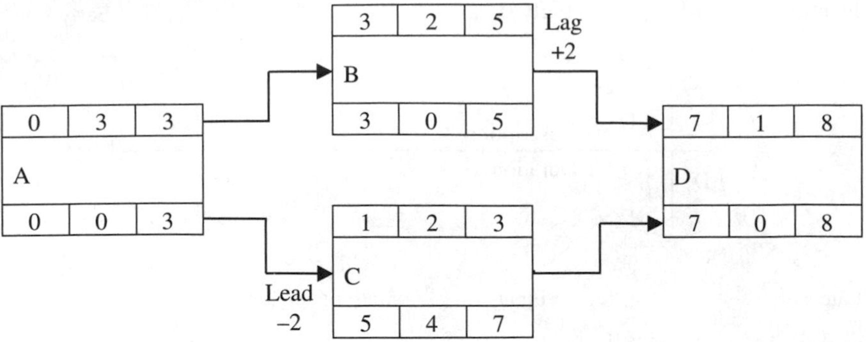

FIGURE 9.11 Network after back pass.

the next activity. The start date of the first activity is taken as day one, Monday morning, rather than zero as I have used above. The overall effect is just to add one to all the start dates you would obtain using the method I have proposed above.

Scheduling the Project

The network only calculates early and late dates. The baseline or scheduled dates must be chosen taking account of other factors. Hopefully they will be between the early and late dates. There are three options:

- Schedule by early start (hard-left): used to motivate the workforce
- Schedule by late finish (hard-right): used to present progress in the best light to the customer
- Schedule in between: done either to smooth resource usage (Sec. 9.4) or to show management the most likely outcome

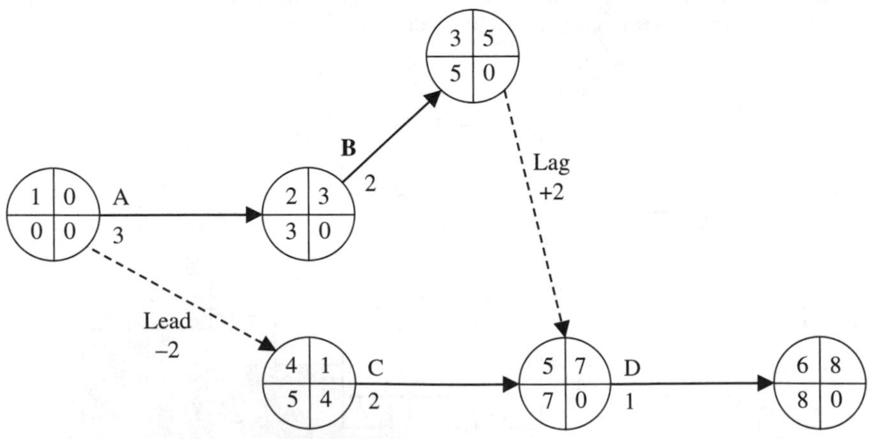

FIGURE 9.12 Arrow network after forward and back pass.

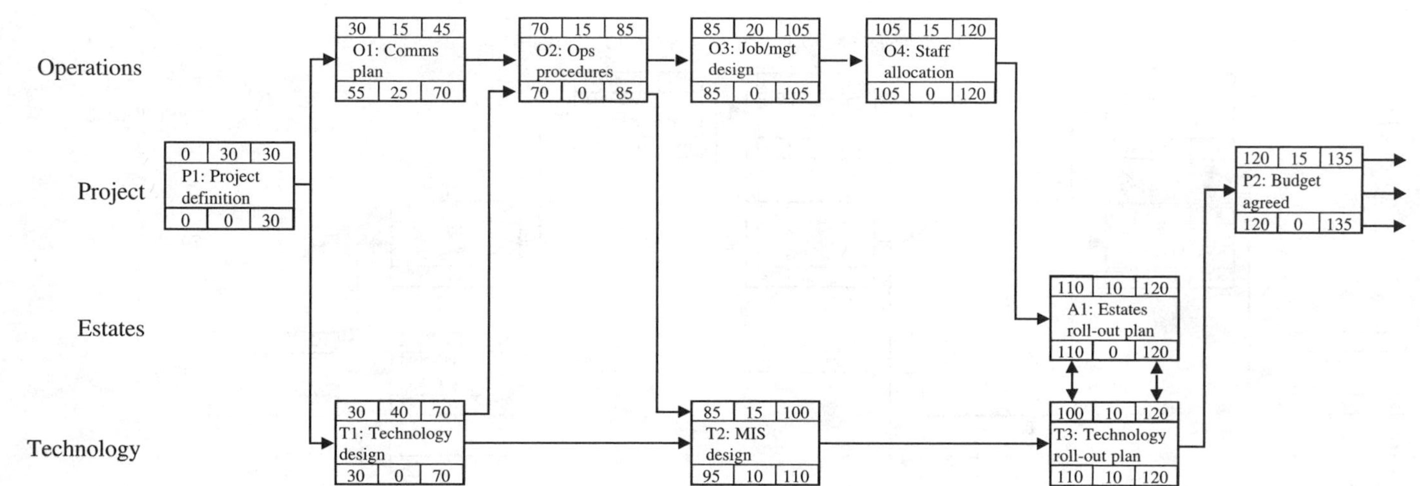

FIGURE 9.13 Precedence network at work-package level for CRMO Rationalization Project.

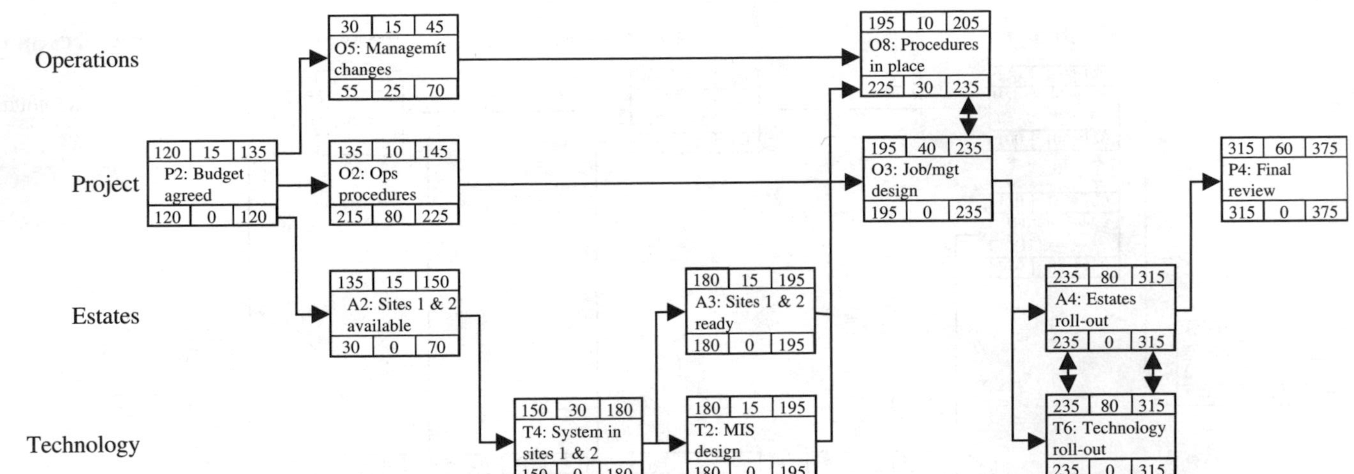

FIGURE 9.13 Precedence network at work-package level for CRMO Rationalization Project. (*Continued*)

Using Networks

Networks are a mathematical tool to be used as appropriate. This does not depend on the size of the project. In Sec. 5.5, I gave examples of multimillion pound projects where they were not used. It depends on the complexity of the interdependencies and resource sharing and the manager's ability to analyze these without computer support. As a mathematical tool, they help the manager calculate the schedule and analyze the impact of changes, what-if analysis. However, networks should not be used to communicate the plan or schedule. Bar charts, activity listings, or responsibility charts should be used for that.

9.4 RESOURCE HISTOGRAMS AND RESOURCE SMOOTHING

Using a network, you can calculate the early and late start and finish for work elements. However, in order to set the baseline or scheduled dates, it is necessary to take account of other constraints. Resource constraints are the most common. If the resource requirements for all activities are known then, once the project has been scheduled, you can calculate a resource profile for the project as a whole. This is known as the resource schedule and is either listed as a table of resource levels with time or is drawn as a resource histogram. This resource schedule can be compared to the known availability of each type of resource, and if the requirement exceeds availability it may be necessary to adjust the schedule to reduce the requirement. It may be possible to do this by consuming some of the float on noncritical activities. Alternatively, it may be necessary to extend the duration of the project.

Table 9.5 is an activity listing for a small project which I will use to illustrate the concept of resource scheduling. There are two resource types: analysts and programmers. Figure 9.14(*a*) shows the bar chart and resource histogram for both resource types with the project scheduled by early start. This produces quite wildly varying resource levels. If there were only one analyst available to the project, he or she would be overloaded during the first two months of the project. One person can work up to 22 days in a month without overtime. To overcome this problem we can try to use the float associated with some of the work elements

TABLE 9.5 Activity Listing for a Project with Resources

TriMagi
Activity Listing

Activity	Duration (mths)	Early start (mth)	Late start (mth)	Early finish (mth)	Late finish (mth)	Resource requirement: Analyst (days)	Resource requirement: Programmer (days)
A	3	0	1	3	4	24	0
B	2	0	2	2	4	24	0
C	2	0	2	2	4	16	16
D	1	3	4	4	5	0	12
E	1	0	3	1	4	0	4
F	4	0	0	4	4	16	0
G	1	4	4	5	5	12	8
H	1	5	5	6	6	4	8

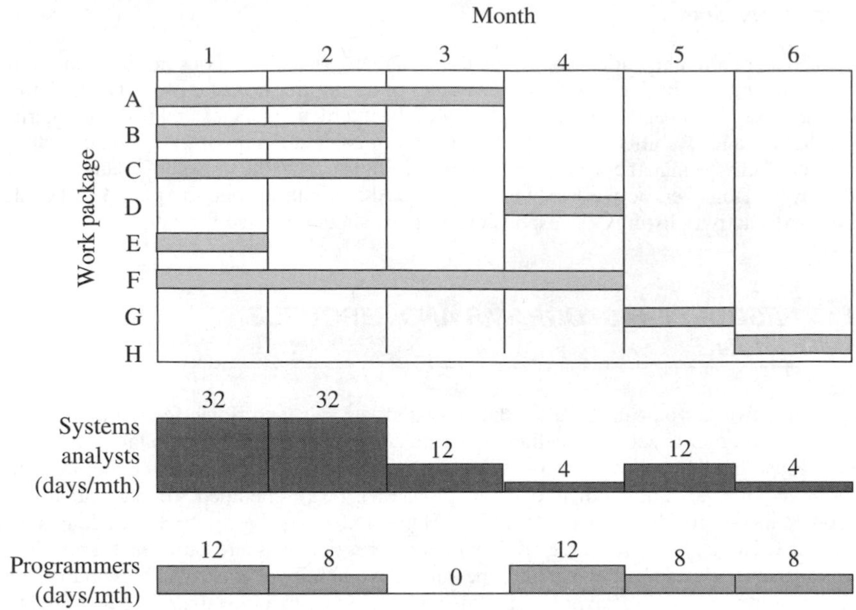

(a) Project scheduled by early start

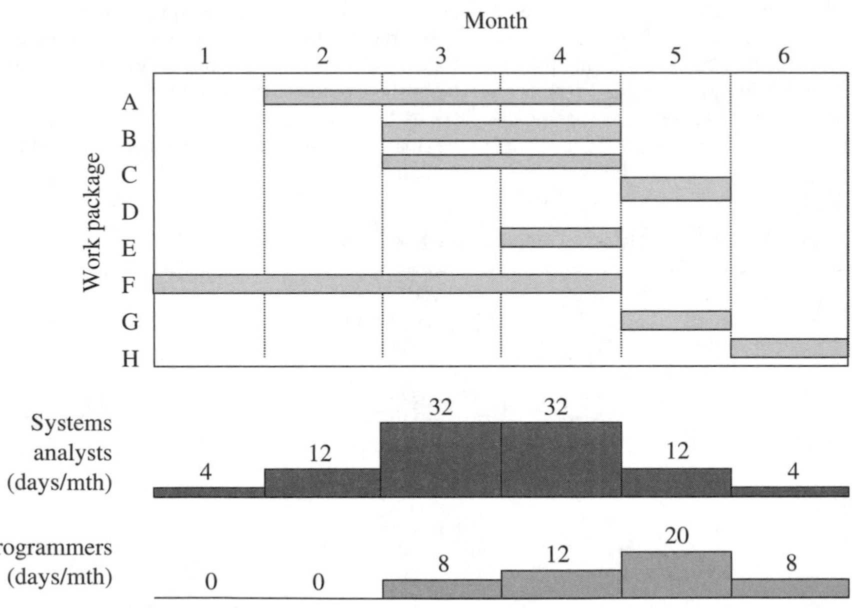

(b) Project scheduled by late start

FIGURE 9.14 Resource smoothing.

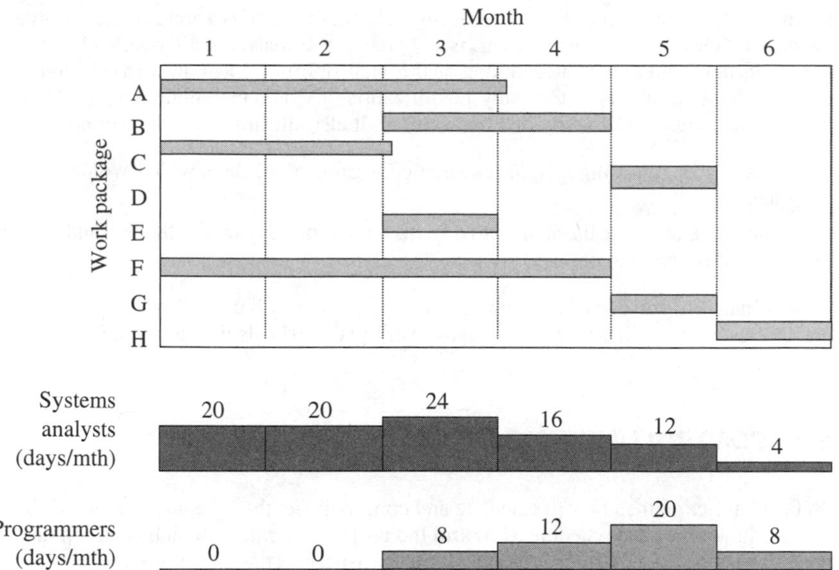

(c) Project scheduled with analyst as priority

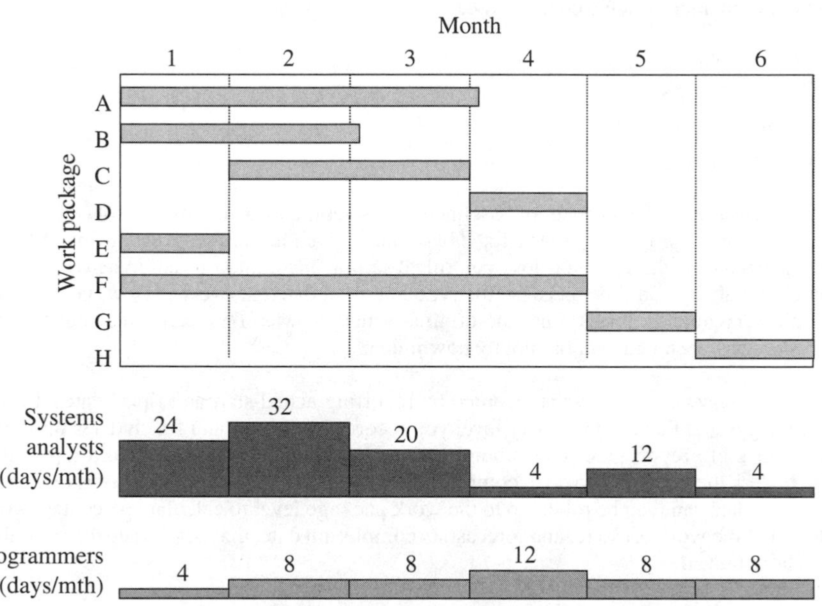

(d) Project scheduled with programmer as priority

FIGURE 9.14 *(Continued)*

to smooth the resource profiles. Figure 9.14(*b*) shows the bar chart and resource profiles for the project scheduled by late start. This is no better as the analyst is still overloaded, but now in months three and four. Concentrating on the analyst, Fig. 9.14(*c*) shows a schedule which gives the least variability of the analyst's utilization, giving a maximum level of 24 days in month three. This can be easily met by overtime. It also illustrates two further points:

- The danger of imposing a rigid resource constraint of 22 days which would delay the project
- The need to encourage the analyst to take his annual holiday in months five and six rather than months one to three

Alternatively you can take the programmer as priority. Figure 9.14(*d*) shows the schedule and resource profiles in that case. However, this overloads the analyst again.

9.5 CONTROLLING TIME

So far I have explained how to calculate and communicate the schedule. I conclude by discussing how to use the schedule to control the project's duration, which is its primary purpose. I describe the control process and tools to visually represent progress.

Control Cycle

There are four steps in the control process (Sec. 7.2):

- Set a measure.
- Record progress.
- Calculate the variance.
- Take remedial action.

Set a Measure. The planned, or baselined, dates set the measure for control of time. It is vital to measure progress against a fixed baseline. If you measure progress against the most recent update of the plan you lose control. It is not uncommon to come across projects which are always on time, because the schedule is updated at every review meeting, and people very quickly forget what the original schedule was; they can remember that the schedule has been updated, but not by how much.

Record Progress. Progress is recorded by reporting actual start and finish dates. In Sec. 5.1, I suggested that at the activity level you record actual start and finish dates only. It is problematic to report percentage completion, although team members should report how much work they have left to do to complete the activity for cost control purposes (Sec. 8.4). Progress data can then be rolled up to the work package level to calculate percentage completion of the work package, and forecast its completion date, that is, the date the milestone will be achieved.

Calculate the Variance. The variance is calculated either in the form of delays to completion of critical, or near critical, work, or as the remaining float of subsequent activities. Forward-looking control (Sec. 8.4) focuses on the remaining float of subsequent work, or on future delays to the start of critical or near critical work. That is what we can do something about. Delays to critical or near critical work have an impact on the remaining float of subsequent work, and when the remaining float of a subsequent work element becomes negative, that extends the forecast completion date of the project. It is important to monitor near critical work, and not focus solely on the critical path. The mathematical exactitude of the network can produce an undue focus on one area of the project, whereas it may be one of several other near critical paths which determines the duration of the project, and it was only estimating error that caused one of these to be identified as the critical path. Indeed, if you focus all your management attention on one path, you can guarantee another will determine the duration.

Where delays occur to bulk work, it will have little effect on the remaining float of future activities, until it has been delayed so much that it is itself critical. Indeed, resources may be switched from bulk work to critical work to maintain progress on the latter. However, if bulk work becomes significantly delayed, resource availability may determine the duration of the project, not the logic of the critical path.

In order to determine the impact of any delays on the project, and any proposals for eliminating them, it is necessary to analyze the effect of each on the overall project. This is a repeat of the what-if analysis described above. If the WBS has been well constructed this analysis can often be conducted manually, by analyzing the effect of the delay on the work package within which it occurs and then the effect of the work package on the overall project. The milestone plan is a powerful tool for determining whether a work package is critical and its effect on the project. This approach gives greater management control. Alternatively, where there are complex interdependencies and multiple shared resources, the analysis can be performed using the network. This provides a more accurate picture of the effect of changes, but it is difficult to determine the appropriate changes in the first place. The network does provide a valuable support to the manual approach, avoiding oversights.

Visual Representation

There are several tools which provide a visual representation of progress on the project.

Tracked Bar Charts. Figure 9.3(*d*) is a tracked bar chart. It shows current progress against baseline. Figure 9.15 shows a tracked bar chart for the CRMO Rationalization Project at a point part way through the project.

Milestone Tracker Charts. A problem with the tracked bar chart is it doesn't show how much the project has slipped since the last report; it just provides a snap shot of current progress. The milestone tracker chart (Fig. 9.16) shows the change since the last report. On the horizontal axis we plot the planned completion date for each milestone. On the vertical axis we plot the report date. So at each report date we can see the current planned completion date of the milestone, and can compare it to the planned date at the last report date, and the original or baseline date shown in the first line.

This makes it very difficult for the project manager to hide what is going on in the project, and there are two things the client managers do not want to see:

1. *Milestones slipping every report date*: If a milestone slips, the client managers want to see the project manager make and hold a commitment to the new date, not have it slip every time.
2. *Early milestones slipping and later ones being shown as not slipping*: Perhaps some early milestones will not be on the critical path, but the nature of milestones is that many of them will be critical. So if early milestones slip the client managers expect the project manager to show the impact on later milestones.

TriMagi Project Responsibility Chart

Project:	Rationalization of the Customer Repair and Maintenance Organization
Project Sponsor:	Steve Kenny
Project Manager:	Rodney Turner

X	e**X**ecutes the work
D	takes **D**ecisions solely/ultimately
d	takes **d**ecisions jointly
P	manages **P**rogress
T	on-the-job **T**raining
I	must be **I**nformed
C	must be **C**onsulted
A	may **A**dvise

No	Milestone name	Regional board	Operations director	CRMO managers	CRMO team leader	CRMO staff	Project manager	Project support office	Estates manager	Estates department	Network manager	Networks department	IS department	Operators	Personnel	Suppliers	End Date
P1	Project defintion	D	D	dX	dX	I	PX	X	X	I	X	I	C	C	C		5-Mar
T1	Technology design					C					PX	X	X	X		A	30-Apr
O1	Communicaton plan	I	D	d	PX												22-Mar
O2	Operational procedures	I	D	d	PX	X											15-May
O3	Job and management design	I	D	d	PX	C									TX		31-May
T2	MIS functional spec	I	D	d	dX						PX		X				31-May
O4	Staff allocation	I	D	d	PX	C									TX		15-Jun
T3	Technical roll-out plan	D	d				C	X	C	I	PX	X	X	I		C	15-Jun
A1	Estates roll-out plan	D	d	CR	X	I	C	X	PX	X	C	I	IS I	I	I	C	15-Jun
P2	Financial approval	D	d	I	CRMO		PX	X	C		C		C	A	A	C	30-Jun
A2	Sites 1 and 2 available		I	I	I				PX	X	I						15-Jul
O5	Management changes	I	DX	X	PX	I											15-Jul
T4	Systems in sites 1 and 2		I	I	I					X	PX	X	X	I		X	31-Aug
O6	Redeployment and training			D	PX										TX		31-Aug
A3	Sites 1 and 2 ready		I	I	X	X	P			X		X	X	I		X	15-Sep
T5	MIS delivered	I	D	I	X						PX		X			X	15-Sep
O7	Procedures implemented	D	D	PX	X							A	A	I	X		30-Sep
P3	Intermediate review	D	d	CR	CRMO		PX	X	A		A		A	A	A		30-Nov
A4	Roll-out implemented	I	D	dX	dX	X	PX	X	I	X	I	X	X	I	X	X	31-Mar
P4	Postcompletion audit	D	d	CR	CRMO		PX	X						C			30-Sep

Project Schedule

Period:	Month	Target end:	30-Jun-02

1	2	3	4	5	6	7	8	9	10	11	12	13	14	15	d
February	March	April	May	June	July	August	September	October	November	December	Jan-Mar	Apr-June	Jul-Sept	Oct-Dec	Duration

Time Now

FIGURE 9.15 Tracked bar chart for the CRMO Rationalization Project.

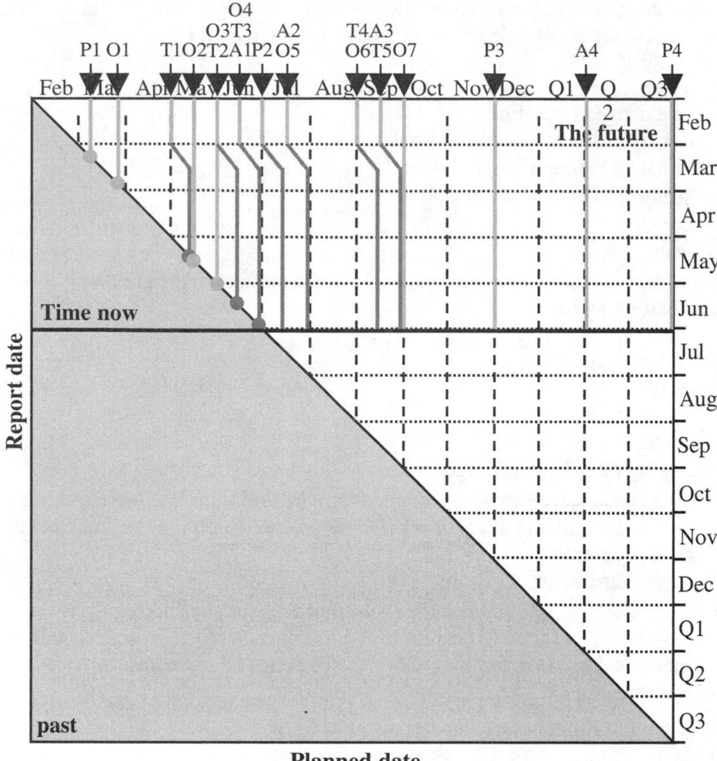

FIGURE 9.16 Milestone tracker chart for the CRMO Rationalization Project.

The milestone tracker chart is mainly used by the project manager to report progress against on the overall project. The team could use it at a lower level of WBS to report progress against activities on the packages of work they are doing.

Figure 9.16 is a milestone tracker chart for the CRMO Rationalization Project at the same reporting period as Fig. 9.14.

S-curves. S-curves, plotted as part of earned value analysis (Sec. 8.4), provide a pictorial representation of whether the project is on average, ahead of or behind schedule. The schedule variance introduced in that section is another time variance, in addition to the remaining float on critical activities.

SUMMARY

1. The purpose of scheduling time on a project is
- To obtain timely benefits which justify the expenditure
- To coordinate resource inputs
- To schedule resource availability
- To assign priority for resources between projects
- To meet a specified end date

2. The schedule specifies the duration, start and finish date, and float of the activities in the project. There are several dates recorded against each activity:
   - Early date
   - Late date and float
   - Baseline date and baseline float
   - Most likely date and remaining float
   - Actual date and remaining duration
3. The schedule can be communicated as:
   - An activity listing
   - Bar charts
4. The duration is calculated by comparing the work content to the number of people available, and allowing for:
   - Lost time
   - Part-time working
   - Interference
   - Communication
   - Lead times
   - Sequencing of tasks within activities
5. The early and late dates can be calculated from the durations and logical sequence of the activities using a critical path network. There are two main types of network:
   - Precedence network
   - Activity-on-arrow network
6. Given the initial schedule and resource requirements for each activity, a resource schedule can be calculated showing the requirements for each type of resource with time. This can be smoothed by delaying bulk work to fill peaks and troughs, or by extending the duration of the project. The resulting schedule is frozen as the baseline.
7. Progress against the schedule can be monitored by:
   - Recording progress on the critical or near critical paths
   - Using tracked bar charts and milestone tracker diagrams
   - Recording progress on S-curves

REFERENCES

1. Morris, P.W.G. and Hough, G.H., *The Anatomy of Major Project: A Study of the Reality of Project Management*, Chichester, U.K.: Wiley, 1987.
2. Williams, T., *Modelling Complex Projects*. Chichester, U.K.: Wiley, 2002.

CHAPTER 10
MANAGING RISK

Over the last five chapters I have described methods, tools, and techniques for the five functions of project management: managing scope, project organization, quality, cost, and time. All five of these require us to make predictions about future performance, but as we all know, we cannot predict the future. We can make informed estimates (guesses), but there always remains some residual uncertainty. I have said several times over the last nine chapters, that the more effort that is put into our estimates, and the more historical information that can be used in guiding them, the more accurate they will be. However, if we put in too much effort, we reach a point where the estimate costs more than the impact of the inherent risk. In a repetitive production environment the uncertainty can be reduced to a very low level, and the emphasis of management becomes to eliminate any variations from the status quo because variations remove certainty and reintroduce risk. In a project environment, because of the essential uniqueness of projects, some uncertainty must always remain, and the emphasis of management becomes to manage the risk. In my view, the essence of project management is risk management.

I describe risk management in this chapter. At the core of risk management is a risk management process. I start by introducing a generic risk management process, and then describing four steps in this process, identifying, assessing, analyzing, and controlling the risk.

10.1 THE RISK MANAGEMENT PROCESS

When describing the management of quality, cost, and time, I repeatedly used the control cycle (Fig. 7.2) to define a process for managing each of them. However, in the case of those three functions, the process informed our thinking, but was not central to the discussion. With risk management, the risk management process (RMP) is central and drives our thinking.[1,2] Table 10.1 shows a generic risk management process, and how it is realized by several global standards. Table 10.2 lists those and several other risk management standards. The steps in Table 10.1 are as follows:

Focus on risk management: You should set the project plan up from the start in a way that facilitates risk management.

Identify risks: You should identify potential risks on your project.

Assess risks (qualitatively): There are two parts to assessing the impact of the risks on the project: the qualitative assessment and the quantitative analysis. Qualitative assessment should be made mandatory. It should be done on almost all projects.

Prioritize risks: You need to select risks to be concentrated on. If you try to deal with all the potential risks on a project you will be swamped with too much information

TABLE 10.1 Generic Risk Management Process

Generic process	APM PRAM[3]	PMI PMBoK[4]	PRINCE[5]	Chapman and Ward[1,2]
Focus on risk	Initiate			Focus
Identify risk	Identify	Identify	Identify	Identify Structure Assign ownership
Assess (qualitatively)	Assess	Assess	Evaluate	Estimate Evaluate
Prioritize				
Analyze (quantitatively)		Analyze		Evaluate
Reduce	Plan response Implement response	Mitigate	Identify response Select response	Harness
Control	Manage response	Manage	Plan & resource Monitor & report	Manage

(Example 10.1). You need to focus on the significant few, and put the insignificant many to one side. Don't forget about them entirely, but let them look after themselves unless they look to become significant. You might ask what if it is a significant many and an insignificant few? The answer is probably don't do your project.

Analyze risks (quantitatively): You can also conduct a quantitative analysis using tools such as three-point estimating and Monte Carlo analysis. This is optional. It takes a lot of effort and so only repays that effort on larger, more complex projects. The qualitative assessment is mandatory, the quantitative analysis is optional. Three-point estimating is easier than full Monte Carlo analysis and so may be used on medium-complexity projects.

Develop a response plan: Having identified the risks, assessed their impact on the project, and selected those for management, the next step is to a plan how to reduce their impact on the project. There are several ways of doing that, depending on the nature of the risk.

Manage the risks: Finally, you manage those risks, and the response plan throughout the project. You monitor whether or not the risk occurs, and take action if it does. Hopefully your response plan reduced the impact if it does occur and makes you better able to respond, further reducing the impact.

TABLE 10.2 Risk Management Standards

Institution	Name	Number
Association for Project Management	*Risk Analysis and Management Guide*	PRAM
Project Management Institute	*Guide to the Project Management Body of Knowledge*	Chapter 11
Office of Governance Commerce		PRINCE2
British Standards Institute	*British Standard for Risk Management*	BS6079
Institution of Civil Engineers	*Risk Analysis and Management for Projects*	RAMP
Australian Standards	*Australian Standard for Risk Management*	AS4340

Example 10.1 Focusing on the significant few risks

I worked with the British Museum that had been developing a new gallery. They commissioned a consultant to do a risk analysis. He identified 100 risks and produced a report which became a doorstop. The British Museum didn't know where to start. On the other hand, on a €300 million project to build a fixed link between Copenhagen and Malmö in the south of Sweden, the team identified just 10 significant risks, and by focusing on those managed to reduce the project's duration by six months.

10.2 IDENTIFYING RISK

I cannot tell you what risks you are likely to encounter on your projects. What I can tell you is how to identify your risks, and how to categorise them, which may help you assess their impact on your projects.

Techniques for Identifying Risks

There are two main techniques for identifying risk: the organic, creative process and the mechanistic process.

Organic, Creative Process. This approach encourages creative, free-flowing thinking to try to identify risks that may not be obvious in the mechanistic process. Brainstorming, which we met in Sec. 5.3, can be used. Brainstorming is a two step process:

a. During the first step the facilitator (project manager) stands at the white board or flip chart, with a pen in hand, and the team members are encouraged to shout out potential risks. They should be encouraged to say whatever comes into their minds. The facilitator should NOT say that a particular idea is stupid, and so won't write it up. The idea is to encourage free-flowing thinking. A stupid idea from one person can stimulate a good idea in somebody else, and if you tell somebody their ideas are stupid, they will shut up for the rest of the process. The emphasis of this step is quantity not quality of ideas.
b. During the second step you try to sort the wheat from the chaff; delete the ideas that are not sensible risks on the project to be left with ideas for further analysis. What we are left with, the wheat, is both the significant few and insignificant many. We cannot distinguish between those until we have started the assessment. What we delete here is things which are just not relevant. The emphasis of this step is quality of ideas.

Mechanistic Process. With the mechanistic approach, you take a version of the project plan, and work through it in a methodical way to identify risks in each element of the plan. You could use the milestone plan (Sec. 5.3) or the responsibility chart (Sec. 6.4). If you are using the milestone plan, you can ask yourself, milestone by milestone, what can go wrong at this milestone, and if it does go wrong what impact will it have on other milestones. Then as part of the risk-reduction process you can ask yourself how you can stop the risk at that milestone, or if you can't, how can you reduce the impact on other milestones. The use of expert judgement, checklists, and people who have done similar projects in the past can help in this process.

Beware, that if a risk occurs in one milestone, it can have an impact on another to which it is not linked to logically, even on a milestone that is already finished. The logic represents the project going well, and risks are the project not going well. For instance, in an early milestone you can make a design assumption, and the logic represents that design assumption flowing through the project. If at a later milestone you find you cannot make that assumption work, it impacts on every milestone dependent on it, even the milestone where it was first made. There is such a link between milestones A2 and O5 in Fig. 5.2, for instance. Both assume we know sites 1 and 2, and if that is the case, they are not linked. However, if at A2 it proves impossible to use the chosen sites, then O5 is affected. The risk-reduction strategy is to try to make the design assumption less dependent on what is done at the later milestone.

Categorizing Risks

Risks can be categorised according to

- The impact they have
- Where control of the risk lies

Impact of the Risk. There are two types of risk under this heading:

1. Business risks
2. Insurable risks

Business risks: These are the risks, or uncertainty, inherent in all our estimates. People tend to treat their project estimates as point-wise correct. However, in reality, our estimates just represent some mid-range value, and they can turn out better or worse than that. (It never ceases to amaze me that in their lives people accept some uncertainty in their estimates of how long things will take, but on their projects they expect their estimates to be exactly correct, Example 10.2.) Business risk is a two-sided risk or uncertainty. Sometimes our projects will turn out better than we expect, when we will make more profit, and sometimes worse, when we will make less profit or even a loss. Table 8.5 analyzes the impact of business risks on a project using three-point estimating.

Example 10.2 Uncertainty of estimates

I did a series of workshops with a consultancy which was having a problem with overruns on its assignments. As a result, it reduced their overruns from an average of 10 percent, twice their annual profit, to about 2 percent. At an early workshop, a director gave a list of overruns. He grouped them by size of overrun in dollars. He started with some nightmares, jobs estimated to cost $40,000 and ending up costing $100,000. His last group were projects with overruns between $2000 and $4000, and the last was a project estimated at $400,000 that overran by just over $2000. I pointed out that the last one only overran by one half of 1 percent, and nobody could expect to estimate better than that. He was not pleased by my contribution.

Insurable risks: These are risks which can only go wrong. There is a hopefully small and random chance that some item of the project will fail. They are called insurable risks, but that is not to say either that an insurance company will want to buy the risk off us, or that we would want them to; see the discussion on risk reduction in Sec. 10.5.

Control of Risk. Risk can also be categorised by where control lies. Control can be internal or external to the project manager's organization, or legal. Internal risks can be technical or nontechnical. External risks can be predictable or unpredictable. Legal risks can fall under the criminal law or civil law, and those under the civil law under the law of contract or the law of tort.

Internal Risks

Internal, technical risks: These risks arise directly from the technology of the work, or the design, construction, or operation of the facility, or the design of the ultimate product. They can arise from changes or from a failure to achieve desired levels of performance.

Internal, nontechnical risks: These usually arise from a failure of the project organization or resources (human, material, or financial) to achieve their expected performance. They may result in schedule delays, cost overruns, or interruption to cash flow.

External Risks

External, predictable but uncertain risks: These are ones with reasonably predictable outcomes, should they occur (like tossing a coin). There are two major types of risk in this category: the first is the activity of markets for raw materials or finished goods, which determines prices, availability, and demand; the second is fiscal policies affecting currency, inflation, and taxation. They also include operational requirements such as maintenance, environmental factors such as the weather, and social impacts.

External, unpredictable risks: These are more ambiguous, with possibly unknown potential outcomes. They arise from the action of government or third parties, acts of God, or from failure to complete the project due to external influences. Government can unexpectedly pass new regulatory requirements. Whether a change of government at an election falls in this or the following category is a moot point. Actions of third parties can include sabotage or war, and acts of God are natural hazards such as an earthquake, flood, or the sinking of a ship. Failure to complete can arise from the failure of third parties through bankruptcy, or a totally inappropriate project design. By their nature, these risks are almost all "insurable" risks.

Turning Internal Risks into External Risks. Before discussing legal risks, I wish to discus a point arising from this issue of internal and external risk. In the 1980s, standard contracting practice was to dump risk down the contract chain. The client passed risk on to the contractor and the contractor on to the subcontractor. What you sometimes did was take a risk that the client could control and do something about reducing, and convert it into a risk external to the contractor, for which they can do nothing but allow a contingency. The client then chooses a contractor via compulsory competitive tendering, and awards the job to the contractor that bids the least amount, that is the contractor that has allowed the least contingency and is therefore most likely to fail (Example 10.3 is an apocryphal story about this). In Table 8.5, do you award the job to the contractor that bids 700, 950, 1000, 1050, 1100, 1200, or 1500 units? If you award the work to the firm that bids 700, and they go bankrupt when you are only half way through the project, you have little recourse to cover your losses, and you may have to start the project again.

In the 1990s, standard practice became to try to assign the risk to the party best able to control it: the client took client risk, contractor A their risk, contractor B theirs, and so on. This did not work either because risks are coupled. What happened was the client tried to reduce their risk and increased contractor A's, contractor A tried to reduce theirs and increased contractor B's, and contractor B increased the client's.

What is now viewed as best practice is where there are risks controlled by multiple parties, you should form an alliance of those parties to manage the risk together. Sometimes the risk is only controlled by the contractors, and then it should be assigned to an alliance of the contractors working under a fixed-price contract. Sometimes it is controlled just by the client, in which case they should keep it. Sometimes it is controlled both by client and contractors and they should then form a partnering arrangement to work together to reduce the risk. Viewing the project as a partnership was a necessary condition for project success suggested in Sec. 3.3.

Example 10.3 Risk sharing

Neil Armstrong was being interviewed about the moon landing and was asked what was the most frightening moment; was it as the Lunar Lander came down and might crash; or was it as he stepped off the ladder; or was it when they came to blast off from the moon and the rockets might not have been powerful enough. No, he said, the most frightening moment was being on the launch pad at Cape Canaveral, and under him were 2000 components, every single one of which had been bought on minimum price tender!!! And one of them did fail in 1986.

Legal Risks. There are three types of legal risk: risks under the criminal law, risks under the law of contract, and risks under the law of tort. (The law of tort is the duty of reasonable care we all have to our fellow citizens. Even where we do not have a contract with somebody, we have a duty to behave responsibly and with reasonable care.) If an employee is killed in an accident at work, you can be prosecuted under health and safety legislation, fined, and potentially sent to jail. You can be sued by his or her estate under the contract of employment, or under the law of tort. If a visitor to your site is killed, you can be prosecuted under the criminal law as above, or the law of tort, but you may have had no contract with the individual. This applies to the software industry as much as the engineering industry with the development of computer control systems to control complex plant (Example 10.4).

Under the criminal law there have been several attempts in the United Kingdom to bring charges of corporate manslaughter. The most recent was when track on a high-speed rail failed, causing a train to derail and killing half a dozen people. In the subsequent enquiry it was discovered that the rail company and their contractors had been reducing maintenance work to save money and so charges were brought. Corporate manslaughter is difficult to prove because one person has to be responsible for the decision that caused the accident, whereas often it is caused by a series of mistakes. In the United Kingdom, the current Labour government is proposing to introduce a charge of corporate killing which could be based on a general culture of sloppiness and irresponsibility, rather than a single incorrect decision.

In the event of a charge being brought, whether as corporate manslaughter or under more general health and safety legislation, or under the law of contract or the law of tort, the case is judged on the basis of what any reasonable professional would have done in the circumstances. Standards improve with time, so you cannot necessarily condemn what somebody did 20 years ago by today's standards, and likewise you cannot excuse a mistake today by the standards of 20 years ago. Examples 10.4 to 10.7 contain four cases, showing how this might apply. The law is not necessarily fair or logical, as Example 10.6 shows. It just tries to be precise.

Example 10.4 Testing a computer control system

Some years ago I was on a course where we were discussing the health and safety legislation, and the duty of care. One of the delegates said he was responsible for testing the control software for a jet fighter used by the Royal Air Force. He said that in a reasonable amount of time they could test 90 percent of all the paths through the software, which would represent 99.9 percent of all the occurrences. However, to test all the paths would take 100 years. His question was what would happen if there was a failure because the control system locked into a path that had not been tested but which had a fault. He was told that he would be judged by what any reasonable professional would have done, and because it was not sensible to test all the paths, he would not be held liable.

A few years later one of that type of aircraft suddenly ejected its pilot over the middle of England and then flew about 150 miles to crash in the Irish Sea. They weren't sure whether the pilot had committed suicide or the computer control system had failed, though the fact that the plane seemed to be on course to crash in the Irish Sea tended to point to the former. However, I thought of the course delegate.

Example 10.5 Seeking damages after 50 years

A woman who worked in an asbestos factory in the late 1930s developed asbestos-related diseases in the 1980s. She sued her former employers claiming they had been negligent in the containment of asbestos in the factory. Her employer had to be judged by the standards of the 1930s, not the 1980s, but was still judged to have been negligent.

Example 10.6 The law is not fair, but scrupulously exact

In the United Kingdom, children with stunted growth are sometimes given growth hormones. Up to 1980, this was made from extracts from the brains of dead people. From July 1978, the government knew this could cause Creutzfeldt–Jakob disease (CJD), the human equivalent of mad cow disease, but did not replace it with a synthetic alternative until 1980. The families of people who had suffered CJD sued the government. The courts ruled that anyone who had been fed the hormone for the first time on or after 1 July 1978 should receive compensation. Anyone who had received it on 30 June 1978 or earlier could not because the government could not have known there was a problem before then. There was one person who had received it for the first time on exactly 30 June 1978, and everyone said this is not fair—not fair but scrupulously exact. (The ruling was subsequently overturned by the Court of Appeal, and all people suffering CJD could claim. People not suffering CJD want to claim now for the fear they have to live with!!!)

Example 10.7 Not judging by today's standards

It was suggested to Sir Winston Churchill in early 1945 that the allies might bomb the railway line leading to Auschwitz, and he said it was not worth the risk. People now react in horror that he could have said such a thing, but they are judging by today's standards. With the technology of 60 years ago they were lucky to drop the bomb within two miles of the target. It saved more lives to use the pilot's life to shorten the war than to go on a fool's errand.

Expecting the Unexpected

Good project managers learn to be risk aware, to expect failure where they least expect it. This is known as *Sod's law* or *Murphy's law*, sometimes stated as: *if something can go wrong it will; if something can't go wrong, it still will*! The value of this attitude is that if you expect things to go wrong, you will be on your guard for problems, and will be able to respond quickly to them. The failures may be ones you had predicted, or ones you least expect. If you anticipate problems, and plan appropriate contingency, you will not be disrupted when those problems occur. If the unexpected then also occurs, you will be able to focus your management effort into the areas that might now cause greatest disruption (Example 10.8). Having said that you must not be so pessimistic you cannot make progress. You need to achieve a balance between blithe optimism and morbid pessimism (see Example 10.9).

Example 10.8 Expecting the unexpected (1)

In the early 1980s, I managed an area of work on the shutdown-overhaul of a petrochemical plant. We were uprating the steam system, and this required us to run a line between the 50-bar and 30-bar steam mains. On the overhaul, all had to do was make the break ins into the two mains at each end of the line. For the 30-bar main this was easy. We made an 8- by 6-inch T-section in advance of the overhaul. In the overhaul we just had to cut the line, which would be completely cold, weld in the T-section, and install an isolation valve. The break-in to the 50-bar main, however, carried greater risk. We had to weld a 6-inch branch onto the 12-inch main just downstream from the main isolation valve, separating the plant main from the factory main. This valve had not been closed in 12 years, and so we did not know if it would shut tight. If it did not, the job would be more difficult, or even impossible. We put considerable effort into drawing up contingency plans in the event of a partial or full leak of the valve. In the event it shut like a dream. However, when we offered up the T-section at the other end, we found it had been made 6 by 6 inch instead of 8 by 6 inch. We therefore had to make a new T-section in a hurry, and an 8-inch pipe of the right pressure rating was not immediately available. That particular job almost extended the duration of the overhaul. However, the time spent planning the other job was not wasted. I knew that so well, I could leave it to run itself and focus my attention on procuring 8-inch pipe.

Example 10.9 Expecting the unexpected (2)

I play bridge as a hobby. We are taught that if the play of the hand looks easy, you should be pessimistic, think about the worst possible layout of the cards and play for that (as long as it doesn't cost you the contract). On the other hand, if the contract looks impossible, you should be blithely optimistic, and play for the only layout of the opposing cards that will enable you to make the contract, no matter how unlikely.

10.3 ASSESSING RISK

Having identified possible sources of risk to the project, we need to determine their impact. First, we assess the impact of individual risks through qualitative assessment. On more complex projects, we can then determine their combined impact through quantitative analysis. What follows applies to insurable risks. The impact of business risks can be determined through three-point estimating as illustrated in Table 8.5.

The Impact of a Single Risk

The impact of a risk factor depends on its likelihood of occurring and the consequence if it does occur (Fig. 10.1):

$$\text{Impact of risk} = \text{Likelihood of risk} \times \text{Consequence of risk}$$

To illustrate this concept, consider the question of whether buildings in the British Isles have earthquake protection. The answer is very few do. Multistory office blocks in London do not. The consequence of an earthquake in London of force 7 on the Richter scale would be severe loss of life. However, the probability of such an earthquake is so small, virtually zero, that it is considered unnecessary to take precautions. However, one type of building which does have earthquake protection is nuclear power stations. The likelihood of an

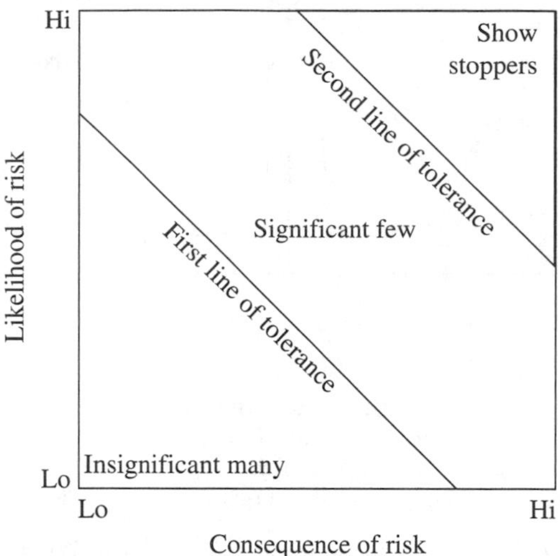

FIGURE 10.1 The likelihood by consequence plot.

earthquake has not changed, but the consequence is now unacceptably high; it would make the surrounding countryside uninhabitable for the next 10,000 years. It has pushed the impact of the risk over a line of tolerance where action has to be taken.

It is suggested that there should be two lines of tolerance in Fig. 10.1. Risks below the first line are the insignificant many, the risks that can almost be allowed to look after themselves. An earthquake of force 7 on the Richter scale is classified as such in the United Kingdom for most building construction. Risks above the first line are the significant few, ones where a risk response is necessary. There is a second line of tolerance of very high likelihood, very high consequence risks. Risks above this line are showstoppers; if the risk cannot be eliminated the project should not go ahead. The consequence of an earthquake under a nuclear power station pushes it here. Earthquake risks in the middle band in the United Kingdom are earthquakes of force 4 or 4.5 on the Richter scale. There is an earthquake force 4 on the Richter scale somewhere in the United Kingdom about three times a year. However, if a building is designed to stand up it is designed to withstand such an earthquake. So action is taken to withstand such an earthquake, but that is to design the building properly according to design regulations. An earthquake force 4.5 will cause more damage, but it is a once in 10 year event. Once every 10 years there will be an earthquake force 4.5 on the Richter scale somewhere in the United Kingdom, and it will damage buildings within about a 2-mile radius, but nobody is likely to be hurt. Buildings are not designed to withstand such an event. Instead they are insured against such an event. It is not cost effective to design every building to withstand the event. Instead it is better to pay a premium to an insurance company, and the insurance companies spread the risk over a large number of buildings over a 10-year period. So we have two insurable risks: one the owner insures by spending money on design and construction; and the other the owner insures by paying a premium to an insurance company.

You see through this discussion that through the two lines of tolerance we have begun to prioritize the risks. There are several further issues that arise.

What Do We Mean by High, Medium, and Low? Figure 10.1 shows us categorizing likelihood and consequence as high, medium, and low. What do we mean by these? The answer

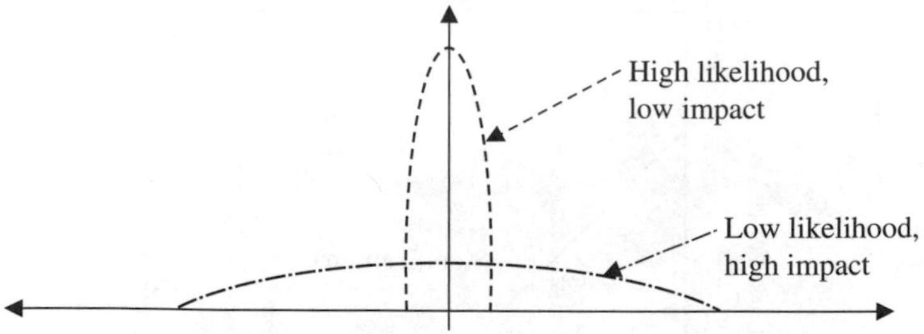

FIGURE 10.2 Low and high likelihood risks.

is whatever is right for your project and your own risk tolerance. You or your organization needs to decide what risk you (it) can tolerate. Project Management Institute (PMI) in their body of knowledge suggests four ranges of figures for likelihood.[4] You can use these, but I do suggest you work out what your or your organization's risk tolerance is.

Are Risk on Lines Parallel to the Lines of Tolerance the Same? The equation above is presented as a multiplication which suggests risks with high likelihood and low consequence are the same as risks with low likelihood and high consequence. Unfortunately they are not. The difference is the spread of possible outcomes (Fig. 10.2). Both have the same expected outcome, but the high likelihood, low consequence risk has a much lower spread of potential outcomes than the low likelihood and high consequence. The former has a very predictable outcome, the latter very unpredictable. Thus with the high likelihood, low consequence risk we can give it to a contractor to manage and they can predict quite closely what its impact will be and allow a cost for it. With latter, we have to insure with an insurance company. They buy a large number of risks, some of which will have no consequence, and some a very larger consequence, and spread their risk over a large number of events. So the mitigation strategy is very different for the two types. We saw this with the earthquakes of force 4 and 4.5 on the Richter scale. The former is medium likelihood, medium consequence and we design the risk out. The latter is low likelihood, high consequence, and we insure it.

The Assessment of Risk Is Often Irrational. Unfortunately the assessment of risk is often irrational (Example 10.10), with people giving huge focus to trivial risks, while ignoring huge ones. Example 10.10 may seem extreme, but it happens in companies that executes focus on trivial issues while ignoring significant risks. You need somehow through the risk identification process to protect against that. Sometimes, however, companies are responding not to their own assessment of the risk, but the public perception (Examples 10.11 and 10.12). Example 10.10 does illustrate one thing. In that case the deaths at the time were running at five per year, but it was not known how high the epidemic would rise. Would the peak number of deaths per year be five, or five thousand, or even five million? This was a very low probability but potentially very high consequence risk. This apparent irrationality in the assessment of risk seems to be an inbuilt mistrust human beings have of low likelihood, high consequence risks, and the associated ambiguity.

Example 10.10 The irrational assessment of risk

A classic example of the irrational perception of risk was the reaction to mad cow disease in Britain. First, the public behaved irrationally, and sales of beef plummeted. The number of deaths from new form CJD, which may, just may, have been caused by bovine spongiform encephalopathy (BSE), was running at 5 per year, about the same number of people that die from allergic reaction to peanuts. But TV reporters would go down to the local supermarket to interview an average shopper, smoking a cigarette (100,000 deaths a year in the United Kingdom from smoking-related diseases), with a trolley load of beer (50,000 deaths a year in the United Kingdom from alcohol-related diseases), and a car in the car park with bald tyres (3000 deaths a year in the United Kingdom from road accidents). "Are you eating beef," asked the reporter. "No," said the shopper, "It's too dangerous."

The public seem to have come to their senses, but then the government began behaving irrationally. For several years they made the selling of a T-bone steak a crime as heinous as the selling of crack cocaine, because it is expected to kill one person every 20 years. The agriculture minister appeared on the TV saying he was concerned about public health!!! If he was concerned about public health he would ban peanuts before a T-bone steak.

Example 10.11 Public perception of risk

In the 1980s, a firm of British engineering design consultants put considerable effort into designing and testing railway wagons for transporting low-level nuclear waste around the country. There were some highly publicised experiments in which a locomotive was slammed into a wagon at 100 miles per hour. In this case, the likelihood of an accident which would result in a release of radiation was small, and the consequence was also small, no immediate deaths, perhaps one or two additional cancer cases resulting in early death several years later. However, this is a highly emotive public issue, and hence the need for indestructible wagons. On the other hand, quite lethal chemicals are transported around in relatively flimsy wagons. In the early 1980s, I worked close to a railway line, along which passed a train towing two wagons filled with cyanide gas twice a day. The consequence of a crash involving a leak in the centre of a city would be instant death to thousands of people, but this is not a public issue. A thousand instant deaths from cyanide gas seem to be more acceptable than two lingering deaths from radiation-induced cancer.

Example 10.12 The value of project opinion and the environment

In the early 1980s, the British government proposed storing medium-level nuclear waste in a redundant mine under the factory I worked near a town called Billingham in the North East England. It may have been one of the safest proposals for storing medium-level waste. The project would apparently cost the company I worked for, ICI, nothing but earn them an income; an attractive project with *no* risk attached. However, ICI would not allow the project to proceed because that was not the way the local community viewed it, and ICI was concerned about local opinion. The ironic thing was ICI used to operate one of the country's largest private nuclear sources on the Billingham site.

It is almost certainly incorrect to say that the project would have "cost ICI nothing." It was causing a loss of goodwill in the local community, and so the cost was whatever value the company put on that goodwill. Clearly they did not think that cost was worth the returns.

Local house prices were falling, so the people who were going to pay for the project were the local community. It was possible to put a price on the environmental impact of the project, and ICI was not willing to bear that, nor let the local community bear it.

Prioritizing Risks

As I have said, having identified the risks, it is necessary to prioritize the risks for further analysis and management attention. Drawing Fig. 10.1 and marking the risks on is one way of doing that. Risks that lie below the first line of tolerance are the insignificant many risks that can almost be allowed to manage themselves. Those between the first and the second line of tolerance are the significant few, for which an active risk-reduction strategy must be found and the risks must be managed using the processes described in Sec. 10.5. Those above the second line of tolerance must either be eliminated or the project not attempted.

Analytical Hierarchical Programming. People are also now recommending a technique known as analytical hierarchical programming[6] (AHP) to prioritize risks. By that technique, pair-wise comparisons are made of the likelihood and consequence of each risk to assign each risk an individual score to be able to rank order their significance, and also determine their relative significance. That can be done with a two-parameter model of likelihood and consequence and a three-parameter model including public perception. The advantage of AHP is you don't need to decide whether each risk scores high, medium, or low against each parameter, just how each risk compares with all the others against that parameter. It is a powerful technique for prioritizing risks.

Table 10.3 shows a simple example, analyzed in Example 10.13. This is a very rough example. If you want to use AHP to prioritize risks, I suggest you read about it more fully.

Example 10.13 Risk prioritization using analytical hierarchical programming

The project in Table 10.3 has three risks, and we do pair-wise comparisons of likelihood and consequence of them, judging each on a scale of 1 to 7 (1 is the same, 7 is an order of magnitude different). We enter the comparisons in a matrix. The diagonal of the matrix is always 1 because we are comparing the risk to itself. In this case, we decide risk 1 is very much more likely than risk 3 and much more likely than risk 2, and risk 2 is much more likely than risk 3, hence the top-half of the matrix. The bottom-half of the matrix is always the reciprocal of the top-half because we are doing the comparison the other way. We add the rows and normalize by dividing by the sum. We repeat the process for consequence. We decide the consequence of risk 3 is much greater than risk 1 and very much greater than risk 2, but the consequence of risk 1 is just greater than risk 2. Thus we enter the comparison of risk 1 and 2 in the top row but the comparison of risk 3 with the other two in the bottom row. We fill in the other cells as the reciprocals

TABLE 10.3 Analytical Hierarchical Programming

Likelihood	Risk 1	Risk 2	Risk 3	Sum	Norm
Risk 1	1.00	3.00	6.00	10.00	0.63
Risk 2	0.33	1.00	3.00	4.33	0.27
Risk 3	0.17	0.33	1.00	1.50	0.10
				15.83	1.00
Consequence	**Risk 1**	**Risk 2**	**Risk 3**	**Sum**	**Norm**
Risk 1	1.00	2.00	0.33	3.33	0.22
Risk 2	0.50	1.00	0.17	1.67	0.11
Risk 3	3.00	6.00	1.00	10.00	0.67
				15.00	1.00

of those we have estimated. We add the rows and normalize again, and then add the two normalized columns. The total for risk 1 is 0.85, risk 2 is 0.38, and risk 3 is 0.77. Thus risks 1 and 3 are almost equivalent and risk 2 half as significant.

Influence Diagrams. Influence diagrams are a tool derived from systems dynamics, which can assist in risk assessment. They show how risks influence one another; some risks reinforce others (+), and some reduce others (–). Figure 10.3 is an example of an influence diagram. The power of the technique is to identify loops of influence. *Vicious cycles* have an even (or zero) number of negative influences and so any disturbance is magnified around the loop. *Virtuous cycles* have an odd number of negative influences, and so a disturbance is attenuated around the loop. In Figure 10.3, loop ADEKLIBA is vicious, and loop ADEGHJIBA is virtuous. Drawing the loop can help with top-down, qualitative assessment, identifying how risks influence each other. However, if used with simulation software, they can be used as part of qualitative analysis as well.[7]

Combining Risks

Having identified and prioritized those risks for further analysis, we should consider how they impact on each other. There are two ways of doing that: a bottom-up approach, which is the subject of the next section; and a top-down approach which continues the qualitative assessment, and is discussed here. The top-down approach was first introduced when

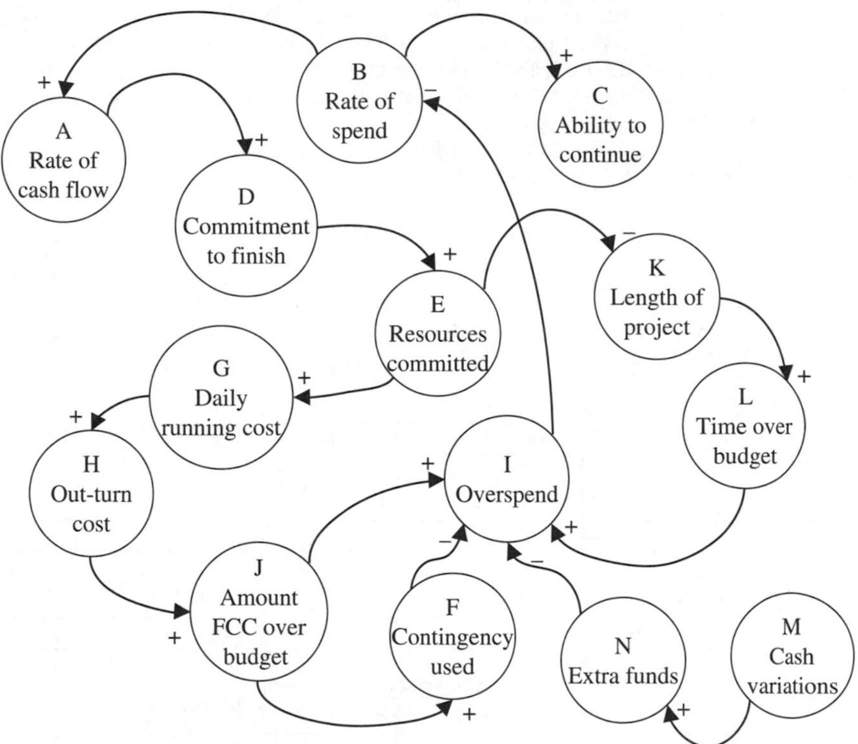

FIGURE 10.3 Influence diagram.

I described the mechanistic approach to identifying risks. You consider how the risks impact on each other, and what you can do to minimize that impact. A word of caution: you do need to be careful with the prioritization above. There may be a risk that on its own appears to be low consequence so you decide not to take it forward for further investigation, but it is a trigger for a much bigger risk. If it is a trigger for a much bigger risk it is not low consequence, but that may not be obvious when looking at it on its own. But you do need to be aware of that when thinking about risks, what they trigger, and what are triggers for them.

The top-down approach can provide the manager with checklists of potential risk factors based on previous experience, and can help them to determine their relative importance. Furthermore, by identifying the controlling relationships at a high level, it enables the project manager to find ways of eliminating the most severe risks from their projects. The approach is to take a component breakdown for the project, and evolve it down to the integrative level with about 20 elements of the breakdown. The component breakdown chosen will depend on what it is that is expected to create risk. It can be the project breakdown structure (PBS), work breakdown structure (WBS), organization breakdown structure (OBS), cost breakdown structure (CBS), or bill of materials (BOM) for the new asset. You then identify the risk associated with each component, and critically the links between the risks. If one risk occurs, does it make another more or less likely. You then concentrate on either eliminating the risk associated with each component, or breaking the links between the risks. If you are successful in breaking the links, you can isolate each risk in the breakdown structure. The reason for limiting yourself to 20 components is if you have a sheet of paper describing each risk and each link, that is, 400 sheets of paper. If there are 30 components you have 900 sheets.

Two tools introduced previously which provide a clear representation of the PBS, WBS, and OBS to an appropriate level are the milestone plan (Sec. 5.3) and the responsibility chart (Sec. 6.4). The milestone plan shows the PBS at the integrative level. The responsibility chart shows the OBS, PBS, and WBS at the integrative level on one document. It also shows how these are influenced by one element of the CBS, the work content, and by the timescale. It is therefore a very powerful document for top-down risk analysis.

Figure 10.4 shows a simple four work-package project to illustrate the top-down approach. The analysis is in Example 10.14.

Example 10.14 Top-down analysis to combine risks in Fig. 10.4

The project in Fig. 10.4 consists of four work-packages to build a warehouse. The duration is seven months. We consider whether it is possible to fast track some of the work and conduct a risk analysis to decide if that is sensible. We decide it is not possible to fast track the lower path, A-C-D, but turn our attention to the top path, A-B-D. We

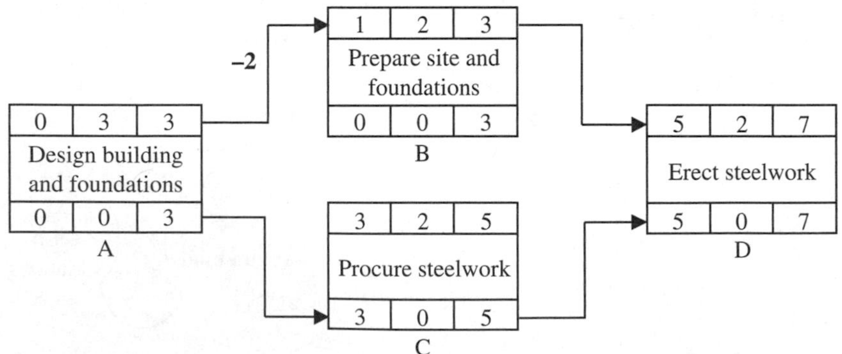

FIGURE 10.4 A four work-package project to construct a warehouse.

decide it is possible to overlap packages A and B; some of the design of the foundations will be finished one month into the design work, and so work on the foundations could start two months before the end of the design work. We show this potential overlap by putting a lead of –2 on the link from A to B. Is there any risk associated with starting the work on the foundations early? The answer is "Yes!" The design of the building may change the design of the foundations and so if we start early there may be potential rework. And we might ask if there is any value in starting early since the duration along the top path is seven months, the same as along the bottom path. Is there any risk in not starting early? The answer is "Yes!" The duration of the top path is much more at risk than the duration of the bottom path, being subject to inclement weather. If, for instance, the design starts on October 1, then without fast tracking the foundation work will be done in January and February, the worst time of the year to be doing it in Britain. Better to try and get it finished in November and December. Having recognized the previous risk, can we reduce the chance of rework? The answer is "Yes!" We could do the foundation work which is unaffected by the design of the building in month two, such as access roads and drains, and the foundation work which is more likely to be affected in month 3, when hopefully the design of the building is more stable (if not the project will be delayed anyway).

10.4 ANALYZING RISK

The bottom-up approach to combining risks is quantitative analysis. As I said in Sec. 6.2, this will only be conducted on larger, more complex projects, because considerable time and effort is required to do it, and so there must be sufficient potential benefits to repay the effort. There are several ways of doing quantitative risk analysis on projects:

Three-Point Estimating. Table 8.5 is an example of three-point estimating. It applies more to business risks than insurable risks, although it is sometimes possible to convert an insurable risk into a three-point estimate. What we do is for each activity or work package estimate the most likely outcome, most optimistic, and most pessimistic. We can do this for cost and time. We then work out the raw estimate, the worst possible outcome and the best possible outcome, though as I said in Sec. 8.3 the chance of actually achieving the best or worst case is small. Using the 1:4:1 formula given in Sec. 8.2, we also calculate the expected outturn for each activity or work package, and from that estimate the expected outturn for the project. Table 10.4 contains three-point estimates for the project in Fig. 10.4. Using the likely values, the critical path is A-C-D and the duration 7 months. Using the optimistic values the critical path is both A-C-D and A-B-D and the duration is 5 months. Using the pessimistic values the critical path is A-B-D and the duration is 12 months. Using the expected values the critical path is A-C-D and the duration is also 7.5 months. (In all calculations I have used the 2-month lead between A and B.) The same process can be applied to the cost of the project, as in Table 8.5.

TABLE 10.4 Three-Point Estimates for Fig. 10.4

Work pack	Optimistic	Likely	Pessimistic	Expected
A	2	3	4	3
B	2	2	6	2.5
C	1	2	3	2
D	2	2	4	2.5

Program Evaluation and Review Technique. The program evaluation and review technique (PERT) applies three-point estimating to critical path analysis (Sec. 9.3). This is effectively what I did above when discussing Table 10.4. You can calculate the schedule and the critical path using optimistic, pessimistic, most likely, and expected values of the duration of each activity. The expected values (calculated using the 1:4:1 formula) are taken as providing the best indication of the schedule.

Monte Carlo Analysis. What three-point estimating shows is there is a range of possible durations and costs for our projects. What we would like to know is what is the probability that the cost or duration will be less than a certain value. I was indicating that in Table 8.5 when I suggested that the sanction value of the project should have an 80 percent chance of being achieved. So how do we calculate these probabilities? With the network in Fig. 10.4 I assumed various probability distributions for the durations of the four packages of work, as follows. The design make take 2, 3, or 4 months with the following probabilities:

2 months: 25 percent
3 months: 50 percent
4 months: 25 percent

The delivery of the steelwork could take 1, 2, or 3 months with the following probabilities:

1 month: 25 percent
2 months: 50 percent
3 months: 25 percent

Because of the chance of snow during December through February, I assumed the duration of the site work could be between 2 and 6 months with the following probabilities:

2 months: 75 percent
3 months: 19 percent
4 months: 3 percent
5 months: 2 percent
6 months: 1 percent

Finally, because of the chance of wind in February, March, and April, the duration of the erection of the steel work could take 2, 3, or 4 months with the following probabilities:

2 months: 75 percent
3 months: 20 percent
4 months: 5 percent

I was then able to calculate the probability of achieving durations between 5 and 12 months. The pointwise and cumulative probability distributions are shown in Table 10.5. You see the most likely duration for the project is 9 months, but there is an even chance it will be finished within 8 months. But the chance of actually achieving the raw estimate, 7 months, is less than 1 in 5. So when this project that was estimated to take 7 months takes 8, because the snow fell in January and the wind blew in March, is the project manager at fault? I think he or she is at fault for not raising awareness of the possible impact of the weather. The project will never last the theoretical minimum or maximum value 5 or 12 months, and has only a small chance of lasting 6, 10, or 11 months. This is compatible with the assumptions I made in Table 8.5.

TABLE 10.5 Pointwise and Cumulative Probability Distributions for the Project in Fig. 10.4

Project duration	Pointwise probability	Cumulative probability
5	0%	0%
6	5%	5%
7	13%	18%
8	32%	50%
9	40%	90%
10	8%	98%
11	2%	100%
12	0%	100%

Calculating these probabilities by hand took an hour for just a four-activity network. It becomes impossible for anything larger, and so we resort to using Monte Carlo analysis.

This is done using a proprietary add-on either to our network scheduling software, or Excel. For every time or cost estimate for which we have a three-point estimate, we choose a probability distribution from standard library. The Monte Carlo software then runs the project model a larger number of times, typically somewhere between 1,000 and 10,000 times. Every time it runs the model, it chooses a value for each estimate for which there is a three-point estimate based on the assigned distribution, and thereby works out the total cost and duration. Each time it runs it will achieve a value somewhere between the theoretical minimum and maximum. The program counts the number of times each duration or cost was calculated, and thereby works out an empirical probability distribution for the duration and cost of the project. These can be used to estimate the cumulative probability distribution for the outturn of the project, as suggested in Table 8.5. Figures 10.5 and 10.6 show pointwise and cumulative distributions respectively, for a project calculated using a standard package, Crystalball.

Another use of the results is they may help us identify risks causing the variability and so help us reduce or eliminate them. It is possible to work out several critical paths through

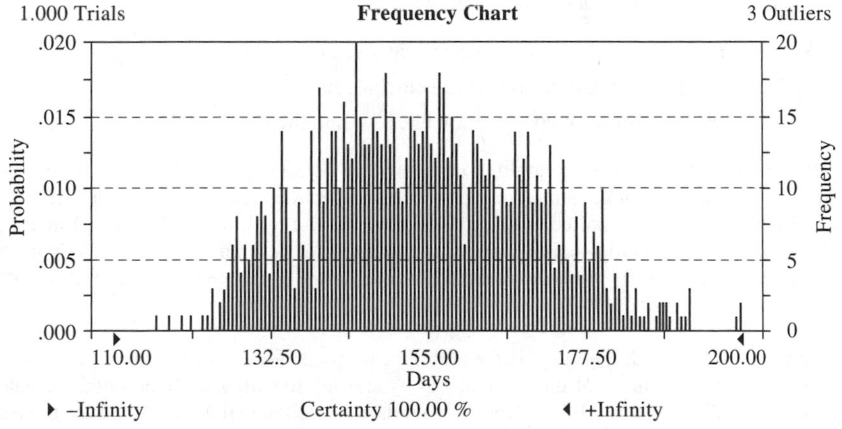

FIGURE 10.5 Pointwise probability distribution calculated using Monte Carlo analysis.

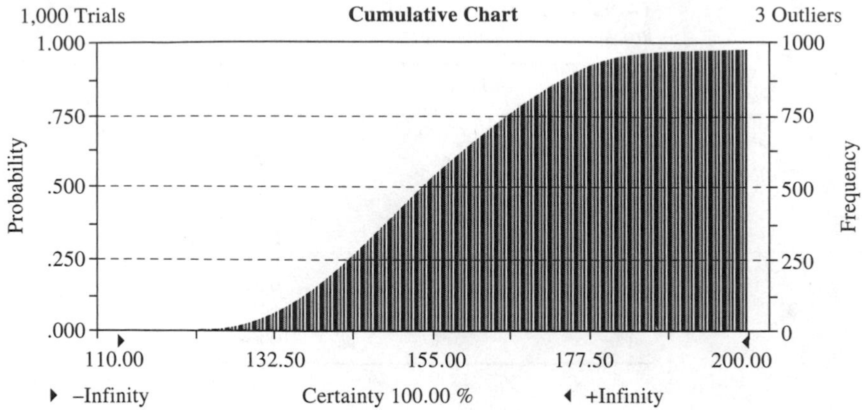

FIGURE 10.6 Cumulative probability distribution calculated using Monte Carlo analysis.

the network, determine what is causing them to be critical, and try to eliminate or reduce those risks. Likewise, you may be able to see several peaks in the costs estimate that again indicates particular risks causing the variability in cost. It is possible to see three or four peaks in Fig. 10.5, which may point to particular risks.

10.5 MANAGING RISK

Having identified and assessed the risk, we now work out how we can reduce the risk and manage it through the project.

Reducing the Risk

There are three basic approaches to reducing the risk:

1. Avoid it: You plan to eliminate or substantially reduce the risk.
2. Deflect it: You try to pass the risk on to someone else.
3. Contingency: You draw up contingency plans should the risk occur.

Pym and Wideman[8] use an analogy of a cowboy being shot at. He can take cover to avoid the bullets; he can get somebody else to stand between him and the bullets (though there is still a chance the bullets will carry on through and hit him); or he can allow them to hit him, hope he gets to hospital in time, and that the damage can be repaired. When you put it that way, plan to fail and hope we don't fail absolutely, it sounds better to use one of the other two approaches.

Avoidance. I showed above on the warehouse project, how to avoid the risk of snow holding up the preparation of the foundations by starting the work early enough, so that it is finished before the snow comes. Under avoidance, you change the plan to reduce the risk or eliminate it entirely.

Deflection. There are three ways of deflecting risk:

1. *Insurance*: by which it is passed on to a third party
2. *Bonding*: by which a security is held against the risk
3. *The contract*: by which it is passed between owner, contractor, and subcontractors

Insurance: A third party accepts an insurable risk (Sec. 10.1) for the payment of a premium. I showed above that this will only be used with risks of low likelihood and high impact. Because they are low likelihood it is not cost effective for the owner to take action, but occurrence of the risk can be devastating. The insurance company spreads the risk over a large number of similar risks, expecting just a small number to occur, but those that do occur will have high impact. Also don't assume what you will insure as an individual your company will insure. You will insure your car for damage and you house against fire, because for you those are low likelihood, high consequence risks. But for a company with a large fleet of cars, they are a medium likelihood medium consequence risk and they will allow a contingency. Likewise, the government doesn't insure buildings for fire damage. The government has so many buildings that it is a high likelihood, medium consequence risk and again it will allow a contingency.

Bonding: One or both parties to a contract deposit money into a secure account so that if they or either party defaults the aggrieved party can take the bond in compensation. This is a way of transferring the risk of one party defaulting to that organization.

Contract: Through contracts, the risk can be shared or passed between owner, contractor, and subcontractors:

a. A risk can be passed to the contractor under two circumstances. The first is where they are the best party to control it. They can then accept any type of risk and apply the techniques described here to control it. Where the contractor cannot control the risk, any risk passed to the contractor should be high likelihood, low consequence risk. Then the impact of the risk is predictable, the contractor can allow a contingency for it, and add a profit margin for accepting it. All the contractor can do is allow a contingency, and the client accepts paying that profit margin. If the contractor cannot control the risk, it would not be sensible to accept anything but a high likelihood, low consequence risk for which a predictable contingency can be made, and it would not be sensible of the client to give it to them because it increases the chance of failure of the project.
b. I said above (Sec. 10.1) that where there are multiple risks controlled by the client and several contractors, it is best to form an alliance and work together to try to reduce the risks. If coupled risks are assigned individually to separate parties they will be fighting each other as each tries to reduce the risk for which they are responsible, to the detriment of the project, and all concerned. Clients need to realize this. So the best arrangement is to form an alliance and work together in partnership to try to reduce the risks. But the client should only be in the alliance if there are risks within their control, and the contractors if there are risks within their control.[9] I said in Sec. 3.3 that a necessary condition for project success is to view it as a partnership.

Contingency. The third response to risk is to make an allowance for the risk, to add a contingency. You can add an allowance to any one of the five functions, scope, organization, quality, cost, and time, but typically there are two main approaches:

1. Make an allowance by increasing the time and/or cost budgets.
2. Plan to change the scope, by drawing up contingency plans should the risks occur.

Time and/or cost: You can either add the allowance as a blanket figure, calculated through a bottom-up approach as above, or you can add it work element by work element. Either way, the project manager should maintain at least two estimates, as described in Secs. 8.2 and 9.1. These are the raw estimate without contingency, and the estimate with contingency. The former, called the baseline, is communicated to the project team for them to work to, and the latter to the owner, for them to provide money and resources. The project manager may also maintain two further estimates, the most likely outturn, the figure to which they are working, and the current estimate, which is the baseline with some contingency already consumed. The reason for giving the project team the baseline or current estimate as the figure to work to is they will seldom come in under the estimate, and will consume contingency if it is given to them. The reason for communicating the estimate with contingency to the owner is they want to budget for the maximum likely time and cost.

Contingency plans: These are alternative methods of achieving a milestone, to be used if a risk occurs. Contingency plans can be of three types:

a. *Purely after the event:* Contingency plans are drawn up to be enacted if the risk occurs.
b. *After the event with essential prior action:* Contingency plans are drawn up to be enacted if the risk occurs. However, some preparation work must be done, such as procurement of long lead items, to speed up the reaction time and so reduce the consequence of the risk. The cowboy above can pay to have an ambulance stood by, with paramedics able to treat him and get him to hospital quicker if he is shot.
c. *After the event with essential mitigating action:* Contingency plans are drawn up, but the design of the facility or work methods changed to reduce the cost of implementing the contingency plan. The upfront cost may be increased to reduce the likelihood of the risk. The cowboy above wears a bulletproof vest.

With (2) and (3) there is a cost associated with the essential or mitigating prior action to reduce the consequence or likelihood of the risk. You can apply the likelihood by consequence formula crudely to see if the action is worthwhile.

$$\mathrm{L} \times \mathrm{C} > l \times \mathrm{c} + \mathrm{cost}$$

If the likelihood (L) by consequence (C) without the action is greater than the likelihood (l) by consequence (c) with the action plus the cost of the action, the action is worthwhile. The comparison is usually clear one way or the other; if it is marginal, don't bother, or do further investigation. In Example 10.8, we drew up alternative plans should the valve shut tight, shut partially, and not shut at all. The latter plans each would have cost more than the first, which is the one we followed, although the second would have only been marginally more expensive.

Contingency plans are the least preferred option. It is better to plan to eliminate the risk than to plan how to overcome it, and it is better to plan how to overcome it than to increase the cost and extend the duration to pay for it.

Summary. Table 10.6 shows where the different response strategies are used in the likelihood by consequence plot of Fig. 10.1.

I have put "no action" against the insignificant many. They should be left on the risk register and watched, but no action taken. You should focus on the significant few, the 20 percent that have 80 percent of the impact.

TABLE 10.6 Risk Response Strategies

	Consequence		
Likelihood	Low	Medium	High
High	Pass to a contractor	Contingency with mitigating prior action	Avoid the risk or don't do the project
Medium	No action	Pure contingency	Contingency with essential prior action
Low	No action	No action	Insure the risk

Controlling Risk

As the project progresses you need to monitor and control the risks. I suggest two documents:

Risk Item–Tracking Form. This is a document prepared for all the identified risks, the significant few and the insignificant many (but not the chaff eliminated after the brainstorming process). It identifies the risk, whether it is a business risk or insurable risk, its likelihood and consequence, and political impact, and the risk-reduction strategy. Table 10.7 is a risk item–tracking form for a risk in the CRMO Rationalization Project.

Risk Regsiter. This is a list of all the risks on the project, with their impact and risk-reduction strategy. They are sorted according to their priority so that the significant few are at the top, where they can be focused on, and the insignificant many are at the bottom, where they can be remembered but they do not occupy too much attention. Table 10.8 is a risk register for the CRMO Rationalization Project.

As the project progresses, you can then monitor the risks as they occur and take action as necessary to recover the plan.

TABLE 10.7 Risk Item–Tracking Form for the CRMO Rationalization Project

TriMagi **Risk Item–Tracking Form**	
Risk item	Loss of team leader of MIS development team
Owner	Rodney Turner
Likelihood	Medium—known to be dissatisfied
Consequence	High—significant delay to work
Rating	1—top risk
Milestones affected	T5
Impact	Loss of expertise Delay in coding
Mitigation strategy	Identify replacements Put them on standby
Progress	Alternative team leader identified

TABLE 10.8 Risk Register for the CRMO Rationalization Project

TriMagi
Risk Register

Rank	Milestone	Risk	Impact	L	C	Strategy
1	T5	Loss of team leader for MIS development team	Loss of expertise Delay in coding	M	H	Identify potential replacement
1	T5	Changes to user interface	H/W and S/W definition Delay delivery	H	M	Ensure user involvement in evaluation of prototype
2	T4	Problems in network diagnostic software	Delay in software completion	M	M	New version of software has fewer faults
2	T4	Availability of work stations for testing	Delay in testing	M	M	Expedite delivery with supplier
2	T3	Testbed interface definitions	Delay	M	M	Expedite definitions
2	T3	Delay in specification of data transmission	Delay in delivery of hardware	M	M	Meeting scheduled to consider alternatives
1	O2	Technical author required	Training & maintenance manuals not available	M	L	Contact agency
1	All	Configuration mgt support required	Poor quality systems and manuals	M	L	Contact agency

SUMMARY

1. There are six steps in risk management:
   - Identify sources of risk
   - Asses impact of individual risks
   - Prioritize risks for further analysis
   - Assess overall impact of risks
   - Develop risk-reduction plans
   - Control the identified risks
2. Techniques for identifying risk include
   - Brainstorming
   - A mechanistic process based on plan decomposition and expert judgement
3. There are two types of risk:
   - Business risk
   - Insurable risk
4. There are five sources of risk
   - External—unpredictable
   - External—predictable

- Internal—technical
- Internal—nontechnical
- Legal

5. The impact of individual risks is a product of the likelihood they will occur, the consequence if they do occur, and the public perception of that consequence.

6. In assessing the combined effect of several risks, you can use
- A top-down approach, based on plan decomposition
- A bottom-up approach and Monte Carlo analysis
- Influence diagrams

7. There are three ways of reducing risk:
- Avoidance
- Deflection, either by insurance or through the contract
- Contingency

8. Risks passed from client to contractor should be high likelihood, low consequence risks. An alliance should be formed to control coupled risks.

9. There are three types of contingency:
- Pure contingency
- Contingency with essential prior action
- Contingency with mitigating prior action

10. The strategy adopted depends on the type of risk.

11. There are four steps in controlling risk:
- Draw up a risk management plan consisting of risk item–tracking forms.
- Monitor progress of the significant few using the risk register.
- Reassess risks at regular intervals, and at key milestones or stage transition.
- Take action to overcome any divergence from plan.

REFERENCES

1. Chapman, C.B. and Ward, S.C., *Project Risk Management: Processes, Techniques and Insights*, 2d ed., Chichester, U.K.: Wiley, 2003.
2. Chapman, C.B. and Ward, S.C., *Managing Risk,* in Turner, J.R. (ed.), *The Gower Handbook of Project Management,* 4th ed., Aldershot, U.K.: Gower, 2007.
3. Association for Project Management, *Project Risk Analysis and Management Guide,* 2d. ed., High Wycombe, UK: Association for Project Management, 2004.
4. Project Management Institute, *A Guide to the Project Management Body of Knowledge,* 3rd ed., Newtown Square, Pa.: PMI, 2004.
5. Office of Government Commerce, *Managing Successful Projects with PRINCE2,* 4th ed., London: The Stationery Office, 2005.
6. Saaty, T.L., *The Fundamentals of Decision Making and Priority Theory with the Analytic Hierarchy Process,* Pittsburgh, Pa.: RWS Publications, 1994.
7. Williams, T., *Modelling Complex Projects.*, Chichester, U.K: Wiley, 2002.
8. Pym, D.V. and Wideman, R.M., "Risk management," in *The Revised Project Management Body of Knowledge,* Newtown Square, Pa.: Project Management Institute, 1987.
9. Turner, J.R., "Farsighted project contract management," in Turner, J.R. (ed.), *Contracting for Project Management*, Aldershot, U.K.: Gower, 2003.

P • A • R • T • 3

MANAGING THE PROCESS

CHAPTER 11
THE PROJECT PROCESS

Part 2 covered the management of the five functions: scope, organization, quality, cost, and time, and the risk inherent in them. We now turn to the second dimension of project management, the management process. In Chap. 1, I suggested that because projects are transient, they have a *life cycle*, going through several stages of development from germination of the idea, to commissioning of the facility, and finally, the metamorphosis into a successful operation. During this life cycle, management emphasise changes; the definition of the project evolves in a controlled way, so the best solution to the owner's requirement is achieved, and money and resources are committed only as uncertainty is reduced. In this part, I consider the management of the project life cycle. In Chaps. 12, 13, and 14, I describe what is done at each of three stages of a simple form of the life cycle: project start-up, execution and control, and closeout.

In this chapter, I start by revisiting the project life cycle and setting it within the context of the product life cycle. I show that projects may run over several stages of the product life cycle, or some projects may be undertaken to deliver individual stages. I then describe two specific stages of the life cycle: feasibility and design, not covered by the Chaps. 12, 13, and 14 of this part. I then consider versions of the life cycle for different types of project, new product development, concurrent engineering, and information systems projects.

11.1 THE PROJECT AND PRODUCT LIFE CYCLE

There is a hierarchy of life cycles, or management processes consisting of:

- The product life cycle
- The project life cycle
- The management processes

In Chap. 1, I introduced two versions of the management process:

1. The one I derived from the work of Henri Fayol:[1] plan, organize, implement, and control.
2. The other from the process recommended by Project Management Institute (PMI) in their PMBoK:[2] initiate, plan, organize, execute, control, and close.

In Chap. 1, I also introduced a generic version of the project life cycle: concept, feasibility, design, execute, and close, which has formed the basis for much of the discussion of this book. In this chapter, I give several other versions of the project life cycle. I also suggested in Chap. 1 that on small projects, especially ones that are part of a program

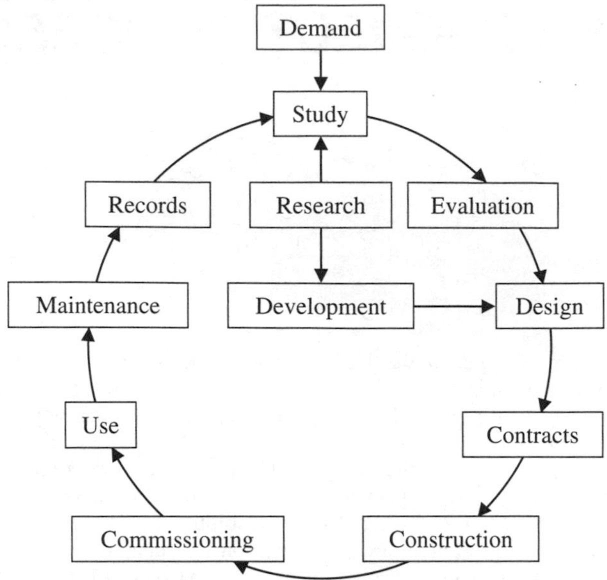

FIGURE 11.1 Wearne's life cycle for industrial projects.

(Chap. 16), there is very little difference between the management process and project life cycle, but on larger projects the two are quite distinct with the management getting repeated at each stage of the life cycle (Fig. 1.8).

In Chaps. 12, 13, and 14, I describe three stages of the life cycle: start-up, execution, and closeout. In reality, the first two relate more to the management process, and the third more to the last stage of the life cycle. In the next two sections, I describe the two stages of the life cycle not covered by these: feasibility and design. (Concept is covered by Chaps. 2 and 12.)

Several versions of the project life cycle set the project within the life cycle of the product made by the facility or asset the project delivers. Stephen Wearne[3] proposed a model (Fig. 11.1), which is essentially a life cycle of the new asset, and is reminiscent of the problem-solving cycle (Fig. 1.6). It starts with a survey of demand for the product produced by the facility. That part of the cycle on or within the circumference describes the life of the facility. The six steps from study to commissioning relate to the three steps of the project life cycle used in the Chaps. 12, 13, and 14. The next three steps extend the life beyond the project to the use of the facility, its maintenance, and monitoring of its performance. The World Bank has a version of the life cycle that is very similar (Table 11.1), as does the European Construction Institute (ECI) (Table 11.2). The World Bank is concerned about the pre- and postproject stages, ensuring the investment decision is sound and the project delivers the benefit postproject. They are not so concerned with the actual construction of the facility which is the responsibility of others. Therefore their version of the life cycle does not detail the actual implementation stages. The ECI's life cycle on the other hand details the pre- and postproject stages, but also details implementation, and is very close to Stephen Wearne's.

Harold Kerzner[4] proposes a model addressing the life cycle of the product produced by the new asset. It is the classic marketing view,[5] (Fig. 11.2). This is the view of projects filling the planning gap (Fig. 2.1) and draws very little distinction between the project and

TABLE 11.1 Project Life Cycle Used by the World Bank

Stage
Identification of project concepts
Preparation of data
Appraisal of data and selection of project solution
Negotiation and mobilisation of project organization
Implementation including detail design and construction
Operation
Postproject review

TABLE 11.2 Life Cycle Proposed by the European Construction Institute

Stage
Concept
Feasibility
Front-end design
Project plan
Specification
Tender and evaluation
Manufacturing
Construction
Commission
Operation and maintenance
Decommission
Disposal

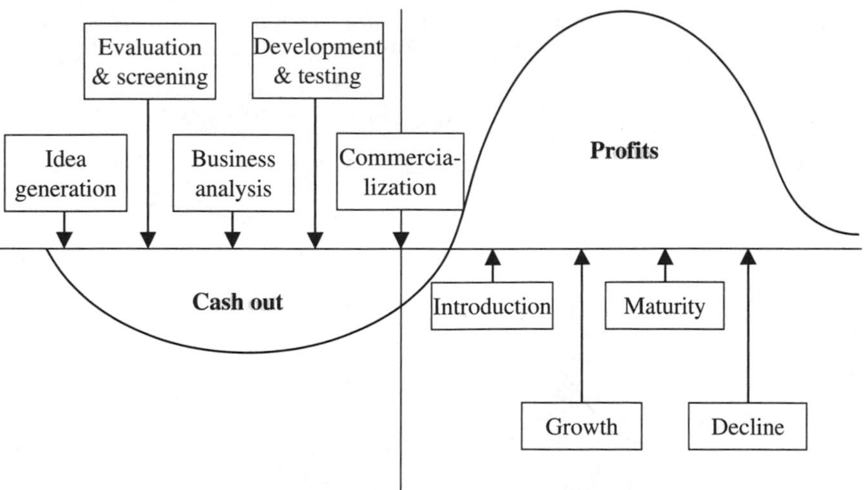

FIGURE 11.2 Classical marketing view of the life cycle of the product.

the product. Some people differentiate between the project and the product life cycles saying the project is the period up to and including commercialization and the product life is the period from introduction of the product until its decline (Fig. 11.2). The project period can last anything from three months in the electronics industry, to ten years in the pharmaceutical industry, to a hundred years for the major infrastructure projects.

There are several implications in Figs. 11.1 and 11.2 which need qualifying:

a. First is an assumption that the facility is an engineering plant that will make a product or perhaps provide a service through its operation, as with an airport for instance. In this book, I have taken a wider view of projects and the facilities delivered. Like an engineering plant, the facility may be a computer system, a design, trained managers, or a set of procedures. With this view, projects can occur at any step in the product life cycle. There are projects to conduct a marketing survey, research and development, and maintenance. From the marketing model, there are projects to launch the new product (Sec. 11.4) and to relaunch the product at deterioration (Fig. 11.3). Projects therefore occur throughout the product life cycle or at any stage in the strategic development of organizations (Sec. 2.5). I describe several of these projects in the following sections.

b. Coupled with this is an assumption that the project is a large engineering endeavour. The project may be a small project taking place as part of a program or portfolio of projects. I discuss program and portfolio management further in Chap. 16. Many change initiatives actually take place as a program comprising several projects, and indeed the life cycles in Figs. 11.1 and 11.2, can be viewed as showing a program of projects (Chap. 16).

c. Finally is the implication that we are talking about the private sector. The Office of Government Commerce (OGC) in the UK has proposed a life cycle for large projects in the public sector, as part of their gateway review process. This shows the project nested within a program, which itself is nested within the formation and implementation of government policy. Again the life cycle tends to focus on the pre- and postproject stages. The assumption with large projects is that the work will be done by external contractors, and so the life cycle does not focus so much on the implementation stage, unlike PRINCE2. OGC's gateway review process has been adopted by the Department of Homeland Security in the United States for the monitoring of information systems projects, and by the federal government in Australia.

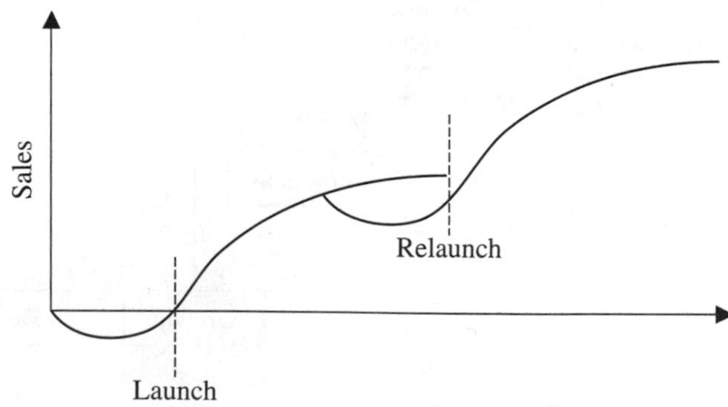

FIGURE 11.3 Life-cycle for government policy.

11.2 THE FEASIBILITY STUDY

After the initial concept stage, it is necessary to develop the definition further, and refine estimates to the level of the second row in Table 8.4, to be able to commit resources to design. This is done through a *feasibility study*. At the concept stage, several solutions may have been proposed. We now assess their technical, commercial, and managerial feasibility, and choose one to take forward to design (assuming at least one is feasible).

Aims of the Feasibility Study

Feasibility studies involve time and money, so it is essential they are well managed, and for this it is important to understand the aims of the study.

Exploring All Possible Options for Implementing the Project. As many ideas as possible should be explored. Each option must be thoroughly reviewed to see whether it can be improved within the limitations of market and technical conditions. The original specification can act as a guide to the study, but it should not stifle imagination and creativity.

Achieving a Clear Understanding of the Issues Involved. The feasibility study must give a clear understanding of the issues. In particular, associated with each option still being considered should be: estimates of costs and revenues; an understanding of the views and objectives of the various sponsors and institutions involved; confirmation of both technical and financial viability; and estimates of the likely economic and financial returns, as described above.

Producing Enough Information to be Able to Rank the Options. The study should produce enough information to rank options. The criteria used are based on the strategic factors described above. Their weighting in the overall ranking of options depends on the sponsor's goals; the public sector will usually give more weight to social and environmental factors than the private sector.

Obtaining a Clear Picture of the Way Forward. The study should result in a clear idea of future stages. It helps to think of the feasibility study as a funnelling and filtering exercise, directing a wide range of possible ideas into a much narrower range of options, with those which clearly fail to meet objectives sifted out. The study should aim to provide a refined specification and a work plan for the next stage, design. It may also result in a draft plan for the design or execution stages (Example 11.1).

Example 11.1 Producing a clear definition of the way forward

I worked on a three-month feasibility study to assess the efficacy of a new process. We launched the study with a two-week workshop. At the end of the workshop, we had a clear objective for the feasibility study but we had produced no plan, not even at the strategic level. Nor did we produce a plan in the yearlong systems design stage which followed. The first plan produced was at the end that year, for the detailed design of the plant.

The Factors Addressed

The study must provide an understanding of factors influencing success, and assess the advantages and disadvantages of each option to enable them to be ranked. These include

Market Conditions. Expectation of returns depends on satisfying demand for the project's product at a certain price level. Usually neither future demand nor future prices can be predicted accurately. If there is a limited portfolio of potential buyers, or the market is volatile, or demand is price-sensitive, as with commodity products, the project is vulnerable to many adverse influences. However, the existing market environment provides a wealth of information on which to base sales forecasts, establish price structures, understand potential purchasers and consumers, evaluate expected trends in demand and the actions of potential competitors, and learn about the expected quality of the product or service.

Supply Considerations. Existing supply conditions are also important sources of information. The feasibility study should assess the cost, quality, and availability of capital equipment, raw materials, and labour. Different technical options should also be explored and specialist technical advice obtained on their feasibility.

Financial Prospects. The profitability of the project can be analyzed by applying evaluation techniques (Sec. 2.4).[6] The financial feasibility also depends on whether the expected return from a project is sufficient to finance debt and provide shareholders with an adequate return to compensate for their risk. Financial feasibility is influenced by economic conditions such as interest and exchange rates prevailing when costs are incurred and income received. The approach differs for projects in the private or public sector. The latter often takes account of nonmonetary benefits and costs, and factors such as environmental impact. Shadow prices are used where the market price is considered not to reflect the economic cost or benefit of a project input or output. The private sector usually places more weight on purely monetary return, although legislation, tax benefits or subsidy, and public relations considerations may encourage it to place value on nonmonetary factors. Adequate consideration must be given to risk and uncertainty (Chap. 10). Risk and uncertainty cannot be eliminated, but they can be managed and reduced by prudent project design and management. You should also remember that the shareholders' evaluation of the project, and hence the share price of the company, depends on their assessment of the risk.

Planning the Study

Appoint an Experienced Manager and Core Team. Their makeup depends on the nature of the project. For the feasibility study, it should include technical, financial, and marketing expertise, and for larger projects may also have economists, legal, and environmental experts, human resources experts, and so on. It is essential that a good balance is struck between specialists, as assessment of the options may be biased if one specialism dominates. For example, if technical experts dominate, they may emphasise technically exciting options, which may not provide the required financial return. It is often helpful to limit the size of the core management team. Compact teams are easier to organize and coordinate than larger groups. The manager of the study will usually not be the project manager for subsequent stages. However, it is usually a good idea for the latter to be a member of the management team for the study, as they will then have greater ownership and commitment to the results of the study, the decisions made and the strategy set.

Scope the Study. Examine the scope of the study to assess the work involved and any constraints imposed. The manager must determine exactly what the decision-makers require to guide them in their choice of the project options, and in what form the information is needed. A work plan with the delivery time and content of interim and final reports should as far as possible be agreed in advance with decision-makers. Remember, project management is fractal management; the study needs planning as much as the implementation of the project.

Plan the Study. Draw up a plan for the study, including a milestone plan and responsibility chart. The milestone plan should identify key stages for the study: interim and final reports, meetings, data collection, and so on. The plan can highlight different lines of enquiry involved and their interdependence, enabling the different aspects of the study to be coordinated. It should be robust, but sufficiently flexible to cope with any unexpected changes. Adequate allowance should be made for the time required to request and collect data as well as processing and interpreting results.

Schedule the Study. Set the timetable and budget for the study. These must be sufficient to enable options to be properly explored and refined, without endangering the feasibility of the whole project. It is important to budget for an adequate exploration of the options without going to the depth of investigation required for the design and appraisal stage.

Managing the Study

Once feasibility study has been planned, we follow the remainder of the management process.

Organization. This involves the adoption of a clearly focused but flexible structure based around the milestone plan. The team should be aware of what is expected, and by when. They should understand how they fit into the study framework, and to whom they should report. Hence, roles and responsibilities must be clearly defined. The responsibility chart is the tool which effectively achieves this.

Implementation. This requires efficient communication within the team. The manager should maintain frequent contact with sponsors to ensure the study remains on target, and any change in requirements is identified. The team should maintain good internal communications to ensure delays are reported, to minimise knock-on effects, to avoid duplication, and to confirm that all information received has been made available to all members of the team. It is particularly important that good communication is maintained between team members in different fields of expertise to ensure any interdependencies are taken into account.

Control. This is the responsibility of the manager who must ensure that milestones are being reached on time, and that the milestones adopted lead to punctual report delivery. Likewise, costs should be monitored to ensure the study remains within budget. Control involves both monitoring of timing and budgets, and rapid and effective corrective action when targets are not met, either by revising targets, or by restructuring present plans within the existing targets.

Completing the Study and Transition to the Next Stage

The feasibility study acts as a spring board for design. The end product should comprise a clear, concise report, the project definition report (Sec. 12.3), which presents the original specification and objectives, with the conclusions and recommendations for use in the next stage. The report should highlight advantages and disadvantages: cost, revenue, strategic considerations, economic benefits, and so on for each of the options which deserve further consideration and the proposed solutions to issues confronting the project. Furthermore, the report should indicate sensitivities to variations from the assumed base case.

11.3 THE DESIGN PHASE

The next stage of the life cycle is design. The primary emphasis of this stage is the development of the project model. The original outline requirements as expressed by the client in the feasibility study are subjected to more rigorous examination to define exactly what is to be done to achieve the project's objectives. A systems design is developed for the new asset, the product it will produce, and the method of building it. This helps define the work (Chap. 5). The project organization is developed, and roles and responsibilities of departments, functions, disciplines, or their managers are described (Chap. 6). The quality specification, cost, time-scale, and risk are all planned and estimated (Chaps. 7, 8, 9, and 10). From this information we determine whether or not the project is viable and represents a good investment at the accuracy of the third line ("Control") in Table 8.4. This appraisal process is vital, as it is the last chance the sponsor has to decide whether to proceed with the project before committing scarce resources to execution. Many of the issues investigated in design are the same as feasibility, but at a greater level of detail.

The design is developed at several levels of the project and stages of the life cycle, Table 11.3 corresponds to different levels of accuracy listed in Table 8.4. It is common to show the life cycle as a serial process, beginning with concept and continuing through feasibility design and execution, until the facility is commissioned and producing the desired output. However, the reality of many projects is different. Design in particular is an iterative process, proceeding through these levels, as our understanding is refined. At each level, the design is checked back to the assumptions set in the project's strategy. Even at one level, there may be several iterations as the design proceeds through several formats. In shipbuilding, the paper design is converted into a plastic model, then into a wooden model from which fabrication jigs are made, before the first vessel of class is made. The ship is thus made four times before it is completed: in paper, plastic, and wood; before being made in metal. The design process is therefore not a single activity, but a set of activities ranging from the outline requirements to the detail design, and these cover all the stages of the life cycle. The computer industry has developed a spiral model of the life cycle (Sec. 11.6), which reflects the reality of the design process, and perhaps has applications elsewhere.

Managing the Design Process

Design involves the production of information to enable a solution to be selected from a series of options and to allow one to be manufactured or constructed. The target for a

TABLE 11.3 Life Cycle of the Design Process

Design stage	Design name	Activities
Definition	Customer requirement	Appointment and problem definition Establishment of solution criteria
Feasibility	Functional design	Evolution of alternative solutions
Appraisal	System design	Evaluation of alternatives Selection of preferred solution
Detailed design	Detail design	Detailed design of selected solution
Delivery	As-built design	Manufacture and assembly Facility construction

good design manager is to produce the right amount of information, using the right people, at the right time, to budget, and to the client's satisfaction, while making a profit for their employer. This balance is not easy to achieve. Engineers are notorious for trying to satisfy the client's requirements, while forgetting the need of their own company to make a profit! The application of good project management procedures to the design process can help to ensure the balance is achieved. It can make the process more flexible, allowing the design to proceed efficiently within a framework of gentle control, in which all designers know what they are doing, why, when it is needed, and what to do if the answer they come up with is not the one originally envisaged. This is not easy. The project manager not only has to deal with the vagaries of their company's management structure but that of the client as well. In a busy commercial environment, they rarely have exclusive use of all the experienced designers they require, competing for the expertise they need. They also often have to deal with heavy pressure from the client to produce action and results. The need for careful planning before quantifiable results are produced is often not understood.

A good design project manager needs to tailor their management style to suit the project. Some projects may be large enough for a task force to be developed with a good working relationship, making communication and management easier. Others may be multidisciplinary, involving short-term input from many different parts of the company which have to be very highly controlled to ensure that the correct product is produced. Yet other projects may be small with very swift programs which have to be fitted in between the longer running projects cutting across other deadlines. The busy project manager will normally have to deal with all these types of project all at the same time and for different clients.

The design process has five stages, Table 11.3. Prior to starting work on any of the stages, the design project manager should consider how the project will be planned and controlled. There are those who say that design as a creative process cannot be controlled. However, to be of value the facility must be obtained by a certain time, Fig. 9.2, and so the process must be managed. Tables 11.4 and 11.5 contain checklists for planning and controlling the design process. The design manager has to be a juggler of resources, costs, and time. In some ways the problem is more complex because the "product" is unique and can change many times before completion. The manager must strike a balance between too much planning and not enough control and too much control and not enough planning.

Managing the Urgency

There is often a tendency to try to shorten the design process to begin work on a project. When discussing the problem-solving cycle, Fig. 1.6, I said people tend to jump from perceiving a problem to selecting a solution, or worse to implementing one. They then never truly determine the cause of the problem and the best method of solving it; they just paper over cracks. It is always important to put adequate time and effort into the design process, and the way to ensure this is to have a proper project plan for the design stage, which measures the progress of the design towards completion against a series of milestones.

There can also be a tendency to overlap implementation and design to make better use of available skilled resources. This is what I described in Sec. 3.4 as *fast build, fast track,* or *concurrency*, which are associated with increasing risk. I cannot stress enough the importance of allowing the design stage to take its course. However, we shall now see that the project manager must guard against the opposing risk, namely the desire of the designers to develop the ideal solution or prolong the design period because of the inherent job interest it offers.

TABLE 11.4 Planning the Design Process

P1:	Examine the problem carefully with the client and if possible with their advisers. Establish what the task is and agree to a fee structure for the work covering various stages of design, taking note of the often highly variable nature of the initial design studies.
P2:	The fee arrangement may also include a collateral warranty. This is commonplace in the construction as a result of case law on the question of latent liabilities. This must also be recognized and dealt with as a milestone because clients frequently cannot get funding released from their backers until the document is signed and completed. The time necessary to complete these procedures is often underestimated, which can cause delays.
P3:	Establish the basis of a planning network, identifying key milestones in design. Plan to do the detailed planning of each phase only when it is necessary, that is, on a rolling basis.
P4:	Confirm the work breakdown and identify packages of design work. Seek for use of the appropriate work package managers and teams from within your company. Select the right people for the right job, and match personalities to the nature of the task. A careful meticulous detailer cannot drive a high-pressured fast-track project forward, but should be used to give support to the innovators and strong managers.
P5:	Assess time and resource requirements for each phase of design work (at the appropriate time) using your own experience combined with that of the work package managers.
P6:	Check each stage of resource allocation against fee available prior to undertaking work. If fee is too small, reevaluate the amount of design and reduce or delay applying resource or renegotiate fee arrangement. Aim to do the right work at the right time.
P7:	Establish that resources available for each stage are sufficient to meet the program, using your master design plan as a basis. Introduce contingency allowances at a fairly high level in the plan so that you can control slippage. Try not to build contingency at each level or else you will never create a workable program.
P8:	Establish, jointly with the department managers in your company, whether your use of their staff (particularly when the project is multidisciplinary) is compatible with their other commitments and schedule resources accordingly. Tie this back to the basic network and evaluate any overall program effect. As far as possible smooth out resource peaks to enable overall company staff planning to be easier and seek to adjust project priorities to suit. A balance always has to be struck.
P9:	Establish work packages, and if possible write down a brief for their managers as clearly as possible. This is often difficult to achieve but is very important because it establishes a firm criteria against which success can be measured in each design package. Ensure this brief is a living document, and that it is continually referred to and updated by mutual agreement of project manager and work package manager as the design evolves.
P10:	Establish the critical path. In theory, your critical path should be determined from the outset by the production of a stable network plan within which variations can take place. In practice this may not be so easy to achieve as there are usually many unforeseeable events which erode your contingency and cause the path to shift. These key activities always dictate whether or not the project is completed on time.
P11:	Information and resource needed is the key to the well being of the project. It is suggested that 80 percent of design problems come from 20 percent of the activities. However an over-preoccupation with activities currently identified as critical can backfire by reducing your awareness of noncritical areas which can become critical. A balance has to be achieved and progress on each facet of the network must be regularly monitored and controlled.

Managing the Users

Throughout design, designer, client, and end user must remain in close dialogue, so the design meets the user's needs. However, it is important that designers and users are not allowed to change the requirements so frequently that no progress is made. Managing the users is vital. The challenge is to ensure essential changes are made, but "nice to haves"

TABLE 11.5 Controlling the Design Process

C1:	Establish communication systems. Decide which level of designer should talk to which level in the client. Ensure you are always in the picture as to progress. Ensure work package managers are aware of their responsibility to control communication. Only correct considered information should be released to avoid incorrect action by outside bodies which destroys confidence in the design team's abilities.
C2:	Establish a Design Review Procedure. Regular (fortnightly) design reviews should take place to ensure the whole package is moving towards its target. An open forum in which package managers can discuss problems should be encouraged. People must not hide major problems but discuss them and to seek help before they get out of hand.
C3:	Establish a design checking procedure to interface the review process. Some projects need a full quality assurance (QA) system. This must be identified at the outset to ensure a quality plan is written and implemented incorporating project management systems. Some projects (eg, bridges designed for the Department of Transport) have checking procedures established for each stage. These may include formal checks by other firms.
C4:	The checks and consequential alterations must be programmed at each stage of the design, and adequate resources and time allowed. If QA is required, you must remember this is only an aid to sound design office procedures and not a substitute.
C5:	Establish regular meetings to interface with client/consultant design team meetings. These meetings may be held instead of or as well as design reviews, depending on the complexity of the job, and should bring together internal design issues and a review of external influences on the design. Following these reviews, a short statement of progress addressing key issues and problems should be prepared for issue to the client. Areas where information is required or where instruction is needed should be identified.
C6:	As changes occur ensure the reasons are communicated if appropriate down to draughtsmen. There is nothing more demoralizing to a draughtsman than facets of the design being repeated when the reason is unclear. Although this may be tedious for the work package managers, the project manager must encourage the team to keep communication lines open. One must always be conscious of the needs and desires of the individual as well the objectives of the project or else neither will be achieved.
C7:	Check expenditure against forecast costs and fee income. Often delays cause an increase in resources to recover the program. Delays may require you to move staff from one project to another to avoid overloading one and under-resourcing another. While plan and control systems should accommodate this, the financial side of your company must not be forgotten. Computerized monthly job cost summaries are out of date before you receive them so ensure you know what the projected cost effects are before they occur.
C8:	Changes to the design brief may be made by the client as design develops. This is part of the design process as the client begins to understand the impact of earlier decisions. Some change should be tolerated, but major changes must be controlled, and additional fees sought before embarking on the extra work. If the client is not aware of the financial implications, they will change their mind without a second thought.
C9:	Establish which outside bodies must be consulted, and their approval obtained prior to commencing. These may be statutory bodies, environmental groups, planning authorities etc. Time to obtain such approvals must be allocated and milestones recognised. You may require input from the consultant team at regular intervals depending on the product. Identify and program this information as a strategic part of the design process.

avoided. Many people suggest freezing user requirements at an early stage. But that can lead to ineffective solutions, as the process of designing the asset and its product can help to clarify user requirements. What is needed is the application of effective configuration management so the design moves steadily forward, until a viable design is produced which meets user's needs.

11.4 NEW PRODUCT DEVELOPMENT

New product development (NPD) can lead to many types of project, including:

- Research and development
- Product design
- Facility design and construction
- Product launch

Encouraging Innovation

NPD plays a key role in organizational competitiveness, yet is it one of the most difficult aspects to manage. Organizations which choose in-house development must create a climate which favours innovation.[7] Top management have a key role in this process, to encourage the establishment of a creative environment, which has three key components.

Climate for Innovation. The innovative climate of an organization and its development policies are inseparable. Product development demands a flexible structure which encourages creativity and entrepreneurship and provides necessary conditions which favour development. However, there can be many pressures within an organization which act to hinder enterprise, and encourage bureaucratic policies and procedures which constrain change.

Innovative Organization. In order to harness innovation, organizations must be versatile and adaptable in their approach to their circumstances. In essence, product development is at its best in organizations which encourage imagination and are organic in nature, rather than those with bureaucratic structures based on routine management processes.

Individual Innovation. Whether bureaucratic or organic organizations consist of people whose personalities and performance affect the success of projects and overall performance. Thus, organizations need to adopt structures which harness individual innovation. This will be reflected in recruitment and selection procedures, opportunities for development, removal of bureaucratic restraint, and rewards to innovators. It is not possible to prescribe the definitive organizational structure to achieve this; much depends on the company's response to its environment.

Planning New Product Development

The nature of product development creates several planning problems, as projects range from modest expenditure to major investments, combined with indeterminate time constraints incompatible with routine reporting cycles. The diverse activities involved in new product programs should move through a logical sequence of events. Though considered contrary to flexibility and creativity, development plans are necessary as they help determine critical components of the project. The sequence, project life cycle, suggested by Kotler[5] is often used to illustrate the new product planning process (Fig. 11.4).

Plans should be used to enhance rather than hinder the development process. Management should not be limited by this logical progression. The sequence outlined

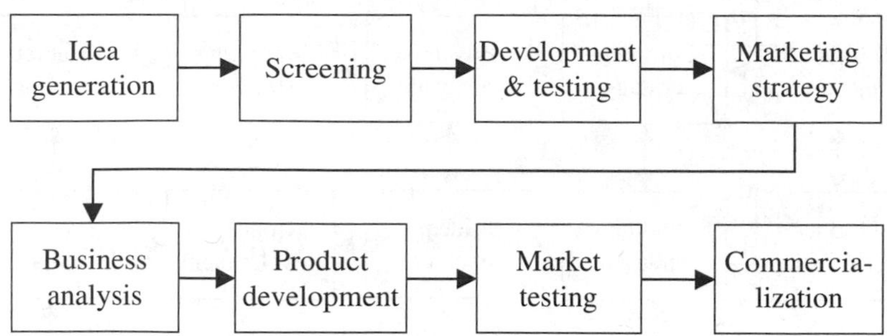

FIGURE 11.4 New product planning process (project life cycle).

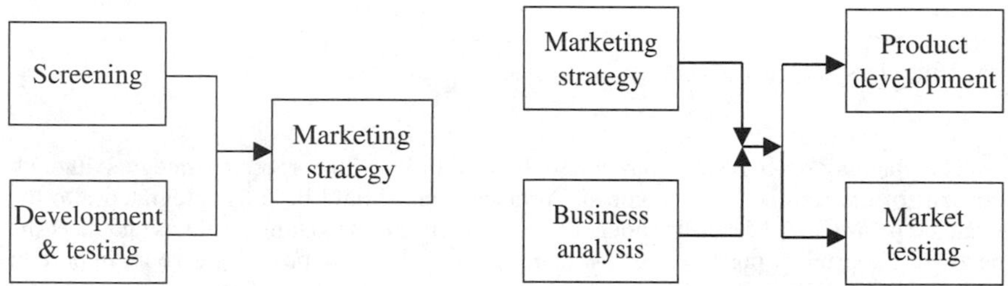

FIGURE 11.5 Revised sequence for ideas generated by users.

FIGURE 11.6 Revised sequence for products not requiring radical change.

is a guideline to help development, not constrain it. Idea generation, for example, does not always automatically occur as part of the formal planning sequence. Ideas may be initiated by users or employees during normal work (Fig. 11.5). Similarly, product development does not always require radical change. Projects may be initiated to modify existing product lines (Fig. 11.6), or may also have several stages running simultaneously (Fig. 11.7).

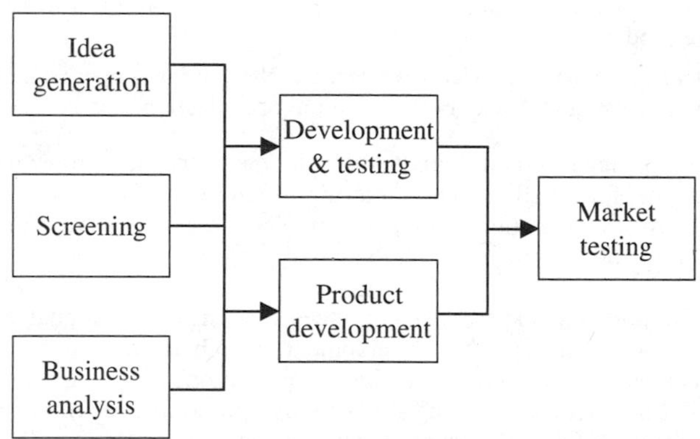

FIGURE 11.7 Sequence with several simultaneous changes.

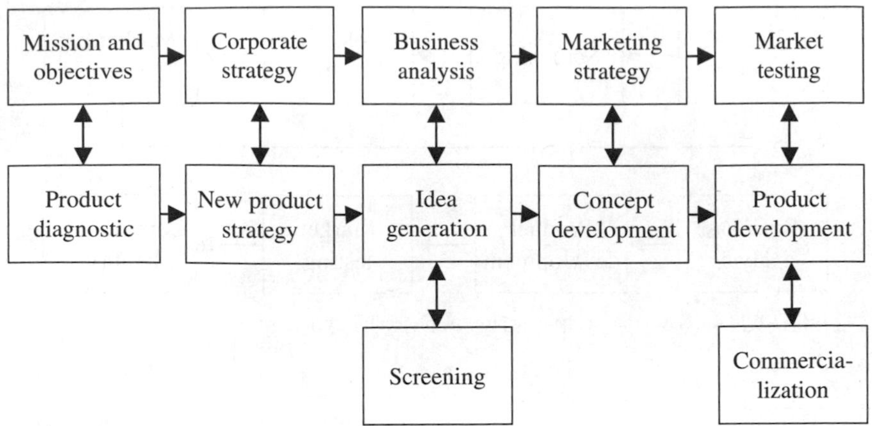

FIGURE 11.8 Revised sequence incorporating strategic focus.

The planning process so far has not established links with corporate strategy. Although not part of the routine of the company, project plans should be fully integrated into the strategic plans (Sec. 2.5). NPD should be complimentary to existing products and meet the needs of the product portfolio against market demands. New products give an important strategic capability for achieving corporate and business objectives. Strategic issues should direct and influence the new product project in three ways: strategic focus, technical criteria, and market acceptance. Figure 11.8 proposes a revised product development planning process which combines strategic focus with the need to combine phases of new product development.

Organizing New Product Development

In a climate which welcomes creativity, the marketing function has two distinct roles:

1. *Routine, operational marketing tasks:* Demanding a structure based on routine activities, planning, and coordination of the marketing mix for products which form part of the existing product line.
2. *Novel projects:* Requiring less defined structure. NPD projects operate in uncertain conditions, and plans require freedom from routine organization.

In order to implement in-house product development, the first problem is to find the right organizational format. By nature, innovation is individualistic, requiring each company to develop their own working arrangements. There are several ways in which a business can organize itself for product development.

New Product Committees. These are committees, meeting on a continual or ad-hoc basis, responsible for coordinating product development. Members are senior functional managers and executives from research, marketing, finance, production, engineering, and so on. Their responsibilities include reviewing and screening proposals, determining policy, planning, and coordination. Often the committee is considered to be the coordinating function which ensures the product maintains its momentum and controls the activities of the multifunctional team developing the product.

Product Managers. Product managers may be given responsibility for developing new products alongside their normal duties of managing existing product lines. There are several reasons for this. As well as monetary benefits, product managers are sympathetic to the customer requirements and considered to be in the best position to ensure synergy with the existing product portfolio. The disadvantages are additional management time required may not be forthcoming, nor can the product managers give this unique activity the specialized attention, resources, and expertise required while maintaining responsibility for routine activities.

New Product Managers. They are given overall responsibility for product development from planning to implementation. Often the new product manager works alongside existing product managers but without their operational responsibility, and can thus turn their attention to the creative role and generate practical new product ideas. Although the establishment of a new product manager formalizes the product development role, there are strong links with existing product lines, leading to minor changes, rather than independent, novel, or radical innovations.

New Product Departments. These are common in large organizations, working alongside new product managers in generating ideas, and evaluating their feasibility. In contrast to other methods, new product departments place the responsibility with a senior manager. The department provides the umbrella for coordination of various functions for continuous project management. It does not have responsibility for operational duties, so may dedicate its efforts to producing quality new products. Sometimes a new product department may be situated within a larger department, such as planning, marketing, research and development, projects, or engineering.

Venture Teams. These are composed of functional specialists working to a closely defined brief, and generally recruited on an ad hoc basis for a short time. While located in the team, the individuals are removed from day to day activities. The team ideally reports to a nonoperating executive.

Task Forces. These groups are organized on an ad hoc basis. Members are seconded from operational duties for the duration of a project, or divide their time between routine activities and project work. The aim of task force management is to ensure continued support from the functions throughout a project. As a project reaches the latter stages, task forces may recruit more members with specialist skills.

Project-Based Product Development. Product development involves individuals with specialist skills, from various functions and managerial levels. The formation of project teams can be effective in solving problems and creating benefits which cannot be achieved in routine ways. However, one structure may not be appropriate at all stages of the project. Just as the activity needs to be fluid and flexible, the organization must also adapt to accommodate the different expertise needed throughout the project.

Controlling New Product Development

Control is an important element of NPD. The application of marketing control systems to an NPD process reduces the risk. Control processes should be integrated into all aspects of the plan and linked to critical components mentioned earlier. A continuous monitoring program provides project teams with valuable information which may determine the successful outcome of projects. The key of any system is the extent to which it allows the manager

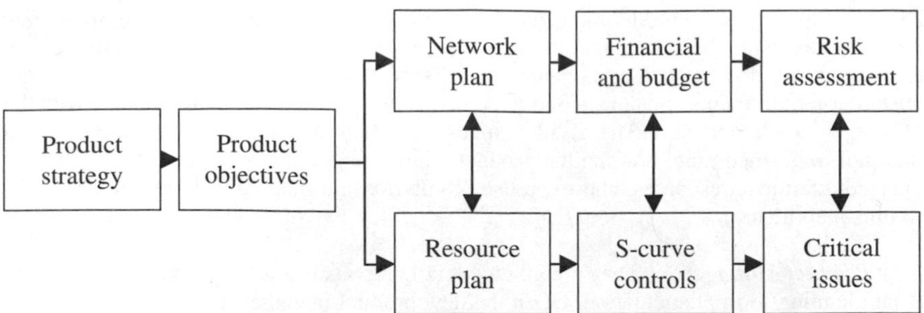

FIGURE 11.9 Schema for a hierarchy of plans for new product development.

to influence the success of the outcome of the venture. Several planning and control techniques may be used to monitor new product projects. Figure 11.9 illustrates how these methods may be combined to monitor progress based on project objectives.

11.5 CONCURRENT ENGINEERING

Figures 11.1, 11.2, and 11.4 show the product development process taking place sequentially. Traditionally, product development took place as a relay race: research, followed by development, followed by product engineering and prototyping, followed by production process engineering. Product development processes became artificially extended, not only as it was insisted that one step was finished before the next started, but inevitably there was a delay between one step and the next. Concurrent engineering attempts to overcome the built-in delays by running the product development process with the steps in parallel, as suggested by Figs. 11.5 to 11.7. The concept was first adopted in the development of fast-moving consumer goods as early as the late 1970s, but subsequently became widely adopted across a range of industries. Concurrent engineering is a systematic approach to the integrated concurrent design of products and their related processes, including manufacture and logistics support. This approach is now used in areas other than manufacturing, including construction and organizational change projects. The objectives of concurrent engineering are to achieve:

- Decreased product development times and hence earlier time to market
- Improved profitability and competitiveness
- Greater control of design and development
- Reduction in product costs
- Improved product quality

Requirements of Concurrent Engineering

Several changes are required within the organization and in its approach to projects, in order to allow this to happen.

A Change in the Organizational Culture. A shift is needed to the flatter, more flexible approaches of project-based management. There needs to be decentralisation of authority, with managers empowered to take decisions without referring them up the line, which builds in delay (Example 11.2). However, this environment creates an almost greater requirement for senior management support, to show their faith and support for the product development process.

Cross-Functional Team Working. The use of cross-functional teams is inherent to concurrent engineering. It is the only way to achieve the necessary parallel development. This requires people to communicate freely across the functional hierarchy, demonstrating the need for the empowerment and support of managers mentioned above (Example 11.2). It also requires teams to work closely with suppliers since their development must take place in parallel. This may require partnering or integrated supply chain management. It also requires advanced contracting methods to allow contracts to be agreed with suppliers long before the closure of design.

Use of Technology. Concurrent engineering only really becomes possible with the use of modern information systems. This includes the use of computer-aided design, engineering, and manufacture systems, (CAD, CAE, and CAM), with access via the intranet, to aid design integration through shared product and process models and databases. It also requires the use of good project management to coordinate the work of the people involved. Configuration management becomes a significant element of concurrent engineering and so the Project Management Information System (PMIS) must be able to perform the status accounting involved.

Techniques. Concurrent engineering requires the extensive use of iterative working techniques to develop all the aspects of the product, process, and logistics design simultaneously.

Example 11.2 Communicating across the hierarchy

Some years ago I did some work with the National Air Traffic Service, responsible for monitoring the U.K.'s air space. One person I interviewed had recently joined from the private sector to help implement a major project. He said he found it difficult doing his project work, because if he needed to communicate with another person working on the project but from another department he had to write a memo. The memo would go to his boss, who would critique it and send it back for revision. It would then go to his boss's boss, who would also critique it, and so on until it reached the lowest common boss, and then it would go to the person he was trying to communicate with. It could take two weeks or more for the memo to get to the person he was trying to communicate with. He said it was impossible because on a project you need to make instant decisions, advancing simultaneously across a number of disciplines from different departments.

The next person I interviewed was an air force officer on secondment. I was reeling from what the previous person had said, and asked the air force officer about it. He said, yes it happens, but what he did was send a draft memo to the person he was trying to communicate with while the official memo went through the cycle. As a military person, used to the need for fast communication (Fig. 1.10), he had found a way around it.

You can understand the concern of middle and senior managers. If a plane crashes because of a decision made by somebody in their department, they want their stamp on the decision trail. But it makes project work impossible. Managers need to be empowered. It is about setting flexible parameters within which they can work, but setting limits where limits matter.

Risks and Pitfalls of Concurrent Engineering

Attitudes of Middle Management. There may be resistance from middle managers (see Example 11.2). Not only does the cross-functional working threaten their influence, it increases their costs initially. Although the increased costs will be repaid through earlier completion times, managers can see an early fall off in the profitability of their departments, and hence a reduction in their bonus in the early years. When told that the higher costs associated with faster development times will be repaid through increased sales over a three-year period, managers may say that they are only in post for two years, and hence do not get the enhanced payback within their period of tenure.

Authorisation of the Concurrent Engineering Project. The first problems to be overcome are those of obtaining sanction for the project and for executing it on a concurrent engineering basis. If this has been done before it should not be a problem, otherwise it may be a long hard battle with all of the organizations and departments involved. The project definition stage is of paramount importance in this respect as it has to satisfy the following major criteria prior to project authorisation and major commitment:

1. The product must be within the organization's aims and objectives.
2. The market need for the product must be established beyond doubt.
3. The supply of raw materials must be shown to be secure.
4. The design of the product must be carried out to a sufficient level to establish its feasibility. (If necessary including models and/or prototypes.)
5. The design of product manufacturing system and its associated support systems must be evaluated, and must be within the organization's intended capability.
6. Economic and financial evaluations must show the product to be viable bearing in mind the predicted life cycle, development, and production costs.

It is usual during project definition to survey the industry, benchmarking to obtain typical implementation costs for similar products (Chap. 17). It may also be possible to use the organization's standard investment appraisal techniques,[6] but these do not always account for combined development and implementation stages. The risk associated with authorisation of a concurrent engineering project is significant as it is an all-or-nothing approach that commits the organization to prosecute the project to completion. (The only factors to prevent its continuation after authorisation would be due to external items such as a dramatic market shift.) In authorising a concurrent engineering project, senior management must be seen to give both the project and the approach their full support. They are not only authorising a technical development, but are authorising radical change and way of life in the organization.

Organizational and Cultural Change. The adoption of a concurrent engineering policy will inevitably involve significant changes to an organization and its culture. Some of the more important of these may be:

- Departmental organization shift to project orientation
- Conventional project-oriented organization shift to concurrent project organization
- Move out of deep hierarchical structure into shallow, multidiscipline teams
- New organizational reporting structures
- Change to business practises and procedures
- Establishing long-term customer/supplier relationships and elimination of counter-productive competitive tendering policies
- Reorientation of accounting policies away from departments and towards projects

The importance of these changes will be determined by many factors including:

- The degree of support from senior management
- The existing organization's size and culture
- Resistance from established functions
- Degree of product novelty and complexity
- Difficulties in implementation

Managing Interfaces. There will be many interfaces to be managed including those between:

- Management and design
- Commercial considerations and design
- Suppliers and design
- The various design functions within the concurrent engineering design team
- New product production and support facilities and those of existing products.

Careful selection of the concurrent engineering team, its working procedures, and the control facilities employed to ensure that these are managed effectively.

Technical Management. The most important function to control is that of design as this largely determines how a product is to be made or implemented and its associated costs. The designer is often only limited by a relatively few critical constraints but his or her work may have great impact on the work of others and on downstream costs. The following aspects of the project are identified at an early date and monitored closely:

1. Differentiation: where there are linkages between highly differentiated departments.
2. Cross-functional requirements: where there is a need to take account of requirements of the other function, particularly those downstream in the development process.
3. Uncertainty: where there is high uncertainty in the use, interpretation, or content of data.
4. Intensity and frequency of two-way flow: where there are major feedback requirements between departments or functions.
5. Complexity: where there is a need to liaise between groups because of the complexity of the product or task.

Standardisation policies help ensure conformity and the extensive use of common electronic data management tools helps to keep all parties working to the same model and standards. The early development of prototypes and prototype testing is a powerful tool used extensively in concurrent engineering. Its major value is that of identifying and forcing problems out into the open at an early stage. These can then be solved before they become too serious.

Cost Control and Release of Finance. The implementation of concurrent engineering requires a significant departure from conventional financial release and cost control methods as it involves

- Initial release of greater fund on more preliminary information in the early stages.
- Negotiation of less well-defined contracts with suppliers or contractors who are to assist in the design process.
- Commitment of greater funding for production facilities at an early stage when parameters are not well defined.

The chief departure from conventional methods is the acceptance of significant financial risk at an early stage and the greater requirement for an effective and continuous cost review procedure that gives early warning of possible cost risk areas. (Bear in mind that the rolling-wave approach to both planning and technical design will have major impact on the increasing confidence in cost estimates).

Risk Controls. More stringent risk management procedures are required with concurrent engineering than with conventional developments. In particular, they have to operate across the whole range of project activities including sales, marketing, personnel, production, and support. They need to impose a consistent risk approach on a continuous basis, covering all items that would otherwise be analysed at stage reviews. As with conventional developments, they fall into the usual categories including technical, commercial, financial, and time. Most of the concurrent engineering techniques and procedures operate with the reduction of risk as a prime motive.

11.6 INFORMATION SYSTEMS PROJECTS

As software has become more complex, managers have a greater need to understand its production, and several models of the software life cycle have been developed to aid this process. Many of the models are applicable to other areas of technology, and to R&D projects. The function of a life-cycle model is to determine the order in which software development should be undertaken, and to establish transition criteria to progress from one stage to the next. Transition criteria include completion criteria for the current stage, and entry criteria for the next. More sophisticated models of software development life cycles have evolved because traditional models discouraged effective approaches to software development such as prototyping and software reuse. This section traces the evolution of the different models, and explains their strengths and weaknesses.

The Code-and-Fix model

The earliest model for software development had two simple stages:

Stage 1: Write some code.

Stage 2: Fix the problems in the code.

Code was written before requirements were fully defined, design done, and test and maintenance procedures described. The strength of this approach was its simplicity, but that is also the source of its weaknesses. There are three main difficulties:

1. *Maintainability:* After a number of fixes, the code becomes so poorly structured that subsequent fixes are very expensive. This reinforces the need for design prior to coding.
2. *User requirements:* Often the software is a poor match to users needs so it is either rejected or requires extensive redevelopment.
3. *Cost:* Code is expensive to fix because of poor preparation for testing. This highlights the need for these stages, as well as planning and preparation for them in early stages.

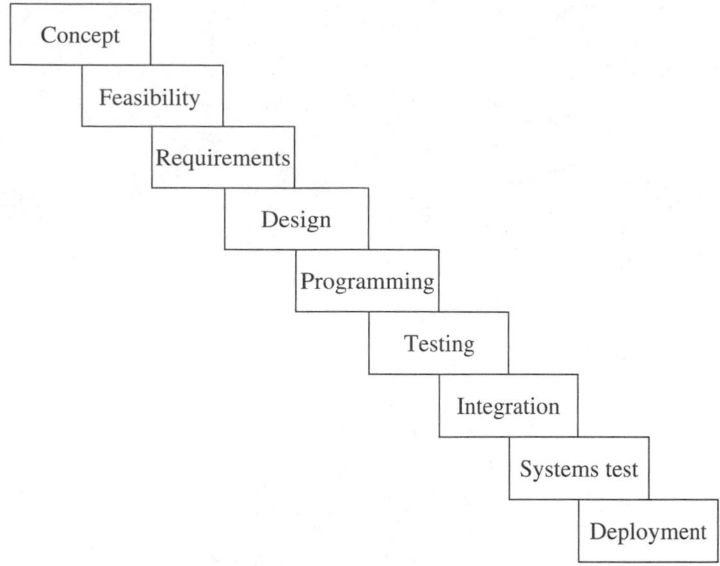

FIGURE 11.10 Stage-wise model of software development.

The Stage-Wise and Waterfall Models

Experience on large software systems as early as the mid-1950s led to the recognition of these problems, which resulted in the development of a stage-wise model. This suggests software should be developed in successive stages (Fig. 11.10). The waterfall model (Figs. 11.11 and 11.12) is a refinement of the stage-wise model from the late 1960s which is still popular. The major enhancement was that it recognized feedback loops between stages, but with a requirement to confine loops back to the previous stage only, to minimise the expensive rework resulting from feedback over several stages. I use the second waterfall model (Fig. 11.12) to illustrate principles common to many of the life cycles, as it can be easily related to other models, although the specific stages and names vary between models.

The second waterfall model is characterized by its V-shape. Down the left-hand side are stages which derive elements of the system, while up the right-hand side is the delivery of the elements to form the system, Table 11.6. Each stage is defined by its outputs, the deliverable, rather than its constituent activities. A tangible output is the only criterion of progress, the only thing which can be assessed objectively. In this way the 95 percent complete syndrome is avoided. The product of each stage represents points on the development path where there is a clear change, where one viewpoint of the design or emerging system is established, and is used as the basis for the next. As such, these intermediate products are natural milestones of the development progression and offer objective visibility of that progression.

To provide management control, the ideas of *baseline* and *configuration management* are adopted. The completion of a stage is determined by the acceptance of the quality of the intermediate products, or deliverables, of that stage. These deliverables form the baseline for the work in the next stage. Thus the deliverables of the next stage are verified against the previous baseline as part of configuration management and quality assessment, before they become the new baseline. Each baseline is documented, and the quality assessment includes

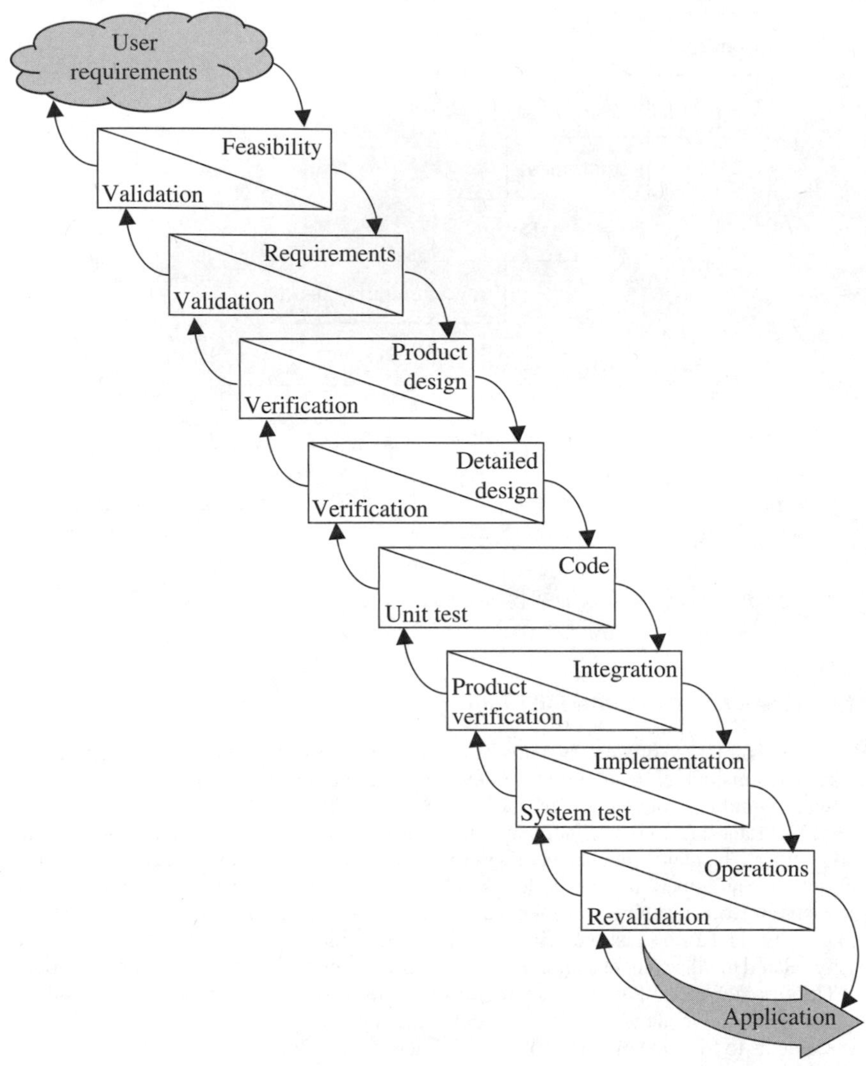

FIGURE 11.11 The water-fall model (1).

reviews of intermediate products by development personnel, other project and company experts, and usually customer and users. We met baselining in Chaps. 8 and 9. However, there the focus was on baselining time and cost, whereas here it is on quality and scope. For these, you cannot baseline the whole project, only one stage at a time, as the definition of scope and quality evolve through the project. This evolution is controlled by configuration management (Sec. 7.3). It is the documentation and reviews which provide the tangible and objective milestones throughout the entire development process. The waterfall model shows how confidence in the project's progress is built on the successive baselines.

This simplistic description of the life cycle could imply that control of software development can only be achieved by rigorous control of the staging, so that no stage is considered

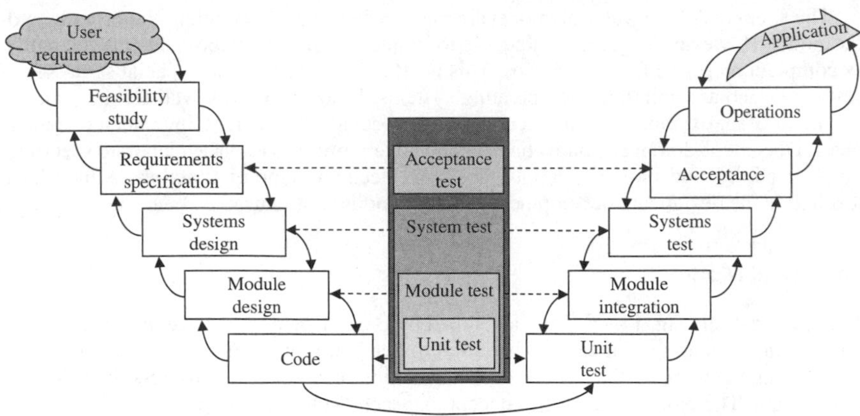

FIGURE 11.12 The water-fall model (2).

complete until all prescribed documents have been completed to specified standards, and no stage can be started until all its input documents are complete, giving nonoverlapping stages. Although the intended rigor of such an approach is commendable, it is unrealistic on a large development project. It is not intended that the life cycle should be interpreted in such as simplistic way.

TABLE 11.6 Stages of the Software Development Life Cycle

Stage	Description
Feasibility study	Production of verified/validated system architecture based on a design study, including allocation of tasks to staff and machines, milestone plan, responsibility chart, schedules of major activities, and outline quality plan
Requirements specification	Production of complete/validated specification of requirements (functional/nonfunctional) the system must satisfy. Produced in close liaison with the end user. Means of system acceptance also agreed with end user
Systems design	Production of complete/verified specification of overall architecture, control structure, and data structure for the system. Production of draft user manuals, and training and test plans for integration
Module design	Production of detailed designs for each module, together with module test plans. This may actually consist of more than one level of design
Code	Module designs are converted into code units in the target language
Unit test	Code units are tested by the programmer. Errors are corrected immediately by the programmer. Once complete, code units are frozen and pass to integration
Module integration (structural testing)	Component units of a module are integrated together, and tested as specified in module test plan. Errors detected are formally documented, and the affected area returns to a stage where the error was introduced
Systems test (functional testing)	Modules are integrated together to form the system, and tested against the system test plan. Errors detected are handled as for module testing
Acceptance test	Client formally witness the exercising of the system against agreed criteria for acceptance
Maintenance	Service life is often grossly underestimated. The cost of development can be small compared to maintenance, but the latter is given little consideration.

The strengths of the waterfall model are that it overcomes the problems in the code-and-fix model. However, its great weakness is its emphasis on fully elaborated documentation as completion criteria for early stages. This is effective only for some specialist classes of software, such as compilers and operating systems. It does not work well for the majority of software, for example, user applications and especially those involving interactive interfaces. Document-driven standards have pushed many projects to write elaborate specifications of poorly understood user interfaces and decision support functions, which have resulted in the design and development of large amounts of unusable code.

The Spiral Model

The spiral model (Fig. 11.13), first developed by Barry Boehm,[8] can accommodate all the previous models as special cases. The radial dimension represents the cumulative cost of undertaking the work to date. The angular dimension represents the progress of each cycle of the spiral. The model reflects the concept that each cycle involves a progression through a repeated sequence of steps for each portion of the product, and for each elaboration from overall concept document to coding of each individual program. Each loop of the spiral passes through four quadrants:

1. Determine objectives, alternatives, and constraints.
2. Evaluate alternatives and identify and resolve risks.

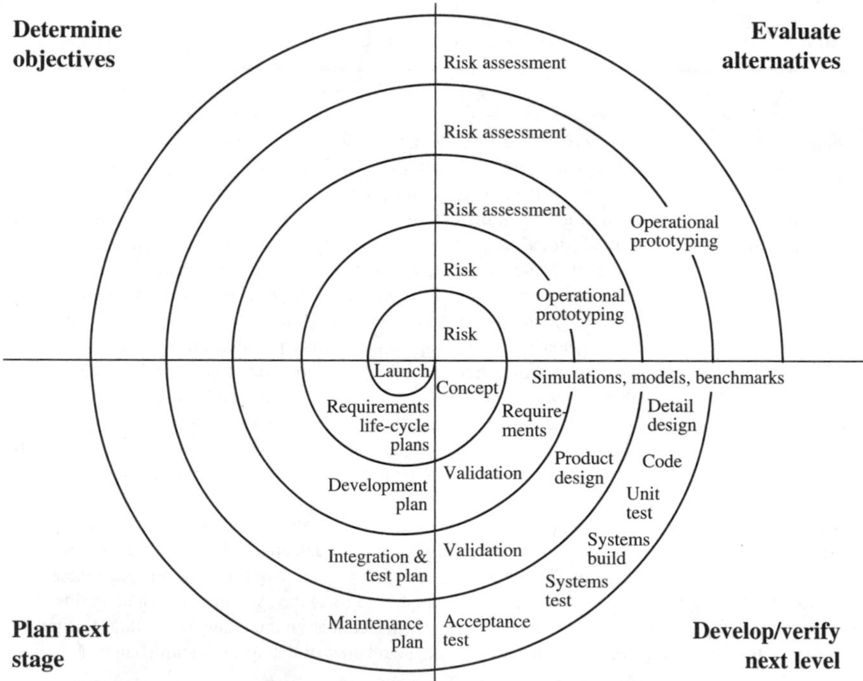

FIGURE 11.13 The spiral model.

3. Develop and verify the next level of product.

4. Plan the next stage.

Determine Objectives, Alternatives, and Constraints. After planning and launching each cycle begins with identification of:

- Objectives of this portion of the product being set, including performance, functionality, ability to accommodate change, and so on
- Alternative means of delivering this portion of the product, including alternative designs, reuse, or buying in
- The constraints imposed on final deliverable by the various alternatives, including cost, schedule, interfaces, and so on

Evaluate, Identify, and Resolve Risks. The next step is to evaluate alternatives against the objectives and constraints. Frequently this process identifies areas of uncertainty which are significant sources of risk. If so, this stage should involve the formulation of a cost-effective strategy for resolving the sources of risk.

Develop and Verify the Next Level of Product. Once the risks are evaluated, the next stage is determined by the relative importance of remaining risks. This risk-driven basis of the spiral model allows it to accommodate any appropriate mixture of different approaches to software development including specification-orientated, prototype-orientated, simulation-orientated, transformation-orientated, and the like. The appropriate mixed strategy is chosen by considering the relative magnitude of the program risks, and the relative effectiveness of the various approaches to resolving risk.

Plan the Next Stage. This completes the cycle. An important feature of the spiral model, as with others, is that each cycle is completed by a review involving the primary parties concerned with the product.

Management and the Spiral Model

There arc four key points.

Initiating and Terminating the Spiral. The spiral is initiated by the hypothesis that a particular operational objective can be improved by a software solution. The spiral evolves as a series of tests of this hypothesis. If at any time the hypothesis fails, the spiral is terminated. Otherwise it terminates with the installation of new or modified software.

Features of the Spiral Model. The model has three essential features:

1. It fosters the development of specifications which need not be uniform, exhaustive, or formal. They defer detailed elaboration of low-risk software elements, and avoid unnecessary breakages in their design until the high-risk elements are stabilised.

2. It incorporates prototyping as a risk-reduction option at any stage of development. Prototyping and the reuse of risk analysis were previously used in going from detailed design to code.

3. It accommodates reworking or a return to earlier stages as more attractive alternatives are identified, or as new risk issues need resolution.

Evaluation. The main advantage of the spiral model is its range of options accommodates the good features of existing software, while its risk-driven approach avoids many difficulties. Other advantages include

- It focuses early attention on options reusing existing software.
- It accommodates evolution, growth, and changes of the product.
- It provides a mechanism for incorporating software quality into product development.
- It eliminates errors and unattractive alternatives early.
- It identifies the required amount of each resource.
- It uses the same approach for software development, enhancement, or maintenance.
- It provides a viable framework for integrated hardware and software system development.

However, there are three areas which must be recognized:

1. *Matching to contract software:* The model works well on internal development projects, but needs further work for contact software. Its adaptability makes it inappropriate for fixed-price contracts.
2. *Relying on risk assessment expertise:* The spiral model places a great deal of reliance on the ability of software developers to identify and manage sources of project risk.
3. *Need for the further elaboration of the stages of the model:* The steps of the model need further elaboration to ensure all software developers are operating in a consistent manner. This includes detailed specification of deliverables and procedures, guidelines, and checklists to identify the most likely sources of project risk, and techniques for the most effective resolution of risk.

Risk Management. Efforts to apply and refine the model have focused on risk management (Chap. 10). A top 10 list of software risk items (Table 11.7) is a result. Table 3.6 also gave a list of 10 success factors developed by Jim Johnson of the Standish Group.[9] The reverse of these can be suggested as risks.

Agile Methods

One of the success factors that Jim Johnson suggests in the adoption of agile, iterative processes.[10,11] These are the basis of Rapid Applications Development (RAD) or Extreme Programming. These two techniques basically follow the spiral model, but with very short cycle times. Some people say that RAD doesn't follow a life cycle, and therefore shows that the life cycle is not an inherent feature of projects. It does, it is just that it goes repeatedly around the life cycle in very short bursts of activity. So as we have seen in this chapter, the life cycle can be a single sequential process, a parallel process as in concurrent engineering, or an iterative process as in the spiral model, rapid applications development, or extreme programming.

The Problems of Real Life

Unfortunately, software development is not quite as simplistic as these models might imply:

- Exploratory work on subsequent stages, including costing, can be required before the current stage is complete, for example, design investigation is almost invariably required before it can be stated that the user requirement can be achieved within a realistic budget.

TABLE 11.7 A Prioritised Top Ten List of Software Risk Items

Risk item	Risk management technique
Personnel shortfalls	Staff with top talent, team/morale building, cross training prescheduling key people
Unrealistic schedules and budgets	Detailed cost and duration estimates, design to cost, incremental development, reuse of software, requirements scrubbing
Developing wrong functionality	Organization/mission analysis, ops concept formulation, user surveys, prototyping. early user manuals
Developing wrong user interface	Task analysis, prototyping, scenarios, user profiles (functionality, style, workload)
Gold plating	Requirements scrubbing, prototyping, cost-benefit analysis, design to cost, value engineering
Continuing changes to requirements	High change threshold, information hiding, incremental development (defer changes to later increment)
Shortfalls in procured components	Benchmarking, inspection, expediting, reference checking, quality auditing, compatibility analysis
Shortfalls in subcontracted tasks	Reference checking, pre-award audits, fixed price contracts, competitive design/prototyping, team building
Shortfalls in real-time performance	Simulation, benchmarking, modelling, prototyping, instrumentation, tuning
Straining the capabilities of computer science	Technical analysis, cost-benefit analysis, prototyping, reference checking

- Problems encountered in later stages may require reworking of earlier stages—failure to recognise this leads to earlier documentation becoming inaccurate and misleading.
- The users' requirement may not remain stable throughout a protracted development process—it is then necessary to consider changed requirements and consequential changes.

It is important that the life cycle is not rigidly imposed. In reality, there are no clearly defined break-points between the stages. Equally, all the stages are composed of several substages or packages of work. Once this is recognized, it leads not to the conclusion that the life cycle model must be discarded, but it represents a valuable model of what is involved in the work of software development. The biggest single problem in the software cycle is communication across the boundary from one stage to the next. At each stage there can be a degradation of the definition of the users' requirements. Quality assurance (Sec. 7.2) and configuration management (Sec. 7.3) play a crucial role in managing this flow of information.

Resourcing the Life Cycle

Many of the names of stages in the life cycle are similar to resource types working in software development. This results in each type becoming primarily associated with a stage: systems analysts with design, programmers with coding. You will hear IT people referring to the work of each resource types as an "activity," and then they confuse the "activity" of the resource, with the work of the stage. Resource types are then not assigned to the project until the work of the stage with which they are associated is about to begin. The result is long lead items are ignored by earlier resource types, resources have no time to prepare before starting, and resources cannot complete their input within the time allotted. In reality, most resource types should work throughout the project. Where the work of one resource type overlaps with a stage that defines a work package. Early work packages are in support of the design process, and in preparation for the stage in which the resource is primarily involved. Later work packages are in support of implementation.

SUMMARY

1. Projects can be classified in many ways. Different types of projects require a different approach or emphasis to their management.
2. In addition to traditional projects to deliver and commission a facility, projects can conduct:
   - Marketing surveys and product development
   - Research and development
   - Maintenance and decommissioning
3. New product development can lead to many projects:
   - Research and development
   - Product design and prototyping
   - Facility design and delivery
   - Product launch
4. New product development can be managed through:
   - New product committees
   - Product managers
   - New product managers
   - New product departments
   - Venture teams
   - Task forces
5. The stages of the product development life cycle include
   - Idea generation and screening
   - Concept development and testing
   - Marketing strategy
   - Business analysis
   - Product development
   - Market testing
   - Commercialization
6. Concurrent engineering is used to overlap stages in the product development cycle, to speed up the delivery of new products.
7. Concurrent engineering requires the adoption of new project management practices, including:
   - A change in organizational culture
   - Cross-functional team working
   - Use of new technology, information systems, and other new techniques
8. There are risks and pitfalls associated with concurrent engineering, including:
   - Attitudes of middle management
   - Authorization of the project
   - Organizational and cultural change
   - Managing interfaces
   - Technical management
   - Cost and risk controls and release of finance

9. There are five types of model of the life cycle for software development projects:
   - Code-and-fix models
   - Stage-wise models
   - Waterfall models
   - Spiral models
   - Agile methods including Rapid Applications Development and Extreme Programming
10. Stages in all the models are identified not by the work done, but by the deliverables, or intermediate products, in which they result. The control process focuses on the quality of these deliverables.
11. Spiral models, which can incorporate any one of the other four as a special case, view the project as moving repeatedly through four quadrants:
   - Plan the forthcoming stage.
   - Determine objectives, alternatives, constraints.
   - Evaluate alternatives and identify and resolve risks.
   - Develop and verify the next level product.

REFERENCES

1. Fayol, H., General and Industrial Management, London:Pitman, 1949.
2. Project Management Institute, A Guide to the Project Management Body of Knowledge, 3d ed., Newtown Square, P.A.: PMI, 2004.
3. Wearne, S.H., Principles of Engineering Organization, London:Edward Arnold, 1973.
4. Kerzner, H., Project Management: A Systems Approach to Planning, Scheduling, and Controlling, 9th ed., Hoboken, N.J.: Wiley, 2006.
5. Kotler, P. and Keller, K.L., Marketing Management, 12th ed., New York: Prentice Hall, 2005.
6. Turner, J.R and Wright, D., The Commercial Management of Projects, Aldershot, U.K.: Gower, 2009.
7. Keegan, A.E. and Turner, J.R., "The management of innovation in project based firms," Long Range Planning, **35**, 367–388, 2002.
8. Boehm, B.W., "A spiral model of software development and enhancement," Computer, May, 61–72: The Institute of Electrical and Electronic Engineers, 1988.
9. Johnson, J., My Life Is Failure: 100 Things You Should Know to Be a Successful Project Leader, Boston:Standish Group International, 2006.
10. Martin, R.C., Agile Software Development: Principles, Patterns, and Practices, New York: Prentice Hall, 2002.
11. Beck, K. and Andres, C., Extreme Programming Explained: Embrace Change, 2d ed. New York: Addison Wesley, 2004.

CHAPTER 12
PROJECT START-UP

In this chapter, I describe the project start-up processes, explaining how to get the project initiated. Some people assume project start-up only occurs at the very start of the project, in the concept stage, but in reality the processes described in this chapter can take place at the start of any of the stages in the project life cycle, even in close out. You may want to conduct the start processes whenever there is a significant change in the project team, either in its composition or structure, or when you believe the project team's attention needs refocusing on the objectives of the stage ahead. In this chapter, I explain the need for the start-up processes, and describe the objectives and methods of start-up. I describe the tools of start-up including the start-up workshop and the project definition report and project manual.

12.1 THE START-UP PROCESS

A project requires the undertaking of a unique task using a novel organization, which must be created at the start of the project. When new teams form, the members take time to learn how to work together before becoming truly effective. Typically, a team goes through four stages of formation in which its effectiveness first falls and then rises[1] (Sec. 4.5):

a. *Forming*: The team members come together. The team members are proud they have been chosen to work on this important project. However, they are uncertain of each other and of their roles and so their effectiveness is only at a medium level. Second worst-case scenario is the team remains in this stage for the duration of the project.
b. *Storming*: They find areas of disagreement. They can't agree on the problem they are trying to solve, nor the objectives of the project. They therefore can't agree to the project plan, who is responsible for what, what work they have to do, nor how to get started. Perhaps they can't even agree upon the project management methodology to use. The team's performance falls, and worst-case scenario is it continues to fall and never recovers.
c. *Norming*: The team finds they can agree on some things and build cooperation around that. They agree upon the problem they are trying to solve, and thus the project objectives. From there they develop a milestone plan and responsibility charts, and define the work they individually have to do. This is called norming because during this stage the team forms norms of behaviour about what it means to be a member of this team, and identify with it more. Part of the norming process is they agree with the project management methodology. During this stage team effectiveness rises.
d. *Performing*: The team maintains its peak performance throughout the project.

A project is subject to time constraints, and so this process must be undertaken in a structured way to ensure it happens quickly. This is called *project start-up*.[2,3] The term *project start-up* is used to differentiate from start; the former is a structured process for team formation; the latter is an action at an instant of time. Morten Fangel draws the analogy with starting the engine of a car, and starting-up the diesel engine in a ship.[2,3] The former is achieved by flicking the ignition switch, the latter by a structured series of activities, a start-up process, which gives the most efficient and economical operation. The same applies to projects.

The start-up process can take two to three days but some people think that they don't have time to devote to it. They have to get on with the project, rather than spend time sitting around working on project plans. But you have to ask yourself, what is more effective, to have the team work at 50 percent efficiency for the duration of the project, or 120 percent efficiency for all but the first three days (Example 12.1). It is now widely accepted that a structured start-up process is an essential part of project management. It is necessary, on a unique, novel, and transient endeavour, to improve the understanding of the project team of the task they face, and how they will approach it, and to get them working effectively as a single unit. There has been an increasing need for effective start-up on projects, and this may be due to:

- The increasing complexity of technologies used
- The use of qualified project management earlier in the life cycle
- The need for team building and cross-cultural cooperation
- The need for increased effectiveness caused by shorter product life cycles
- Changes in the way projects are managed, including goal-directed approaches, which reinforce the setting of objectives, the use of group methods for building cooperation, and the management of the team through the use of a clear and common mission

Example 12.1 The power of start-up

I worked with a project team in British Telecom, a U.K. telephone company. The team of six people had seven weeks to complete a highly critical project. The project manager's manager was a great believer in the start-up process and insisted the team do no work in the first week. They were to spend the first week planning, three of those five days being spent in a start-up workshop with me. At the start of week two, the team hit the ground running and went on to have an extremely successful project. You have to ask which is better:

- To work for seven weeks at 50 percent efficiency
- Or to work for six weeks at 120 percent efficiency

My view is obvious.

The Objectives of Start-Up

For start-up to be successful, the participants must understand what the objectives of the process are at any stage, and must be aware of what specific outputs are needed to achieve the necessary level of understanding. These objectives can include

- To create a shared vision for the project, by identifying its context, the desired performance improvement, and its objectives in terms of desired output and outcome.
- To gain acceptance of the plans by defining the scope of work, project organization, constraints of quality, cost and time, and the planned response to the inherent risks.

TABLE 12.1 Shift of the Start-Up Objectives through the Life Cycle

Objective	Concept	Feasibility	Design	Execution	Close-out
Context and objectives	Draft	Main	Review		Monitor
The project model	Initiate	Draft	Main	Review	
The management approach		Initiate	Draft	Main	Review
Commission and handover			Initiate	Draft	Main

- To get the team working, agreeing its mode of operation and channels of communication.
- To refocus the project team onto the purpose of the project and the method of achieving it.

The first three objectives correspond to Parts 1, 2, and 3 of this book, respectively; the fourth runs throughout. As they move through the project, the team's understanding of these develops in turn. During concept, the emphasis is on identifying the project's context, the desired performance improvement, and the changes needed achieve that, and from there to developing the shared vision and project strategy. During feasibility, the emphasis shifts to developing the project model (Fig. 1.11) and determining the feasibility of that model in terms of the ability of the project to deliver the desired output and the operability of that output to achieve the desired outcome and performance improvement. During design, the emphasis now is on formalizing the model into a plan for implementation. In execution, we deliver the desired output, including the detail design of the project's output and work methods, and actually doing the work. Finally, as the new asset is commissioned, and handed over to the client, the emphasis changes back to the purpose of, the benefit expected from the new asset, and the product it produces, to ensure the team are aware of what is actually required and so are better able to achieve it. Hence, the objectives of project start-up will be different at each stage of the life cycle (Table 12.1). Although, as you move from one stage to the next you may review the objectives of the previous stage and look forward to those of the next.

Below each of the 4 objectives are 15 subsidiary objectives (Table 12.2). These in turn may influence the emphasis of the work of the project team depending on the type of activity undertaken and decisions taken. The emphasis of the team's work may be:

- Analysis: of the project's context, previous plans, future tasks, and management routines
- Planning: of objectives, scope of work, organization, and routines
- Communication: between participants of the results of the analysis and plans
- Motivation: of participants to carry out work or make decisions

Table 12.2 relates the emphasis of the team's work to the 15 subsidiary objectives. When linked to Table 12.1, this shows that during the life cycle the emphasis shifts from analysis and planning to communication and motivation until the end when it switches back to analysis, which I think will match the experience of most people.

The Methods of Start-Up

Another requirement of a systematic approach to project start-up is the use of appropriate methods. There are three standard methods of start-up:

- Project, stage, or even milestone launch workshops: to develop project plans in a joint team-building process
- Start-up or stage review reports: to collate the results of analysis undertaken during start-up or from a previous stage in accessible form for use during the subsequent stage
- The use of ad hoc assistance: to support and guide the project team

TABLE 12.2 Ten Subsidiary Start-Up Objectives and Their Effect on the Working of the Team

Subsidiary objectives	Analyze	Plan	Communicate	Motivate
Context & Objectives				
Impact of context	A		C	M
Business purpose	A	P		
Objectives of project		P	C	M
Project Model				
Milestone plan	A	P		
Responsibility chart		P	C	M
Detail work plans		P	C	M
Resource allocation		P	C	M
Management System				
Management system		P		
Principles of cooperation			C	M
Control processes		P	C	
Commission				
Timely, efficient end		P	C	M
Disband team		P	C	M
Handover to client		P	C	
Obtain benefits		P	C	
Record data	A		C	

These techniques may be used individually or in combination. The choice depends on several factors. First, the different methods require varying amounts of time, so you must ensure key team members are willing to devote it. Secondly, the methods have different efficacy in achieving the objectives in Tables 12.1 and 12.2. Table 12.3 shows the different impact of each method. Thirdly, through project start-up you should try to build as much historical experience into the project definition as possible, to minimise the uncertainty. You should choose a method which does that for the case in hand. Other methods of start-up include case studies, study tours, social events, education programmes, and other media, such as videos.

Launch Workshops. A launch workshop held at the start of the concept or feasibility stages may be called a *project definition workshop*, and at the start of design or execution an *initiation* or *kick-off meeting*. The objectives of the workshop, the agenda, and the people invited depend on the stage being launched, and are discussed more fully in Sec. 12.3.

Stage Review Report. A start-up or stage review report can be prepared at the end of any stage to launch the next. A report produced at the end of the concept stage for launching the

TABLE 12.3 Effectiveness of the Techniques for Project Start-Up

Start-up technique	Analyze	Plan	Communicate	Motivate
Launch workshop	High	Medium	High	High
Review report	Low	High	High	Medium
Ad-hoc assistance	Medium	High	Low	Medium

feasibility study may be a one- or two-page *project scope statement* (Sec. 5.3 and Table 5.1). During the feasibility study, this is expanded into a *project definition report* or *client requirements statement*, used to launch design. At the end of that stage, a full *project manual* or *project requirements statement* may be produced in support of the design package, and that used to launch execution. The contents of each of these reports depend on the stage being reviewed, and are described in Sec. 12.4.

Ad hoc Assistance. This may be from:

- Internal professionals, such as the project support office
- External consultants
- Team members from similar or earlier projects
- Organizational behaviour experts helping manage the team dynamic

External professionals can fill one of two roles. They may be there to facilitate the team dynamics, to initiate the storming and the forming. Or they may be invited to bring specific technical expertise or bring experience from having worked on similar projects in the past. This can provide additional resources with special skills, which may motivate key people. Having someone to share ideas with can be stimulating. A disadvantage is there can be some confusion over responsibilities, which can lead to wasted effort.

Start-Up and the Type of Project

In Sec. 1.4, I introduced the goals and methods matrix, Figure 1.13. This defined four types of project. The emphasis of start-up differs for each type.

Type 1 Projects. The goals and methods of delivery are both well defined, and the team can move quickly into activity-based planning. These projects are usually very similar to ones done in the past, (not so unique and novel, runners or repeaters, Sec. 1.2). The emphasis of start-up is therefore on briefing the team on the standard techniques. External facilitators who have done similar projects in the past may be used to brief the team. The role of the project manager is something of a conductor, leading the team through the predefined score, but putting his or her own interpretation on it.

Type 2 Projects. The goals are well defined, but the methods of achieving them are not. The start-up workshop develops a milestone plan for the project, where the milestones represent the known products. It then develops a responsibility chart to define who is going to take responsibility for determining how to achieve the milestones. The workshop requires a broad cross-section of disciplines to be represented, including all the people who may have a contribution to make on how best to achieve the project. A facilitator may be used to norm the team's behaviour, gaining agreement to the milestones and responsibility chart. The role of the project manager is that of a coach. There is a clear objective of getting the ball in the goal as many times as possible in the next 90 minutes. The coach trains the team in standard plays, but leaving them to put them together as the game unfolds.

Type 3 Projects. The goals are not well understood, but the project will follow a standard life cycle, and the definition of the goals will be refined as the project proceeds, using configuration management. The emphasis of start-up will be on agreeing to the purpose of the project, the nature of the goals, if not their precise definition, and the life cycle and review points to be followed in reaching a better understanding. A facilitator may be used to help negotiate agreement on these. The role of the project manager is that of a sculptor, starting with a shapeless block of clay or marble. Somewhere in there is a statue. He or she will use standard techniques to cut away the clay or marble. However, they will need to avoid flaws, and so the precise nature of the statue will not be evident until it is finished. (I hope the sculptor is more like Michelangelo using an army of apprentices than a hermit in an attic.)

Type 4 Projects. Now neither the goals, nor the method of achieving them is known. The emphasis start-up is very much on agreeing to the purpose of the project. In the early stages of the project, the team works on defining first the goals (turning the project into a Type 2) and then the methods of achieving them. Planning will be by defining a series of gateways (review points) that the project must pass through before they close forever. A facilitator may again be used to negotiate agreement, and then help with team formation as the project converts to Type 2. The project manager must now take the role of an eagle. He or she must be able to hover above the project and see how it fits into the overall context of the organization, but also be able to identify small problems (a mouse) and go down and deal with them. They must then be able to rise back above the project again, before going down to deal with another mouse.

Scheduling Start-Up

A schedule of the start-up activities helps to focus attention on the process, and acts as a means of implementing the chosen techniques. The schedule may take the form of a responsibility chart (Fig. 12.1) with both a definition of roles and responsibilities, and a timescale. The schedule for starting-up design and appraisal may be included in the project definition report.

12.2 START-UP WORKSHOPS

Project start-up workshops can be used to start a project, or a stage of the project. Indeed, mini workshops may be held to start a work package, by the rolling-wave principle. A workshop held at the start of the concept or feasibility stage is called a *project definition workshop* and at the start of design or execution stage an *initiation* or *kick-off meeting*.

Workshop Objectives

The main objectives of the workshop are

1. *Gain commitment and build team spirit*: This is the primary objective of a workshop. Many of the others can be achieved by people working alone or meeting in smaller groups. By coming together, they may develop a common understanding, and resolve items of confusion, disagreement, or conflict through discussion. If people are briefed after a meeting (presented with a fait accompli) they may nod their heads in agreement, but you often find they do not truly accept what they are told. If people agree to a course of action in a meeting, you usually find they have internalized that agreement, but if they have not, it is difficult for them to avoid their commitments later because several people have heard them make them.
2. *Ratify earlier project definition*: Whatever stage is being launched, it is vital for the team to agree what the current level of definition entails, and that it truly represents user requirements.

TriMagi Procedural Responsibility Chart

Project: Procedure for project start
Project Sponsor: Steve Kenny
Project Manager: Rodney Turner

X	eXecutes the work
D	takes Decisions solely/ultimately
d	takes decisions jointly
P	manages Progress
T	on-the-job Training
I	must be Informed
C	must be Consulted
A	may Advise

		Project management tasks											Responsibilities					Period:			Week				Target end:								
No	Milestone Name	Background	Purpose	Objectives	Work breakdown	Responsibilities	Detailed plans	Resource allocation	PM system	Cooperation	Information		Steering committee	Project manager	Project office	Participants	Consultants	1	2	3	4	5	6	7	8	9	10	11	12	13	14	15	d (Duration)
	Project description	S	M	M	M	M							DX	C	C	I																	
	Definition workshop	S	S	S	S	S	S	S	S	S	S			PX	X	X	T																
	Draft issue of plan	M	S	S	S	S	M	M	M	M	M		I	PX	X	I																	
	Meeting consultants			S		S				S				PX	X	X	A																
	First issue of plan	M	M	M	M	M	S	S	S	S	S		I	PX	X	I																	
	Invitation to start-up workshop														X																		
	Start-up workshop																																
	Part 1	M	M	M	M	M	S	S					DX	PX	X	X	T																
	Part 2	S	S	S	S	S	M	M	M	M	M		DX	PX	X	X	T																
	Final editing of plan	M	M	M	M	M	M	M	M	M	M		I	PX	X	I																	
	Approval of plan	M	M	M	M	M	S	S	S	S	S		D	PX	I	I																	
	Issue of definition report	M	M	M	M	M	M	M	M	M	M		D	PX	X	C																	
	M = Main issue																																
	S = Subsidiary issue																																

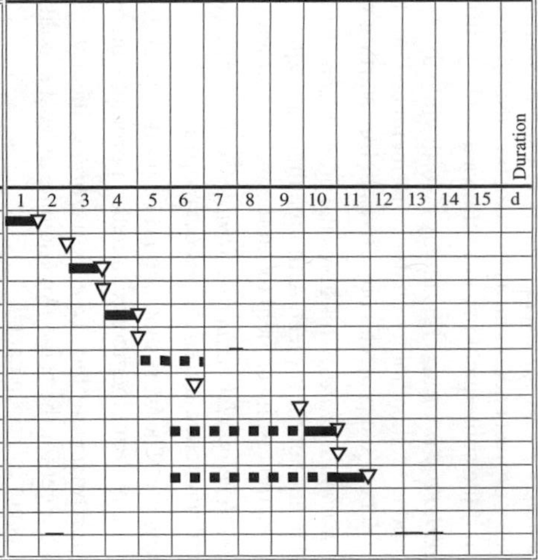

FIGURE 12.1 Responsibility chart used as a start-up schedule.

3. *Plan the current stage*: The workshop is used to launch the current stage and so producing a plan for the stage is key. This should at least consist of a milestone plan and responsibility chart.
4. *Prepare preliminary plans for execution*: It is usually worthwhile to prepare a draft milestone plan for project execution, as this can be a useful basis for the feasibility study or design, even if the subsequent project follows a slightly different course.
5. *Conduct a stakeholder analysis:* It is always worthwhile to identify the project's stakeholders and conduct a stakeholder analysis. Remember you should gain agreement of all the stakeholders to the project's objectives before starting work.
6. *Prepare preliminary estimates*: This gives the project team some idea of the expectation of the cost and benefit of the project. Although their subsequent work should not be constrained by the estimates, it can help to set the basic parameters.
7. *Access risk and develop risk reduction strategies*: Preliminary risk analysis should be undertaken, and risk reduction strategies developed.
8. *Start work promptly*: The workshop should be used to plan the initial work of the current stage so that the team members can make a prompt start.
9. *Agree a date for reviewing the stage deliverables*: Ideally, the plan should contain a timescale and budget for the stage. An end date at least should be set for completion of the stage so it is not left open ended.

Apart from the first, these are the objectives of start-up given above, but at a more detailed level.

Workshop Attendees

The workshops should be attended by key managers, including:

- The project sponsor
- The manager of the current stage
- The manager designate of future stages, especially execution
- Key functional managers whose groups are impacted by the project, including technical managers, user managers, and resource providers
- A project support office manager
- A facilitator

The sponsor may attend the definition workshop, but not later ones. Possible attendees for a project definition workshop on the CRMO Rationalization Project are given in Table 12.4.

Workshop Agenda

A typical agenda for a workshop is

1. Review the current project definition
    - Purpose, scope, and outputs of the project
2. Define the objectives of the current stage
3. Determine the success criteria of the project and the current stage
    - Set a project mission

TABLE 12.4 Project Definition Workshop Attendees for the CRMO Rationalisation Project

Role	Possible person
Sponsor	Regional managing director, or Regional operations director, or Regional financial director, or Regional technical director
Manager of the feasibility study	Regional operations director, or Regional financial director, or Regional technical director
Project manager designate	Customer services manager, or Network manager, or IT manager
Key functional managers	Estates manager, and Finance manager, and Sales and marketing manager
Project support office	May already exist, otherwise a temporary one may be created for this project

4. Prepare a milestone plan for the current stage
5. Prepare a responsibility chart against the plan
6. Estimate work content and durations for the work packages
7. Schedule the work packages
8. Define the quality objectives of the current stage
9. Assess risk and develop reduction strategies
10. Prepare initial activity plans
11. Prepare a management and control plan

Most effort goes into the milestone plan and responsibility chart, as that is the most effective use of group work. Sections 5.4 and 6.3 described how to develop them using whiteboards, flip charts, Post-its, and a data projector. Involving everyone present around a whiteboard, gains their commitment to the plans produced. Working around a table with pen and paper can isolate members of the team from the working process. Estimates and schedules are best agreed through a process of negotiation immediately after the workshop. The initial activity schedules are prepared so that the team members know what to do immediately following the meeting; it is an initiation meeting. The *management and control plan* agrees the approach to be used in managing the project and the mechanisms, priorities, and frequency of the control process. It may be the basis of the management approach outlined in the project manual (Sec. 12.3).

Workshop Timetable

A workshop typically lasts one to three days. I usually allow two hours per item, except items 4 for which I allow four hours. However, it is important not to stick rigidly to a timetable, but to allow discussion to come to a natural conclusion, as people reach agreement and a common understanding. I sometimes include project management training as part of the timetable, which extends the duration by about a day. I find it useful to schedule a break in the middle of agenda item 4. When developing a milestone plan, people often reach a blank; the plan will just not make sense. However, when left for a while, it just seems to fall into place.

12.3 PROJECT DEFINITION REPORT AND MANUAL

Stage review reports gather the results from the work of one stage and are then used to launch the next stage. Three reports may be produced at the end of each of the first three stages of the project (Table 12.5). I describe the project definition report and project manual here. The project scope statement is described in Sec. 5.3. Table 12.5 also shows the names used for equivalent documents by the PRINCE2 process.[4]

Project Definition Report

Objectives of the Project Definition Report. The project definition report gathers the results of the feasibility study into a readily accessible document. It is a handbook for the management, design, and execution teams, which defines what the owner expects from the project, and the reasoning behind the chosen options and strategies. This reasoning can always be open to questioning. It is healthy that the teams involved in later stages question earlier decisions. However, by having earlier reasoning recorded, the project teams can avoid repeating work, and more importantly avoid following previous blind alleys. The project definition report will also be used to launch the design stage, and may be the input to a kick-off meeting at the start of that stage. Hence, the objectives of the project definition report are

- To provide sufficient definition, including costs and benefits, to allow the business to commit resources to the design stage
- To provide a basis for the design stage
- To provide senior management with an overview of the project's priority alongside day-to-day operations and other projects, both proposed and on-going
- To communicate the project's requirements throughout the business
- To define the commitment of the business to the project

Most of these objectives look forward; the report is not produced as a bureaucratic exercise to record the feasibility study, but as a basis for the future stages.

Contents of the Project Definition Report. The suggested contents of the report are

1. *Background:* Sets the context of the project, describing the problem or opportunity which creates the need for the desired performance improvement. It may describe the purpose of the higher level program of which the project is a part (See Examples 2.3 and 2.6).
2. *Purpose, scope, and objectives:* The reason for undertaking the project with expected benefits, the sort of work needed to achieve that, and the product to be produced by the project in order to achieve the returns (See Table 5.1.).

TABLE 12.5 Stage Reports

Report	Produced in	To launch	See	PRINCE2 names
Project scope statement	Concept	Feasibility	Table 5.1	Project mandate
Project definition report	Feasibility	Design	App. A	Project brief
Project manual	Design	Execution	Sec. 12.3	Project initiation document

3. *Success criteria and project mission:* A statement of how the project will be judged to be successful, and what the project aims to achieve. This may include a statement of different stakeholders' aspirations (See Table 3.3).
4. *Work breakdown structure:* Initiates work breakdown, starting with areas of work, and including a milestone plan, with a list of milestones in each area of work. It may also include milestone scope statements (See Fig. 5.3 and Table 5.2 and 5.3).
5. *Project organization:* Defines the type of project organization, including:
   - Organizational units within the business involved in the project
   - Their involvement in different areas of work
   - Managerial responsibility for different areas of work
   - The type of project organization to be used
   - The location of project resources
   - The source of the project manager
   - Their source and limits of authority

   and describes the responsibilities of key managers and groups in the business, including:
   - Project sponsor, champion, and manager
   - Work-area and work-package managers
   - Project steering board
   - Quality assurance board and project support office manager

   It may be necessary to include a tentative resource schedule, so the project can be assigned priority. This schedule is derived from high-level assumptions applied to areas of work or work packages. It should not be based on a detailed definition of work except in areas of high risk because that requires an investment in planning resource before the business has agreed to commit it (Fig. 6.5).
6. *Stakeholder register:* Shows the stakeholders their personal expectations of the project, their influence on the project, and the planned communication strategy to win their support for the project (Table 4.3).
7. *Quality plan:* The milestone scope statements will show the measures of achievement of each milestone (Table 5.3).
8. *Schedule:* The bar chart or the responsibility chart shows the planned schedule for the project (Fig. 6.5).
9. *Cost estimates:* The cost estimate at the project level may be included showing the cost-breakdown milestone by milestone (See Fig. 8.4).
10. *Risks register:* The project definition report should include the risk register (Table 10.8) but not the individual risk item-tracking forms (Table 10.7). The latter will be included in the risk log.
11. *Initial activity plans:* Initial activity plans will be included for early milestones to show how work on the next stage of the project will begin. These may include cost estimates at the activity level (Fig. 8.5).
12. *Project appraisal:* Initial statements of cost and benefit and associated payback may be included. This will justify the commitment of resources to the design and appraisal stage.
13. *Project management system:* Tools and techniques for planning and controlling the project and supporting computer system may be described. This may include preliminary quality plans and control procedures.

The report is typically 10 to 40 pages long, depending on the size and complexity of the project, and its impact on the organization. It is developed throughout the feasibility stage.

However, once ratified by senior management at the end of that stage, it should be sacrosanct, and only modified by formal change control. Appendix A contains the project definition report for the CRMO Rationalization project, incorporating the figures and tables introduced throughout the previous chapters into single document.

Project Manual

Objectives of the Project Manual. The results of the design stage are recorded in a project manual. This is a definitive document which explains how the owner's requirements set out in the project definition report are to be delivered by describing the objectives and scope and management strategy for the project as they are defined at the end of the stage. It is used as the briefing document for all people joining the project team during execution.

The manual is developed progressively by the project manager from the project definition report throughout the design stage. The draft manual is reviewed by the owner and project manager together, until it is signed off at the end of the stage as reflecting their mutual understanding of how the owner's requirements are to be delivered. When the manual is signed off, the project manager must accept responsibility for delivering the project as defined in the manual, and from that point on changes to the manual can only be made through strict change control. The development of the project manual and the master plan it includes often represents the largest proportion of the project manager's efforts during design after the management of the actual design process itself.

Through execution, the manual is extended down to the work-package level as part of the start-up of individual work packages. The manuals at the work package level must be derived from the project manual but they may highlight the need for modifying the project manual.

Contents of the Project Manual. The contents of the manual may include the following items:

1. *Project description and objectives:* These summarise the project definition report, as modified by the design process (Example 12.2).
2. *Master project plan:* This forms the major part of the manual. The design process results in this master plan. The contents of this plan, which cover the definition of scope, organization, quality, cost, and time in the project model are summarized in Table 1.5.
3. *Management plan:* This describes how the project is planned, organized, implemented, and controlled, although the first two only need to be done at lower levels of work breakdown.
4. *Performance specification:* This defines the required levels of performance of the facility and its product. This is one of the major elements of the quality specification of the project and would have been developed and refined during the design process.
5. *Functional specification:* This explains the technology to be used in the development of the facility, and how that will function to deliver the required output.
6. *Acceptance tests and acceptance criteria:* These are derived from the previous two and are an important part of the manual. They must be defined before work starts for two reasons. They must be independent. The project team members must not be allowed to develop testing procedures which match the facility built. Secondly, the project team must know how they are to be judged, if they are to deliver a quality product. In Chap. 8, quality was defined as meeting customer's requirements. These must be defined in advance so the team members know what their objective is, and they do not produce a product which is either over- or underspecified.

7. *Project constraints:* These are derived throughout the first three stages, and so must be recorded for all people joining the project at a later date.
8. *Risks and assumptions:* These must also be recorded for two reasons. So people joining the project later know what has been addressed, and so others, especially owners, sponsors, financiers, and auditors can see they have been properly addressed and that adequate weight has been given to them.

Example 12.2 Summarising the project definition report in the project manual

I had a discussion with managers attending a course at Henley Management College about whether the manual would contain the definition report in its entirety, or whether it would be summarised into a single section as a background. We decided on the latter for two reasons. First, it is the job of management to summarise the instructions from the level above when passing them on to the level below, so the next level down can focus on those things which enable them to do their jobs effectively. You inform people on the next level on a need-to-know basis. This does not mean you need to be excessively secretive. You tell the next level enough to motivate them, and make them feel part of the overall management team, without overburdening them with unnecessary information. Secondly, taken to the extreme, you would include the entire corporate plans in the briefing documents to every project.

SUMMARY

1. There are four stages of team formation:
   - Forming
   - Storming
   - Norming
   - Performing
2. Project Start-up is a structured way to moving the project team quickly and effectively through these four stages, so as to:
   - Define the project's context and objectives
   - Develop the project model
   - Define the management approach
   - Commission the facility and hand it over
3. The methods of project start-up include
   - Stage launch workshops
   - Start-up reports
   - Ad-hoc assistance
4. A stage launch workshop may be held with the objectives:
   - To gain commitment and build the team spirit
   - To ratify the project definition as produced in the previous stage
   - To plan the current stage of the project
   - To prepare preliminary plans for the execution stage
   - To prepare preliminary estimates for the project
   - To ensure work starts promptly
   - To agree a date for review of the stage deliverables
5. A project definition report may be prepared with the objectives:
   - To commit resources to design
   - To provide a basis for design

- To set the project's priority
- To inform all those effected by the project
- To gain commitment

6. The contents of the report may include
- Background
- Purpose, scope, and objectives
- Project success and mission
- Work breakdown structure
- Project organization
- Stakeholder register
- Quality plan
- Cost estimates
- Schedule
- Risk register
- Initial activity plans
- Project appraisal
- Project management system

7. The systems design produced during the design and appraisal stage may be summarized in a project manual, which may have as its contents:
- Project description and objectives
- Master project plan
- Management plan
- Performance specification
- Technical specification
- Acceptance tests and criteria for acceptance
- Project constraints
- Risks and assumptions

REFERENCES

1. Tuckman, B.W., "Development sequence in small groups," *Psychology Bulletin,* 1965.
2. Fangel, M., "The essence of project start-up: The concept, timing, results, methods, schedule and application," in the *Handbook of Project Start-up: How to Launch Projects Effectively*, Fangel, M. (ed.), Zurich: International Project Management Association, 1987.
3. Fangel, M., "To start or to start-up?—That is the key question of project initiation," *International Journal of Project Management*, **9**(1), 1991.
4. Office of Government Commerce, *Managing Successful Projects with PRINCE2,* 4th ed., London: The Stationery Office, 2005.

CHAPTER 13
PROJECT EXECUTION AND CONTROL

During execution, most of the work to deliver the objectives (build the new asset) is done, and thus most of the expenditure made. The stage starts with completion of detailed design. At the previous stage sufficient design (systems design) has been done to prove the concept and obtain financing. From the detail design a cost estimate corresponding to the fourth line of Table 8.4 ("Control") is developed. This design may require three or four times as much effort as the systems design developed at the previous stage, but it is only done after the project has been proven and finance raised. Work can now begin. Resources are selected, and they plan the detail work on a rolling-wave basis. Work is authorized by the project manager and allocated to teams or individuals. As work is done, progress is measured to ensure the desired results are achieved, that is, the new asset is delivered within the constraints of quality, cost, and time, and it will achieve the required benefit. If there is a shortfall, appropriate recovery action is taken. This may mean doing nothing because the variances are small, replanning the work to recover the original plan, or revising the plan to accept the current situation. In extremis, it may mean terminating the project.

In this chapter, I describe the management of the execution stage. I start by explaining the selection of resources, implementation planning, and the allocation of work. I then describe the requirements for effective control, how to monitor progress and analyze variances to forecast completion, and how to take action to respond to deviations from plan.

13.1 RESOURCING A PROJECT

One of the recurrent questions of project management is: *do you assign work to people or people to work?*

In one approach—assigning work to people—you form a project team, they decide how best to achieve the project's objectives, and assign work to themselves. The risk is you will find the skills of the people in the team are inappropriate for the work you have to do. In the other approach—assigning people to work—you define the scope of work and then form a project team of appropriate skills. The risk is that the project manager will not be a technical expert, and so will be dictating to experts how they should undertake the task.

To overcome this dilemma, you develop the definition of the work and the organization in parallel down the breakdown structure. During the concept stage, you define the areas of work and the functional areas of the organization involved. During the feasibility stage, you work with functional managers to develop the milestone plan and responsibility chart at the strategic level. From the responsibility chart, you determine the skill types required and form a team. The team determine how they think the work should be done and so define the work at the activity level. The project manager and work-package manager agree and authorize the work and assign it to the team. Hence, the people to do the work are selected from a resource pool, which is identified by planning the work at the strategic level in the project hierarchy.

The process of resourcing a project includes the following steps:

1. *Identify what is to be achieved*: through the milestone plan.
2. *Identify the skills and skill types required to do the work*: through the responsibility chart. The skills required include technical, craft, professional and functional skills, or managerial knowledge.
3. *Identify the people available*: through discussion with the resource providers. It is important to obtain people with the correct skills. There is a danger, especially with a fixed project team, of selecting somebody to do work because they are available, not because they have the right skills, or that the resource provider may try to provide their least competent people, and retain their best individuals within their own sphere. You should take account of people's true availability. A person may only be available to a project part time and be retained for the remainder of their time on their normal duties.
4. *Assess the competence of the people available*: to identify any shortfall in skills. Even after selecting people of the correct skills, there may not be a perfect match to requirements.
5. *Identify any training required*: to overcome the deficiency in skill levels. Training may be in the form of open or bespoke courses, or on the job coaching.
6. *Negotiate with the resource providers*: Throughout this process, you must negotiate with the line managers of the people who will do the work, so they willingly release their people. If the resource providers will not cooperate, the manager can bring pressure to bear via the sponsor. However, even then they may not cooperate and block their people working on the project, so it is best to win the resource providers' support. This can be done by gaining their commitment to the project's goals, and by helping them understand how the project is of benefit to them.
7. *Ensure appropriate facilities and equipment are available*: Facilities may include office space, meeting rooms, security arrangements, and transport. Equipment may include computers, computer software (including word processing, spreadsheets, and project management information systems), telephone, internet access, and e-mail.

13.2 IMPLEMENTATION PLANNING

Having identified the people to do the work, the team can then define the details of the work to be done and assign work to themselves for execution. The detail work should be planned on a rolling-wave basis, as it is only when you are about to start the work that you have all the information required to plan activities in detail. In this way, you can also allow people to plan their own work. I did suggest in Sec. 5.4 that you can create a preliminary activity definition through work-package scope statements for early estimating.

Planning and Scheduling Activities

There are five steps in planning and scheduling activities:

1. *Define activities required to reach a milestone*: When selecting activities, the team should choose ones which are controllable, that is, they should
   - *Produce a measurable result*: It must be possible to determine when an activity is finished. It is no good dividing a work package into five activities each equal to 20 percent of the work. In those circumstances the last activity often takes 80 percent of the effort.
   - *Have average duration roughly equal to the frequency of review* (Sec. 5.1).
2. *Ratify the people involved*: The people to do the work have been chosen as described above. However, once the activities have been defined it may be necessary to review the team to ensure it contains all the necessary skills and no redundant skills.
3. *Define roles and responsibilities*: The involvement of each team member in the activities is then identified. A responsibility chart can be a useful tool for this (Fig. 6.6).
4. *Estimate work content and durations*: The work content and durations are estimated by applying the processes used on the work-package level.
5. *Schedule the activities within the work package*: Finally the activities are scheduled within the work package to deliver the milestone on time. This can be done manually, or by building the activities into a nested network (bar chart), as illustrated in Figs. 13.1 and 13.2.

If you adopt rolling-wave planning, estimates of work content and duration at the activity level will be made at a later stage than those at the work-package level, after sanction has been obtained. Some people are uncomfortable with this, fearing that the activity estimates will turn out to be different from—usually higher than—the work-package estimates.

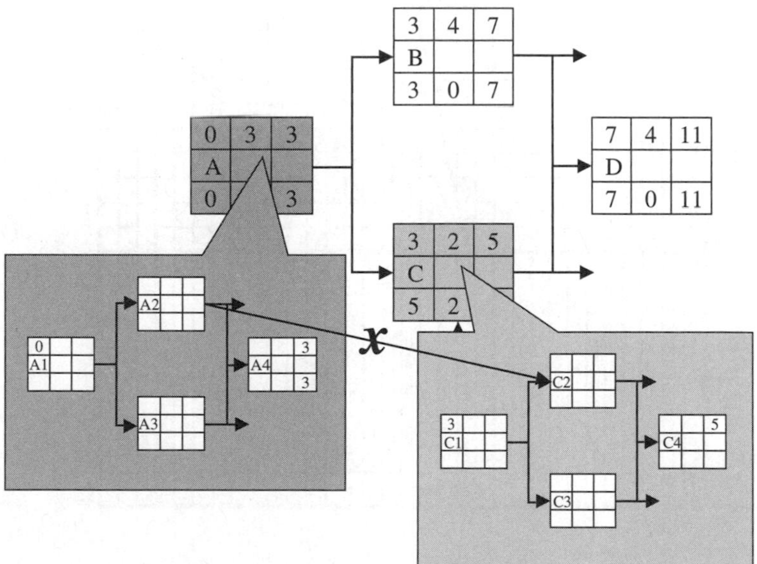

FIGURE 13.1 Nested networks and bar charts.

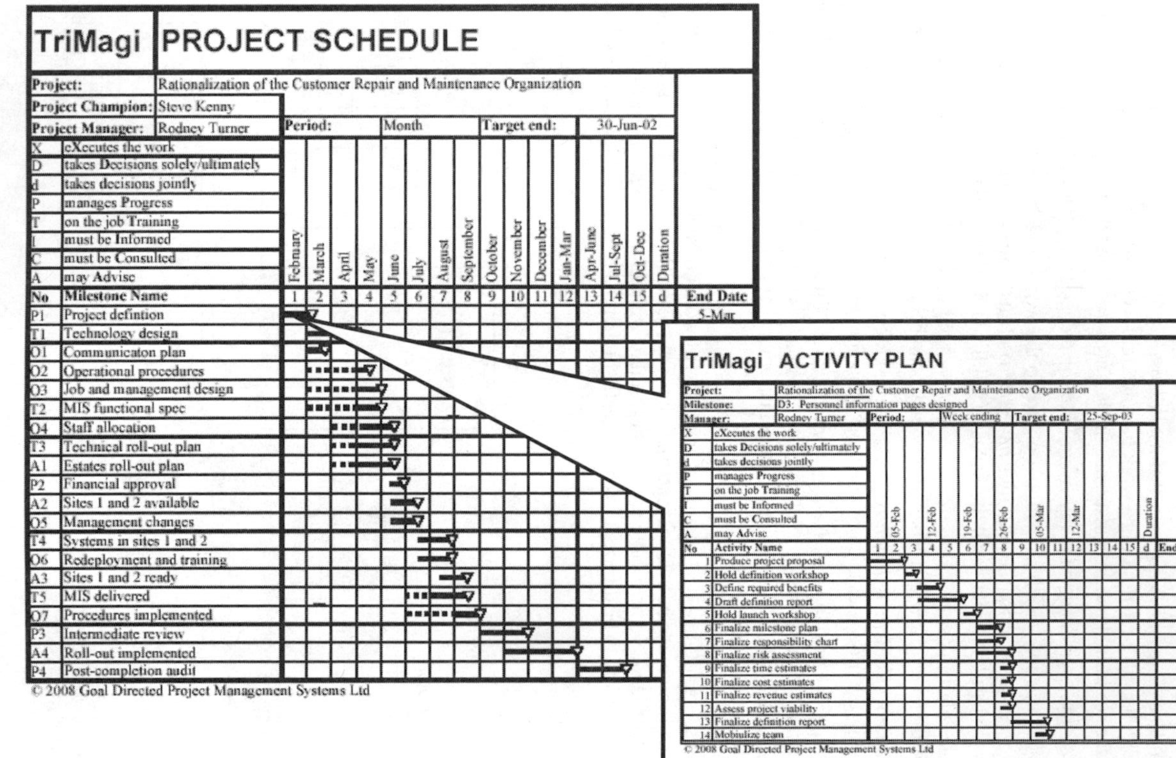

FIGURE 13.2 Nested networks and bar charts.

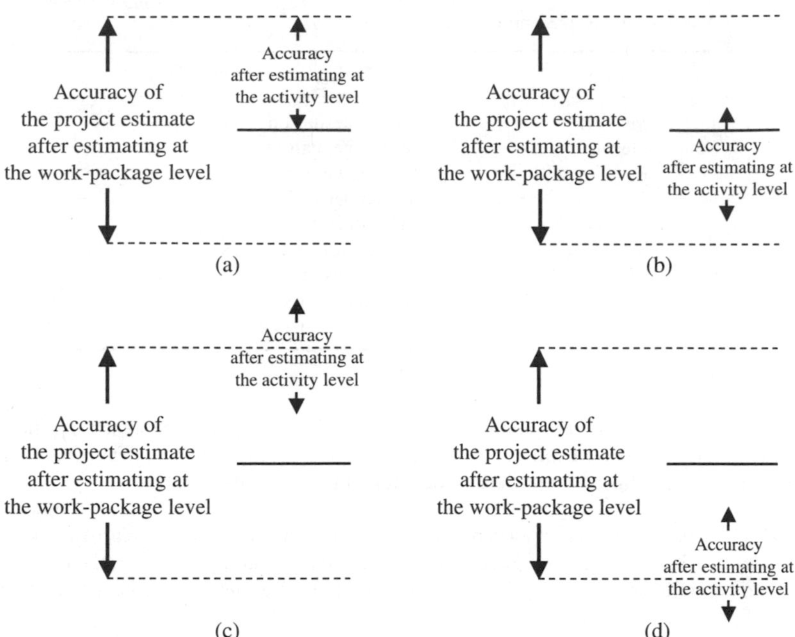

FIGURE 13.3 Comparison of total project estimates following estimating at the work-package level and activity level.

What should happen, of course, is that the range of possible outturns for the total project after activity estimating should fall within the range after work-package estimating. Table 8.2 shows that the range of accuracy for the project after estimating at the work-package level may be of the order of 10 percent and after estimating at the activity level may be 5 percent. Figures 13.3(*a*) and (*b*) show acceptable activity estimates and Figs. 13.3(*c*) and (*d*) unacceptable ones. Figure 13.3(*d*) is unacceptable because overestimating can lead to viable projects being cancelled, or capital being tied up and becoming unavailable to fund other worthwhile projects. If the estimates consistently fall outside the allowable range of those prepared at the work-package level then the estimating data used for the latter needs improving. It is therefore important to feed the results back to the estimators so that they can improve their data.

If it is not possible to schedule the activities to deliver the milestone on time (subject to Fig. 13.3) then the delay to the plan must be subjected to change control. The change can be to declare a variance between the current schedule and the baseline, or if the delay is severe to update the baseline.

Representing the Activity Schedule

There are several ways of representing the activity schedule.

Responsibility Charts. The responsibility chart provides a complete picture of the schedule of activities which make up a work package. Figure 6.6 shows the activity schedule for Milestone P1 in the CRMO Rationalization Project.

TABLE 13.1 Representing the Plan

Level of breakdown	Representation of plan
Integrative level	Project definition report
Strategic level	Milestone plan
	Project responsibility chart
	Master network
	Master bar chart
Detail level	Responsibility chart as activity schedule
	Work-package estimating sheets
	Nested network
	Nested bar chart

Estimating Sheets. The responsibility chart is an effective tool for representing the people involved, but weak for estimating work content. An estimating sheet (Table 9.3) can be used for the latter. This is often usefully developed in a spreadsheet.

Nested Networks. You can draw a network of the activities which make up a work package. Figure 13.1 illustrates nested networks. If you are using computer tools, the nested network can be included as hammocked networks in the master network or kept as a separate subnetwork linked to the master network.

Nested Bar Charts. Similarly you can draw a bar chart of the activities which make up a work package. Figure 13.2 shows nested bar charts. Responsibility charts at several levels are nested bar charts.

Representing the Overall Plan

The activity schedules represent the plan at the lowest level of breakdown. The plan will be represented in many different ways depending on the level of breakdown and for whom it is being prepared. Table 13.1 summarizes the different representations of the plan.

13.3 ALLOCATING WORK

When work is being done, it is allocated to the team via work-to lists. A *work-to list* is a list of activities to which a person or resource is assigned. The activities may be listed by:

1. *Work package*: The person or resource is given the activity schedules for all work packages on which they are working, as a responsibility chart or estimating sheet.
2. *Time Period*: They are given a listing of the activities they are assigned to for a given period of time from all work packages they are working on. The period is typically the current control period and one or two periods into the future. The work-to list contains
   - All activities started but not finished
   - All activities due to start in the period

The work-to list may be in the form of a responsibility chart (Fig. 6.6) or output from a computer system (Table 13.2). It is now quite common for the output to be sent from the

TABLE 13.2 Computer-Generated Work-to List and Time Sheet

Act no.	Description	Orig dur (*d*)	Rem dur (*d*)	Sched start	Sched finish	Actual start	Actual finish	M (h)	Tu (h)	W (h)	Th (h)	F (h)	WE (h)	Work rem (h)
P1A	Project proposal	5.0		01 Feb	05 Feb	01 Feb	05 Feb							
P1B	Definition workshop	2.0		08 Feb	09 Feb	08 Feb	09 Feb							
P1C	Define benefits	3.0	3.0	10 Feb	12 Feb									
P1D	Draft definition report	6.0	6.0	10 Feb	17 Feb									
P1E	Launch workshop	2.0	2.0	18 Feb	19 Feb									
P1F	Milestone plan	3.0	3.0	22 Feb	24 Feb									
P1G	Responsibility chart	3.0	3.0	22 Feb	24 Feb									
P1H	Assess risks	3.0	3.0	22 Feb	26 Feb									
P1I	Estimate time	2.0	2.0	25 Feb	26 Feb									
P1J	Estimate cost	2.0	2.0	25 Feb	26 Feb									
P1K	Estimate revenue	2.0	2.0	25 Feb	26 Feb									
P1L	Assess viability	2.0	2.0	25 Feb	26 Feb									
P1M	Finalize definition report	3.0	3.0	01 Mar	05 Mar									
P1N	Mobilize team	2.0	2.0	04 Mar	05 Mar									

master plan to the individual's work station electronically, and they will subsequently become their time sheet. The work-to list may contain the following information:

- Activity number and name
- Baselined dates and duration
- Current estimate/actual dates and duration
- Estimated work content
- Work to date
- Bar chart of baselined dates
- Bar chart of current estimate/actual dates

Some people also include float, but I do not agree with this. The team do not need to know the float and they will invariably consume it. This strongly held view causes some controversy. Some people say I do not trust the team. I am afraid I believe in Parkinson's law: work done expands to fill the time available. I also know that most people are working to tight deadlines on much of their work, and they will put off anything with large float (I do). For this reason, I also do not think people should be given work with very large float; you should wait until much of the float has been consumed. It is inconsiderate to give busy people nonurgent work, and work which may actually need to change as other work is done. However, I do agree that the team should be told whether work is critical or not. If they need to consume float, they should negotiate that with the project manager, not assume that the float is there for them to consume by right because it is shown on the work-to list.

At the end of the control period the work-to list will become a *turnaround document* (Sec. 13.5) through which the project team reports progress. The processes of drawing up the activity schedules, including them in the master plan, and issuing work-to lists are shown in the procedure for monitoring and control (Fig. 6.5).

The equivalent list for gathering materials on a project is called a *kit-marshalling list*. This lists all the materials required for an activity and the date they are required by. If the materials are held in a store, then the list may be issued in the reporting period before that activity, so that the materials can be collected together (marshalled) to a central point ready for use. If they need to be procured, clearly the list must be issued earlier still. The planning system needs to record the lead time. A computer system can be very useful for this.

13.4 REQUIREMENTS FOR EFFECTIVE CONTROL

Everything I have covered up to this point has brought us to the point where we are doing work. As the work is done we must ensure we achieve the planned results; that we deliver the new asset and desired changes and performance improvement to the desired specification, and within the cost and time at which it was thought to be worthwhile. Furthermore, as the facility is commissioned, we must ensure it delivers the expected benefits which were used to justify the money spent. We can be sure that this will not occur in a haphazard fashion. The structured process by which we check progress and take action to overcome any deviations from plan is control. As we have seen several times up to now, there are four essential steps to the control process (Fig. 7.2):

1. Plan future work and estimate performance.
2. Monitor and report results.
3. Compare results to the plan and forecast future results.
4. Plan and take effective action to recover the original plan, or to minimize the variance.

The book so far has dealt with the first step. In the remainder of this chapter, I deal with the other three steps in turn. In this section, I start by explaining the requirements for effective control. For control to be effective, each step in this four-step process must be effective.

Effective Plans

I have discussed the requirements of effective planning throughout the book. In particular, the plans must be comprehensive and frozen into a baseline to provide a fixed measure for control. If the plans are updated frequently, without the application of strict change control, then there will be no measure for control. The project will always be on time, because the plans have just been updated. Team members may develop new activity schedules, but the project manager must authorize them before they are included in the master plan. Work is done against current work-to lists, issued regularly.

Effective Reporting

Effective reporting mechanisms should satisfy the requirements discussed in the following sections.

Reports Should Be Made Against the Plan. To ensure people are interpreting the reports in the same way, they should be made against the plan. Example 3.6 describes a case in which the project manager and team members were working on different plans. The team members were making verbal reports and reporting satisfactory progress. The project manager could not understand why they were not achieving his milestones. Turnaround documents are a tool for reporting against the plan.

There Should Be Defined Criteria for Control. Likewise it is important to have defined criteria. If people are asked to make ad hoc reports, they tend to report the good news and hide the bad news. If asked to report against a set of closed questions, they will usually answer honestly. If they report dishonestly, it will become obvious at the second or third reporting period. Defined criteria are given in Sec. 13.5.

The Control Tools Should Be Simple and Friendly. Team members should spend as little time as possible filling in reports. If submitting reports takes an excessive time, people rightly complain they are being distracted from productive work (Example 13.1). Simple friendly tools means single-page reporting nested in the work-breakdown structure (WBS) and reports against the plan with defined criteria requiring numeric or yes/no answers. Reports are often filed against work-to lists. These are *turnaround documents*. The work-to list contains space for the report, and is returned at the end of the reporting period.

Example 13.1 Simple friendly tools

I used to work on ammonia plant overhauls, each a four-week project. Every day, supervisors came to an one-hour control meeting in the morning, a two-hour meeting in the afternoon, and spent one hour after work completing daily returns. They complained that they should spend more time on the patch motivating their men.

Reports Should Be Made at Defined Intervals. Just as it is necessary to report against defined criteria, it is also necessary to report at defined intervals. You should not ask people to report only when there is something to discuss. People hate to volunteer failure, so they will not ask for help until it is too late to recover. If people know that they must report both good news and bad at defined intervals, then they will report more freely and accurately. The frequency of the reporting period depends on:

- The length of the project
- The stage in the project
- The risk and consequence of failure

On a yearlong project, you may report fortnightly at the activity level. In areas of high risk you may report more often. Towards the end you may report weekly or even daily.

Reports Should Be Discussed at Formal Meetings. To be effective the reports must be made and discussed at formal meetings. Passing the time of day at the coffee machine is part of effective team building, but not of effective control. To keep the meetings short and effective, the discussion should focus on identifying problems, and responsibility for solving them, but the meeting should not attempt to solve the problem.

The Reports Should Stimulate Creative Discussions. To link into the next steps of control the reports must generate creative discussion, so the team can identify where variances are occurring and ways of taking effective timely action.

Effective Reviews

Using the data the team determines whether the project is behaving as predicted and if not calculate the size and impact of the variances. The two quantitative measures of progress are cost and time, and so receive most attention. The team uses the reports to forecast time and cost at completion, and calculates any differences between these figures and the baseline. It may simply be that work is taking longer and costing more than predicted; or delays or additional effort may be caused by variances in quality, people failing to fulfil their responsibility, externally imposed delays, or changes in scope. Therefore the variances in time and cost can point to a need to control one or more of the five functions of project management. The defined criteria, formal meetings, and creative discussions are key to this process.

Effective Action

To close the control loop, the team must take effective action to overcome any variances. This may mean revising the plan to reflect the variances, but hopefully it means taking timely effective action to stop them getting worse and preferably reducing or eliminating them.

13.5 GATHERING DATA AND CALCULATING PROGRESS

Gathering Data

The first step in the control process is to gather data on progress. Suggested data are given in Table 13.3. These are usually collected at the activity level but may be collected at the work-packages or task levels. When collected at a lower level they can be aggregated to report at a higher level. The use of these data in the control process is described in Sec. 13.6.

I said above that data is most effectively gathered against defined criteria using turnaround documents, work-to lists issued at the start of the reporting period and used at the end of the period to gather data. Turnaround documents provide reports against the plan,

TABLE 13.3 Criteria for Control and Required Data

Criteria for control	Quantitative data	Qualitative data
Time and cost	Actual start/finish Forecast start/finish Effort to date Effort remaining Other costs to date Other costs remaining	
Quality		Achievement of milestone and activity measures Problems encountered
Organization		Responsibility chart kept
Scope		Issues Changes Special problems

defined criteria, and simple friendly tools. They can also be used as the focus for formal meetings. I find it effective to use a data projector to project the turnaround document onto a white board. The team can fill in the document on the board in a group meeting. This process encourages creative discussions to identify any problems, but also enables the meeting to be kept short. Figure 13.4 is a manual turnaround document encompassing the activity schedule from Fig. 6.6. Table 13.4 contains a computer-generated turnaround document. Figure 13.5 is a turnaround document at the milestone level.

The required time and cost data can be gathered using people's time sheets. At the start of the week the individual is given a blank time sheet listing the activities they should be working on day by day; they enter the amount of time spent working on each activity, and at the end of the time period they enter the amount of time left to work on each activity:

- The first time he or she books time against an activity is actual start.
- Total time booked gives effort to date.
- Effort remaining is entered at the end of the period.
- When this is zero you can look back to find the last time he or she entered time against the activity for actual completion.
- Forecast completion can be extrapolated from effort to date and effort remaining.

Calculating Progress

The data gathered is used to calculate progress on all five project management functions: time, cost, quality, project organization, and scope. In particular, with the first two we try to forecast the final outturn, the time and cost to completion, as this gives better control than reporting the actual time and cost to date. This concept is part of the forward-looking control. This is an important principle of project management—you can only control the future, not the past. The only value of the past is to give you information to help you control the future. But you cannot undo the past; all you can influence is the future and so it is the future you need to control. Thus we try to forecast time and cost to completion.

Forecasting Time to Completion. Time is the simplest function to monitor and that is perhaps why it receives the greatest attention. All you have to do is schedule the rest of the project in the same way you scheduled it initially. Tracked bar charts (Fig. 9.15) and milestone

TriMagi Activity Plan	
Project:	Rationalization of the Customer Repair and Maintenance Organization
Milestone:	D3: Personnel information pages designed
Manager:	Rodney Turner

X	eXecutes the work
D	takes **D**ecisions solely/ultimately
d	takes **d**ecisions jointly
P	manages **P**rogress
T	on-the-job **T**raining
I	must be **I**nformed
C	must be **C**onsulted
A	may **A**dvise

Activity Schedule

Period:	**Week ending**	**Target end:**	**5-Mar-08**

Report

Author:	Rodney Turner
Date:	13 February
Checked	Ian Simmons

No	Activity Name	Regional board	Operations director	CRMO managers	CRMO team leader	Project manager	Project support office	Estates manager	Estates department	Network manager	Networks department	IS department	Operators	Personnel	Labour $K	Material $K	Duration d	End Date	Estimated Completion	Work done d	Work to do d	Quality acceptable	Responsibilities kept	Change required	External delay	Special problem
1	Produce project proposal				0.5	1.0	1.0								2.5			5-Feb								
2	Hold definition workshop				0.5	1.0	0.5								2.0			9-Feb								
3	Define required benefits		1.5		0.5	2.0									4.0			12-Feb	12-Feb	3	0					
4	Draft definition report					2.0	3.0								5.0			17-Feb	14-Feb	4	1					
5	Hold launch workshop		1.5	3.0	0.8	1.5	0.8	1.5		1.5					10.5	15.0		19-Feb	19-Feb							
6	Finalize milestone plan					1.0	0.5								1.5			24-Feb								
7	Finalize responsibility chart					1.0	0.5								1.5			24-Feb								
8	Finalize risk assessment				0.5	1.0	0.5								2.0			26-Feb								
9	Finalize time estimates						1.0								1.0			26-Feb								
10	Finalize cost estimates						1.0								1.0			26-Feb								
11	Finalize revenue estimates						0.5								0.5			26-Feb								
12	Assess project viability					1.0									1.0			26-Feb								
13	Finalize definition report				0.5	2.0	1.5								4.0			5-Mar								
14	Mobilize team				0.3	0.5	0.3								1.0	10.0		5-Mar								
	Contingency				0.5	1.0	0.3								2.5											
			3	3	4	15	12	2		2					40	25										

Week numbers: 1 2 3 4 5 6 7 8 9 10 11 12 13 14 15; week ending 5-Feb, 12-Feb, 19-Feb, 26-Feb, 5-Mar, 12-Mar

Time Now

© 2008 Goal Directed Project Management System

FIGURE 13.4 Manual turnaround document encompassing the activity schedule.

TABLE 13.4 Computer-Generated Turnaround Document and Time Sheet

Act No	Description	Orig dur (d)	Rem dur (d)	Sched start	Sched finish	Actual start	Actual finish	M (h)	Tu (h)	W (h)	Th (h)	F (h)	WE (h)	Work Rem (h)
P1A	Project proposal	5.0		01 Feb	05 Feb	01 Feb	05 Feb							
P1B	Definition workshop	2.0		08 Feb	09 Feb	08 Feb	09 Feb							
P1C	Define benefits	3.0	3.0	10 Feb	12 Feb	10 Feb	12 Feb	6	6	2				
P1D	Draft definition report	6.0	6.0	10 Feb	17 Feb	10 Feb	14 Feb			4	6	6		
P1E	Launch workshop	2.0	2.0	18 Feb	19 Feb									
P1F	Milestone plan	3.0	3.0	22 Feb	24 Feb									
P1G	Responsibility chart	3.0	3.0	22 Feb	24 Feb									
P1H	Assess risks	3.0	3.0	22 Feb	26 Feb									
P1I	Estimate time	2.0	2.0	25 Feb	26 Feb									
P1J	Estimate cost	2.0	2.0	25 Feb	26 Feb									
P1K	Estimate revenue	2.0	2.0	25 Feb	26 Feb									
P1L	Assess viability	2.0	2.0	25 Feb	26 Feb									
P1M	Finalize definition report	3.0	3.0	01 Mar	05 Mar									
P1N	Mobilize team	2.0	2.0	04 Mar	05 Mar									

TriMagi	**Milestone Plan**		**Milestone Report**
Project:	Rationalization of the Customer Repair and Maintenance Organization		
Project Sponsor:	Steve Kenny		
Project Manager:	Rodney Turner		

Date	P	O	A	T	Milestone Name	Report Date	Milestone Report
5-Mar	P1				When the project definition is complete, including benefits map, milestone plan and responsibility chart	5-Mar	Project definition complete
30-Apr				T1	When the technical solution, including appropriate networking and switching technology has been designed and agreed	7-May	Technological design required additional wok, completed a week late
22-Mar		O1			When a plan for communicating the changes to the CRM Orgaization has been agreed	22-Mar	Communication plan published
15-May		O2			When the operational procedures in the CRM Offices has been agreed	15-May	Completed. Not delayed by technical design
31-May		O3			When the job design and management design is complete and agreed	31-May	Completed. Not delayed by technical design
31-May				T2	When the functional specification for the supportiong management information system (MIS) has been agreed	7-Jun	Completed a week late. Delayed by the design
15-Jun		O4			When the allocation of staff to the new offices, and recruitment and redeployment requirements have been designed and agreed	22-Jun	Completed
15-Jun				T3	When the technical roll-out stratgey has been defined and agreed	22-Jun	Completed
15-Jun			A1		When the estates roll-out stratgey has been designed and agreed	22-Jun	Completed. Work on sites 1 & 2 will be delayed a week, but)7 can be completed on time.
30-Jun	P2				When the budget for implementation has been determined and provisional fianancial authority obtained	30-Jun	Depayed by 1 week by delay of technical design.
15-Jul			A2		When sites 1 and 2 are available	30-Jun	Work started 22 Jun
15-Jul		O5			When the management changes for sites 1 and 2 are in place (first call receipt and first diagnostic offices)	30-Jun	Work started 22 Jun
31-Aug				T4	When the system is ready for service in sites 1 and 2	30-Jun	Work expected to start 22 Jul
31-Aug		O6			When a minimum number of staff have been recruited and redeployed and their training is complete	30-Jun	Work expected to start 22 Jul
15-Sep			A3		When sites 1 and 2 are ready for occupation		
15-Sep				T5	When the MIS system has been delivered		
30-Sep		O7			When sites 1 and 2 are operational and procedures implemented		
30-Nov	P3				When a successful intermediate review has been conducted and roll-out plans revised and agreed		
31-Mar			A4		When the last site is operational and procedures fully implemented		
30-Sep	P4				When it has been shown, through a post-implementation audit that all benefit criteria have been met		

FIGURE 13.5 Turnaround document at the milestone level.

tracker charts (Fig. 9.16) help in this process. If critical milestones have been delayed, or if the critical path has been delayed (and no other path has become "more critical"), then it is likely that the project has been delayed by that amount. If the team has maintained an up-to-date network for the project, that can be used to forecast the completion date for the project in exactly the same way it was used to predict the end date initially. The record of effort to date versus effort remaining can also be used to control time in one of three ways:

By revising estimates of duration: If there is a consistent estimating error, this will be indicated by a trend. The estimates of duration can be revised accordingly.

By indicating the cause of delays: Table 13.3 shows four possible outcomes of duration and effort. Both may be on (or under) budget, in which case all is well. The project may be on time but effort over budget, in which case there may be minor errors but the team is coping perhaps by working unplanned overtime. The project may be late but no additional effort has been expended. Then the cause of the delay must be due to external factors perhaps other people failing to fulfil their responsibilities, late delivery of some materials, or perhaps the project team have been occupied on work of higher priority (to them). The qualitative control data (Table 13.2) may help to indicate the cause. If both time and effort are overbudget, then the cause may be serious estimating errors, rework due to poor quality, or rework due to change. A trend will indicate the first as described above, and so you will need to monitor effort and duration over several reports. The qualitative control data (Table 13.3) will indicate the second or third cause. You can see from Table 13.5 how the complete set of control data can help initiate discussion over the likely causes of delays, and help in their elimination.

Through the earned value calculation: The schedule variance, calculated as part of the cost control process, will indicate whether the project is on average ahead or behind schedule.

Forecasting Cost to Completion. I showed in Sec. 8.4 how to use the cost data gathered to forecast the cost at completion:

Forecast cost at completion = actual cost to date + forecast cost remaining

This adheres to the principle of forward-looking control. You see in Table 13.2 that we are gathering data about the actual cost to date and the forecast cost remaining of work in progress. For work not yet started you can either use the original estimate, or update based on experience so far (that is where the past is useful). The forecast cost remaining on work in progress relies on people being both reliable and honest but it is the best we have. The S-curve (Fig. 8.8) provides a highly visual representation of the progress to date.

Controlling Quality. Data gathered can show where deviations have occurred from the specification. These quality variances may be identified as part of the quality control

TABLE 13.5 Determining the Cause of Delays by Comparing Effort and Completion Dates

Effort	Duration on time	Late
As predicted	No problem	External delays Responsibilities not fulfilled
Overbudget	Minor estimating errors Minor changes	Estimating errors Major changes Major quality problems

process or may be noticed by team members. The impact of quality problems on time and cost is indicated by Table 13.3.

Controlling Organization. Similarly, the data gathered may indicate where the project organization is not performing as planned. This may specifically be caused by people not fulfilling their roles or responsibilities as agreed in the responsibility chart. Table 13.3 also shows how the control process can indicate the impact of these organizational delays on time and cost.

Controlling Scope. Finally, the data gathered can indicate that changes in scope have occurred. These especially will have an impact on the time and cost of a project (Table 13.3). Changes in scope are usually inevitable. However, they should be rigidly controlled and this requires a change control procedure. Change control is a six step process:

1. Log the change.
2. Define the change.
3. Assess the impact of the change. Seemingly simple changes can have far-reaching consequences.
4. Calculate the cost of the change. This is not just the direct cost but the cost of the impact.
5. Define the benefit of the change. This may be financial or nonfinancial. The latter includes safety.
6. Accept or reject the change based on marginal investment criteria. A return of 40 percent per annum is possible for marginal criteria compared to 20 percent for the project as a whole.

If this procedure is applied rigorously, many changes do not get past step 3. Table 13.6 is a form to aid this process.

13.6 TAKING ACTION

Once we have identified that a project is deviating from plan, we must take appropriate action. The earlier action is taken the better, because it is cheaper to recover or abort the project should it prove nonviable.

TABLE 13.6 Change Control Form

Project: Milestone: Activity:			
Description of change:			
Impact of change:			
Cost of change: Value of change:			
	Name	Signature	Date
Proposed by:	....	....	....
Checked by:	....	....	....
Approved by:	....	....	....

Recovering a Project

The response to the variances can be easily manageable or unmanageable and reactive. The most effective approach depends on the circumstances. There are cases which demand an immediate response. However, in most cases there is time to reflect and recoup. A structured approach to problem solving (Fig. 1.6) is the best means of recovery. Here, I describe a six step version for planning recovery:

Stop: Regardless of the size of the variance and its impact, everyone should pause. Unfortunately, the most common reaction is to seek an instant remedy. Some common solutions such as adding more resources or sacking the project manager may do more harm than good. While this reaction is understandable, it is often wrong because of the emotional state of the team. Terry Williams[1] has shown that sometimes these stock responses can initiate feedback loops which actually make the situation worse. Keep cool, calm, and collected. Remember the first law of holes: if you find yourself in a hole, stop digging (Example 13.2).

Look, listen, and learn: It is important to undertake a thorough review with all team members and the client present. Effective recovery must be based on a clear understanding of the cause of the divergence, and possible ways of overcoming it. Seeking views on what went wrong and what action the team proposes is important in rebuilding commitment.

Develop options and select a likely course: By exploring every avenue and developing a range of solutions. Establish decision criteria so options can be evaluated against agreed conditions. If necessary return to the original financial evaluations, recost and retime each option, air them with the client, and then select one which meets the decision criteria.

Win support for the chosen option: It is important that there is support from all those involved. There is hard work ahead and uncommitted team members will falter at the first hurdle.

Act: Once the agreed course of action has been accepted every effort must be made to implement it. Deviations from the agreed plan will only add to the confusion and make the situation worse.

Continue to monitor: Monitor the impact of any actions to ensure they have the desired effect. If not then the recovery process must be repeated.

Example 13.2 If you find yourself in a hole, stop digging

I was working for an IT vendor and came across a group of salesmen who seemed to be running around like headless chickens. So I asked them what the problem was. They said they were trying to prepare a bid for an order which, if they won it, would make them the largest supplier for a certain line of equipment in Europe. But they said they were making no progress and looked as though they would not complete the bid on time. I suggested they treat the bid like a project and spend three days in a start-up workshop with me. They said they had no time to spend in a start-up workshop with me; they had a bid to prepare. I asked them if they were going to be successful. They said at the rate they were going, no! So I asked what had they to lose, and in fact they may save themselves heart attacks. So they came and spent three days with me in a start-up workshop, successfully won the bid, and became the largest supplier in Europe of that line of equipment.

Options for Action

There are five basic options for taking action.

Find an Alternative Solution. This is by far the best solution. The plan is recast to recover the project's objectives in a way which has no impact on the quality, cost, time, or scope. It may be that two activities were planned sequentially, because they share the same scarce resource. If the first is delayed for other reasons, it may be possible to do the second activity first, and hopefully when it is complete it will then be possible to do the other.

Compromise Cost. This means adding additional resource either as overtime or additional people, machines, or material to recover the lost time. This is usually the instant reaction to project delays. However remember the discussion in Sec. 9.2, when I was describing how to calculate durations: doubling the number of people on a project usually does not double the rate of work. Brooks's law[2] states

> Adding resource to a late software project makes it later still

So actually adding people can have the opposite of the desired effect.[1] The rationale is existing people must take time out to bring the new people up to speed. If you want to add people you need to carve out a bit of the project and give it to them, not increase number of people working on a bit of the project already underway.

Compromise Time. This means allowing the dates to slip. This may be preferable, depending on whether cost or time is the more important constraint on the project. This decision should have been made during the feasibility and communicated to the project team as part of the project strategy.

Compromise Scope. This means reducing the amount of work done, which in turn means taking less on time to achieve some benefit. Notice I did not say compromising the quality. The latter is very risky once the initial specification has been set, and should therefore be discouraged.

Abort the Project. This is a difficult decision. However, it must be taken if the future costs on the project are not justified by the expected benefits. Project teams are often puzzled that their recommendation to terminate a project is ignored; a decision which seems obvious is avoided and good money is poured after bad, depriving other projects. It takes courage to abort a project. During their lives, projects absorb champions and supporters. Senior people may have become associated with its success and feel if the project fails it may damage their reputation. There is often a feeling that "with a little more money and a bit of luck the project can be turned round." The fact is that once an organization makes an emotional commitment to a project it finds it very hard to abandon. Another argument often put forward to support a failing project is that "as we have already spent so much on it we should finish it." Unfortunately, this argument is fallacious: future costs must be justified by the expected benefit no matter how much has been spent so far. If the project's outcome is still important to the organization it may be more effective to abort a project, learn from it, and start afresh.

SUMMARY

1. The process of resourcing a project includes the following steps:
   - Identify what is to be achieved.
   - Identify the skills and skill types required.
   - Identify the people available.
   - Assess their competence.

- Identify any training required.
- Negotiate with the resource provider.
- Ensure appropriate project facilities are available.

2. The five steps of activity planning are
   - Define the activities to achieve a milestone or work package.
   - Ratify the people involved.
   - Define their roles and responsibilities.
   - Estimate work content and durations.
   - Schedule activities within a work package.
3. After creation of the activity schedule, it is entered into the master plan, and at appropriate intervals work allocated to people. Both steps must be authorized by the project manger.
4. Activity schedules may be represented by:
   - Responsibility charts
   - Estimating sheets
   - Nested networks
   - Nested bar charts
5. Work is allocated to people via work-to lists, by:
   - Time period
   - Work package
6. The four steps in the control cycle are
   - Plan future performance.
   - Monitor achievement against plan.
   - Calculate variances and forecast outturn.
   - Take action to overcome variances.
7. For control to be effective, each step in this cycle must be effective. Requirements for effective planning have already been described, and in particular are stated in the five principles of project management at the end of Chap. 3.
8. Requirements for effective reporting include:
   - Reports against the plan
   - Defined criteria for control
   - Simple, friendly tools
   - Reporting at defined intervals
   - Formal review meetings
   - Creative discussions
9. This can be achieved by gathering data using turnaround documents, which can be used to gather data to control the five objectives:
   - Time
   - Cost
   - Quality
   - Organization
   - Scope
10. Time is controlled by recording progress on the critical or near critical paths, or by comparing the cost of work actually completed to that planned to have been completed. In order to do this, the following progress data is collected:
    - Actual start/finish
    - Revised start/finish
    - Effort to date
    - Effort remaining
    - Costs to date
    - Costs remaining

11. Cost is controlled by comparing costs incurred to the planned cost of work actually completed. In order to do this, the same data is required. Costs are said to be incurred when the expenditure is committed, not when the invoices are paid, because at that time the plan can still be recovered.

12. When the divergence of achievement from the plan becomes too great, the project must be recovered. The ten-step problem-solving cycle can be applied to find the solution to plan recovery. Possible courses of action include
- Rearranging the plan
- Compromising time
- Compromising cost
- Compromising scope
- Aborting the project

REFERENCES

1. Williams, T, *Modelling Complex Project,* Chichester, U.K.: Wiley, 2002.
2. Brooks, FP, *The Mythical Man-Month,* 25th anniversary ed., New York: Addison-Wesley, 1999.

CHAPTER 14
PROJECT CLOSE OUT

The last stage of the life cycle is close out. During the closing stages, the team must maintain their vigilance to ensure the work is completed and it is completed in a timely and efficient manner. It is easy for the good effort of execution to be lost as some team members look forward to the next project and others become demob happy or demob unhappy. During this stage, the team's focus must switch back to the project's purpose. During execution, the team concentrates on doing the work within time, cost, and specification. Now they must remember why they are undertaking the project; they are not doing the work for its own sake, but to deliver beneficial change. It is easy to complete the work within the constraints and think that it is a successful project, while failing to use the new asset to deliver the change and obtain the expected benefits, which justified the money spent. There are projects where the new asset is not used properly and no benefit obtained. There are many more where it is used to less than full capacity and the project team does not see it as their responsibility to ensure it is. They are more interested in their next project. At this time, the project team must also remember that it may be the closing stage of the project but it is the start of the operational life of the asset. Adequate mechanisms must be put in place to support it through its life.

As the project comes to an end the team disbands. If the project is completed efficiently, the team may be rundown over some time. As this happens, it is important to ensure it is done in a caring way. Team members may have made significant contributions, or even sacrifices, to the success of the project. If this is not recognized, at best the project will end on an anticlimax, and at worst it will leave lasting resentment which will roll over into the next project. You must ensure the team members are given due reward for their contribution and that the end of the project is marked appropriately.

Finally, there is data to be recorded or lessons learnt for the operation of the facility, or for the design, planning, estimating, and management of future projects. Because the team's attention may be focused on completing the task and looking forward to the next project, this often remains undone. Completing records of the last project is a distraction from the next. It also costs money, spent after the asset has been commissioned, so it provides no immediate benefit. It is money which can be easily saved, especially if the project is overspent. However, it is precisely when a project is overspent that it is important to find out why that happened, so the information can be used for the planning and estimating of future projects.

In this chapter, I describe how to bring the project to a timely and efficient completion. I then explain how to hand the facility over to the users, while ensuring that it is fully commissioned to obtain the benefits and that a proper support mechanism is put in place as it moves into its operation phase. I describe how to disband the project team in a caring way and identify the key data to be recorded at the end of the project, how it is obtained and the purpose for which it is used. This is the second most critical stage of a project. Nobody remembers effective start-up, but everyone remember ineffective close out; the consequences are left to be seen for a long time.

14.1 TIMELY AND EFFICIENT COMPLETION

As the project draws to a close, the team must ensure all work is completed in a timely and efficient manner. The following can aid this process:

- Producing checklists of outstanding work
- Planning and controlling at lower levels of work breakdown to provide tighter control
- Holding more frequent control meetings to ensure problems are identified and solved early
- Planning the rundown of the project team as the work runs down to ensure people are released for other work
- Creating a task force with special responsibility for completing outstanding work
- Closing contracts with suppliers and subcontractors to ensure that no unnecessary costs are booked
- Supporting the project manager by a deputy with finishing skills

Planning and Control at a Lower Level of Breakdown. As you near the end, you begin to look at what needs to be done to complete outstanding work. Instead of waiting two weeks to find out what work has been achieved, you create daily lists of work to complete the asset, hand it over to the client and commission it. This leads naturally to planning at a lower level of work breakdown and holding more frequent control meetings. Towards the end, the risk of delay becomes greater and so it becomes necessary to review progress more frequently, weekly, daily, or even twice daily. I suggested in Chap. 5 that whatever the frequency of control, that should be the average duration of activities, and hence you plan against shorter tasks. The checklists are just more detailed plans.

Disbanding the Team and Forming Task Forces. As the project nears its end, you require fewer resources. This is what gives the S-curve its shape. To ensure the most efficient completion, you must plan the release of resources in advance. You do not want them turning up one day, and sitting around until you realize they are not needed, because that is inefficient for both the project and the organization as a whole. You tell people one or two weeks in advance they will not be required on a certain date. You also tell their line managers, so they can make full use of their people when they are released.

As the teams run down it becomes essential to combine the members into task forces to retain natural *hunting packs*. David Frame describes task forces created at the end of a project as *surgical teams*.[1] The surgical team is a cross discipline team able to work together to complete the daily task lists and tie up other lose ends.

Closing Contracts. Closing contracts with suppliers and subcontractors is another way of planning the rundown of the project team (see Example 14.1).

Manager with Finishing Skills. The skills required to finish a project can be different to those required to start it up and run it. Therefore it may be appropriate to change managers in the final stage (Sec. 6.3). However, if this change is to be seamless the new manager ought to be a former deputy who has been involved for some time. One approach is to have a single project manager for design execution and close out with a deputy for each of design, execution, and close out, a design manager, execution manager, and commissioning manager, respectively.

Example 14.1 Closing account numbers and contracts

A delegate on a course told me he had been telephoned by the accounts department two years after a project had been finished to be told it was overspent. He was asked what he was doing about it. He asked how this could be because the project had finished two years previously, it was underspent then, and no further work had been done. The accounts department said people were still charging their time. The project manager said that there was nothing he could do to stop people charging time and asked the account department why didn't it close the account numbers. They said it was against company policy and it was his duty to make people stop charging their time!!! It wouldn't happen to Dilbert.

14.2 TRANSFERRING THE ASSET TO THE USERS

Key issues in transferring the new asset to the users are discussed in following sections.

Planning for the Transition. There must be a clear understanding of how responsibility for the facility is to transfer from the project manager to the operations manager. Often from a safety perspective, it must be clear when this handover takes place. This will happen during the commissioning process, which should be planned at a lower level of work breakdown than its fabrication.

Ensuring that the Users Accept the Product. I spoke in Chap. 4 about involving users in the decision-taking processes. That will win their acceptance of the specification of the asset. At the end of the project the users must be given the opportunity to agree the facility meets that specification. On a strict contractual relationship, the owner should sign completion certificates to accept the product. But when I worked in Imperial Chemical Industries (ICI), the operating works signed completion certificates even when the plant was built by internal resources.

Training the Users in the Operation of the Facility. The users will usually not be experts in the operation of the facility and so will require training in its use. This should be planned as part of the project. It is probably too late if it is not addressed until close out. However, it is in the transition stage that much of the training should take place. Training will be in the use of the facility, but may also include simple maintenance procedures. Training can be a significant proportion of the cost of a project. When converting a factory to robotic manufacture IBM spent 25 percent of the budget on training.

Ensuring a Definite Cutover. The planned transition and signed completion certificates should result in a definite cutover, at which responsibility is transferred, and final payments made (Example 14.2). It can be important from a safety point of view that there is clear ownership of the facility and that transfer of responsibility is precise.

I also believe that if the new asset is replacing an old system, there must be definitive cutover from use of the old system to use of the new, and this will probably take place as ownership is transferred from the project manager to operations manager. Some people say there should be parallel running as the new system proves itself, but if there is parallel running the new system will never prove itself as the users use the system they are familiar with. The new system should be properly tested, and that may include a period of parallel running on live data while under the control of the project manager. But when transfer occurs the old system should be decommissioned, cold turkey.

Example 14.2 Signing-off completion certificates

I conducted an audit in a company which had taken 18 months to complete a contract, but they had not obtained sign-off three years later. The client was always finding fault, and had effectively had three years free maintenance. At this point, the contractor threatened to switch the equipment off (they effectively still had ownership) and very quickly agreed a final snagging list and obtained sign-off.

Recording the As-Built Design. To ensure ongoing efficient and safe operation of the new asset, it is important the as-built design is recorded. This requires the incorporation of all design changes into the final configuration. This is part of the process of configuration management (Chap. 7). This is effectively now a legal requirement in Europe under a safety directive issued by the European Union. If an accident were to occur because the users were operating a design other than the one built, it would be viewed very seriously by the authorities.

Ensuring Continuing Service or Maintenance of the Facility. The users may be able to undertake simple service or maintenance and operating manuals may help them. However, it is usually ineffective, if not impossible, for them to become experts in the technology of the asset, and so it is necessary to ensure appropriate mechanisms are in place to provide backup. This requires channels of communication and logistics support between owner and contractor throughout the life of the asset. These channels should be defined as part of the handover.

14.3 EMBEDDING THE CHANGE AND OBTAINING BENEFIT

Many people view the project as over when the asset is handed over to the users. However, obtaining the benefit from the project is the final step in the control process. Who is responsible for this final control step depends on the circumstances: it may be the owner, a subordinate (the business change manager), or the project manager. But it should be agreed at the start as part of the project strategy. It will probably be a senior user working for the owner who is ultimately accountable if the sponsor and owner do not receive adequate return on their investment, and so the onus rests there to ensure it happens.

As we have repeatedly seen there are four steps in any control process:

- Create a measure of the desired performance improvement.
- Monitor achievement of that measure.
- Calculate variances between the measure and achievement.
- Take action to remove those variances.

Create a Measure. From the start there must be a clear definition of the project's purpose and the benefits expected from the change introduced and the operation of the new asset. The desired benefit should be defined during the start-up process (Chap. 12), refined during concept, feasibility, and design. It will be defined by the desired performance improvement (Chaps. 2 and 5). However, the desired performance improvement will not be achieved on the day the project ends and the new asset is commissioned (Chap. 2). It may take sometime for it to work through, as illustrated by the benefits map, Figure 2.6. The benefits map provides a measure not just of the ultimate benefit desired, but intermediate steps to the achievement of that, intermediate measures of performance improvement on the way to the desired end result.

Monitor Achievement. Following commissioning, the expected benefits are measured. That is not just the end benefit but the intermediate points as indicated in the benefits map. You might be expecting to take 9 to 12 months for the full benefit to work through, but it can be a risk waiting that time and finding nothing has happened. Thus the benefits map provides essential intermediate review points to measure progress towards the ultimate goal.

If the facility is a computer system, you will be checking to see whether it is delivering the expected returns. For example, if it is a manufacturing planning system, you will check to see whether the inventory is falling, the work in progress is falling, and the lead times are being reduced as predicted. If the facility is a new product, you will be checking to see whether the predicted levels of sales are being achieved. If it is a management training program, you will be checking to see if there is any noticeable improvement in management performance. (The last of these is the most difficult to measure.)

Calculate Variances. Determine the cause of any difference between the expected benefits and those obtained. The cause may be the users are not using the asset to its full capacity, either deliberately or inadvertently. The benefits map will show what should be happening; how the change should be implemented; and how the new asset will support the change. By measuring progress against the benefits map, you can ensure the change is happening as desired, that it is being fully embedded. It is unfortunately the case that people may resist the change, and implement it more in fiction than fact. The benefits map is the tool to track the embedding of the change and make sure it is truly happening.

Take Action. Hopefully a small amount of fine-tuning of the design of the new asset, or a small amount of additional training of users is all that is required to achieve the actual benefit. Projects involve considerable risk because they are novel and unique and so it is quite likely that the design carried some small imperfections which can be easily corrected. In the worst case, improvement will require another project. (That was why the problem-solving cycle, Fig. 1.6, was drawn as a circle.)

14.4 DISBANDING THE TEAM

Over Secs. 14.1 through 14.3, I concentrated on work-related and strategic issues of close out. However, the team members may face the end with mixed feelings. The cycle of team formation and maintenance (Secs. 4.4 and 12.1) suggests that at the end the team may go into mourning with their performance dropping. It can rise as the team look forward to new opportunities or drop if they face unemployment. It is the task of both line and project managers to manage this emotional response so staff is retained and reintroduced into the normal work environment. When considering the motivation of project staff, it must be remembered they belong permanently to the organization and only temporarily to the project. This means that while the project may not need the staff once their contribution has finished, the organization for which they work may value the team members even more because of new skills they learned and may want them for future projects. Retaining project team members is vital and so the process of disbanding the project team must be managed in a caring way. Key elements in this process are discussed in following sections.

Planning the Rundown of Resources in Advance. This, as explained above, is important to achieve an efficient end to the project. It is also important for the motivation of the project team. People feel more motivated to complete their work if they know they are to be transferred immediately to new work. That is only possible if their release has been planned, so that their line managers have been able to plan their future work.

Returning Resources Promptly to Their Line Managers. The organization gets the optimum use of its resources if they are returned promptly to normal duties after completion of the project. Line managers of people seconded to the project are more likely to treat future requests for resources favourably if those people are used efficiently, which means releasing them at the earliest possible opportunity.

Debriefing Meeting. As well as being a learning opportunity for the organization, a project close out meeting can be as important as a launch meeting, as part of the life cycle of the project team. It marks the end of the period of working together, and allows people to show their grief, frustration, or pleasure at having been a member of the project (Example 14.3).

Example 14.3 Releasing frustration at debriefing meetings

I worked on the overhaul of ammonia plants in the early 1980s. We held a debriefing meeting after each overhaul. They served a useful purpose of allowing us to let off steam in advance of the next overhaul. For the four weeks of each overhaul, we used to suspend our feelings, to allow work to progress. We would talk to each other bluntly about what work we wanted done, what it would take, and how we felt about having been let down. It was necessary to make progress in the intensity of the overhaul. In the process, feathers got ruffled, but we had to bite our tongues and get on with it. At the debriefing meeting, it all came out; we said all the things we had bottled up for four weeks. It was all laid bare, forgiven, and we were ready to start afresh on the next overhaul.

End-of-Project Party. The use of "festivals" is an important motivator on projects. They should be used to mark important project milestones, especially the end of the project. In Germany, during the building of a house, festivals are important, especially the completion of the roof. In shipbuilding the launch is a major festival, partway through the project. I worked recently with NASA for whom festivals are very important: the launch, the landing on Mars, the gathering of first scientific data. You should do the same for your projects. Sometimes you can combine the end-of-project party with the debriefing meeting (see Example 14.4). The difficulty is choosing the timing of a party so the maximum number of people can come before being dispersed to new jobs but when they actually have something to celebrate.

Example 14.4 Combining the debriefing meeting with the end-of-project party

I did a series of three start-up workshops with a client, three in total at nine month intervals. Nine months after the last, the project manager of the second rang to say she and the project sponsor wanted to hold a close-out meeting. They wanted meeting to last three days. On the first day (Monday) they wanted to debrief the entire project leadership team, which included the sponsor, the project manager, her two lieutenants, two more people from my client organization and four contractors. Then on Tuesday they wanted to further debrief the project manager and her two lieutenants. On Wednesday they wanted a mini-start-up meeting for another project. The close-out meeting was arranged, and the manager told me it was to be held in an extremely expensive hotel. We were holding the meeting in January, so the rooms were *only* \$300 a night, not the full summer rate of \$700. We started the meeting with dinner on Sunday night. At the dinner, the project sponsor was opening champagne at \$200 a bottle. Finally, I could take no more, so I asked him, "Martin, what's going on? A very expensive hotel, good dinner, and champagne at \$200 a bottle?" He said the project had been budgeted to cost \$18 million and ended up costing just \$16 million and he couldn't give the money back. But the cost of the party was a small amount to pay to say thank you to the team in the overall scheme of things.

Rewarding Achievement. The team members are likely to react favourably to future requests to work for the project manager if their contribution is suitably appraised and rewarded. End-of-project festivals are part of that. However, it is equally important that a person's achievement is recognised by those who matter, especially the manager who is to write the individual's annual appraisal, and determine their annual bonus, so that the person's contribution to the business can be recognised. An important part of this process is winning the appraising manager's commitment to the project, so that they view a contribution to the project as an important achievement during the year. Anne Keegan, Martina Huemann, and I discuss the need to conduct project appraisals, and the link between project appraisals and appraisals in the line.[2]

Disciplining under Achievement. It is also important to discipline poor performance on a project so that good performers do not feel their effort was in vain and that the poor performers know how to improve in the future. For this latter reason the disciplining process should be treated positively, guiding people how to perform better in the future. Of course, it might be possible to take corrective action during the project, so the earlier this is done the better.

Career Review. The fifth stage of team development is mourning, as the project fades into history and the team with it. This is not very good for the self-esteem of the team members who find that overnight they are reduced to "has-beens" unless they go immediately to another project or line job. The end of a long project can be a useful point for all staff to have an interview with their line manager or personnel to review their current position:

- It offers a chance for the individual to review career objectives.
- It offers scope for skills consolidation in the form of theoretical training to supplement the practical experience.
- It shows caring by the organization, which is perhaps the key factor in the whole exercise.

Recalling the case in Example 6.2, perhaps the individuals should have had such an interview well before the end of the two-year period, planning their reentry into the line organization. The individuals could then have taken responsibility for their own career development and perhaps have found opportunities for themselves within the organization where their new skills would have been of great value.

14.5 POSTCOMPLETION REVIEWS

Finally, it is worthwhile to gather data about the project performance for a number of

- To record the as-built design (final configuration)
- To obtain a comparison of final costs and benefits for feeding back to the estimating and the selection of future projects
- To record the technical achievement on the project for feeding back to the design and selection of future projects
- To review the successes and failures of the project and the lessons learned for feeding back to the management of future projects

There are several ways of reviewing the success and failures of projects, two of which include debriefing meetings and postcompletion audits.

Debriefing Meetings. I have already described the role of these in disbanding the project team. It is worthwhile on most projects to hold a meeting of all people who attended the project launch workshop to review the assumptions made. This meeting may last from two hours to a day depending on the size of the project. On particularly large projects they may amalgamate up from a low level, reversing the cascade of project launch workshops.

Postcompletion Audits. On large projects, it may also be worthwhile to conduct a postcompletion audit. This is a formal review of the project against a checklist. An audit is often conducted by external consultants. It is also common only to audit projects which have gone radically wrong. However, better lessons are often learned from successes, so it can be useful to audit projects which have gone well. I describe the holding of audits more fully in Chap. 18.

SUMMARY

1. The key requirements for effective project close out are
   - Finishing the work in a timely and efficient manner
   - Transferring the product to the users
   - Obtaining the benefits
   - Disbanding the team
   - Reviewing progress
2. The work must be finished in a timely, efficient manner. The following can aid this:
   - Checklists of outstanding work
   - Planning and controlling at lower levels of work breakdown
   - More frequent control meetings
   - Planned rundown of the project team
   - Use of task forces
   - Changing the project manager
   - Closing contracts with suppliers
3. Effective transfer of the product to the users is facilitated by:
   - Planning the transition
   - Ensuring user acceptance
   - Training the users
   - Obtaining definite cutover
   - Recording the as-built design
   - Ensuring maintenance of the facility
4. The facility must be commissioned to obtain the required benefit, and this can be controlled by:
   - Defining the desired benefit and desired performance improvement and by drawing a benefits map
   - Monitoring performance improvement and tracking progress against the benefits map
   - Identifying shortfalls
   - Taking action to overcome the shortfalls
5. The project team must be disbanded in an efficient manner, and yet in a way that takes care of their motivational needs. This can be achieved by:
   - Planning the rundown
   - Returning resources promptly to line managers
   - Holding a debriefing meeting
   - Holding an end-of-project party
   - Rewarding achievement

- Disciplining under achievement
- Counselling staff

6. Postcompletion reviews must be held to:
- Record the as-built design
- Compare achievement to plan
- Record technical data
- Learn successes and failures for the future

REFERENCES

1. Frame, J.D., *Managing Projects in Organizations*, 3d ed., San Francisco: Jossey-Bass, 2003.
2. Turner, J.R., Huemann, M., and Keegan, A.E., *Human Resource Management in the Project-Oriented Organization,* Newtown Square, PA.: Project Management Institute, 2008.

P • A • R • T • 4

GOVERNANCE OF PROJECT-BASED MANAGEMENT

CHAPTER 15
PROJECT GOVERNANCE

So far I have focused on the individual. I showed that projects are the means by which organizations introduce change to achieve performance improvement, and described how to manage scope, project organization, quality, cost, time, risk, and the project process to deliver the identified change and achieve the desired performance improvement. Now my attention switches to the support the parent organization can give to the project to facilitate its management and increase its chance of success. In the first and second edition of this book, I presented these as project management procedures and administrative techniques the parent organization can create to support projects. I now see this as part of a wider issue of governance in the project-based organization.

In this chapter, I give an introduction to my model of governance in the project-based organization describing the structures and the roles they imply. I then describe the principal-agency relationship between the project sponsor and project manager, and what that suggests about the communication between them.

In the remaining chapters of this part I describe other governance issues. In Chap. 16, I describe program and project portfolio management. This is governance of the context of the organization, and linking project objectives to corporate strategic objectives to ensure the right projects are done. Then I describe how organizations can develop their project management capability. This is about ensuring that the organizations have the skills to do projects right. In Chap. 18, I describe how the company's board should and can take an interest in projects being undertaken in the organization. Under modern corporate governance regimes, such as those imposed by the Sarbanes-Oxley Act, boards of directors have responsibilities to their shareholders that make it essential for them to be able to forecast the outturn costs and expected revenue benefits of major projects being undertaken by, and so it is essential that boards of directors should take an interest in projects being undertaken by the organization. I describe an organizational governance model, including the use of end-of-stage reviews and auditing. Finally, in this part, I describe the management of international projects. This is not strictly a governance issue apart from the fact that organizations doing international projects should ensure they are done right.

15.1 GOVERNANCE

Clarke defines corporate governance using a definition developed by the Organization for Economic Coopertaive Development, OECD:[1]

> Corporate governance involves a set of relationships between a company's management, its Board (or management team), its shareholders, and other stakeholders.
>
> It provides the structure through which the objectives of the company are set, and the means of attaining those objectives and monitoring performance are determined.

There are two elements to this definition. The second element says what governance is about. It is about defining the objectives of the organization, defining the means of obtaining those objectives, and then monitoring progress to ensure they are achieved. At the organizational level, this means the governance committee is responsible for ensuring the structures exist to define the objectives for routine operations and projects, that is, the top levels in the cascades shown in Figs. 1.9 and 2.8. They are also responsible for ensuring that appropriate structures exist for delivering the objectives and ensuring that progress is tracked towards their achievement. The governance committee does this on behalf of the stakeholders of the corporation, and that is what the first part of the definition deals with. Clarke suggests there are two schools of governance:[1]

1. One school says the sole responsibility of the governance committee of the corporation (the company's board) is to maximize returns to shareholders. But even then there is a tension between achieving quick profits in the short term and growth in the long term.
2. The other school says the board must govern the company on behalf of a wider set of stakeholders, which includes the shareholders, but also includes staff, customers, and the local community.

Within the project-based organization there are three levels of governance:

1. First there is the level at which the board operates and the extent that they take an interest in projects. Under modern governance regimes, boards of directors should take a much greater interest in projects being undertaken in the business than they have in the past. I discuss this level in Chap. 18.
2. There is the context within which projects take place. Part of creating the means of achieving the objectives in the project-based organization is to ensure the organizational infrastructure exists to undertake projects effectively. There are two components of this. The first is creating an infrastructure of program and portfolio management to link projects to corporate strategy. This is ensuring the right projects are done (Chap. 16). The second is ensuring the capability exists within the organization to deliver projects successfully so that projects are done right (Chap. 17).
3. Finally, there is the level of the individual project. The project itself is a temporary organization and therefore needs governing, and so under the principle of fractal management, governance structures should exist at the level of the individual project. This is the topic of the rest of this chapter.

15.2 GOVERNANCE OF THE PROJECT

In the other chapters of this part, I describe the governance support the parent organization gives to the project, and what that means in terms of various administrative routines the parent organization implements. For the remainder of this chapter, I describe the governance of the individual project. The project is a temporary organization and so it too needs governance. With a little adaptation the definition of governance in the previous section can be applied to a project:

> The governance of a project involves a set of relationships between the project's management, its sponsor (or executive board), its owner, and other stakeholders.
>
> It provides the structure through which the objectives of the project are set, and the means of attaining those objectives and monitoring performance are determined.

As with the company, the project can be governed to maximise the returns to the owner only, or to all the stakeholders of the project. But with a project we find it is more important that all the stakeholders should feel they are winners. Ralf Müller[2] and I have identified a necessary condition of project success is that all the project participants should view the project as a partnership, within which their objectives are aligned, and which is managed to achieve the best result for all. I discussed this in Chap. 3 when I said you should aim to balance the objectives of as many stakeholders as possible. Somehow with projects, they are more prone to failure if some of the stakeholders are not committed to the project's outcomes. Projects are coupled systems (Sec. 1.2) and so you must work to optimize the whole project and not individual elements of it. With a routine organization you ought to take care of your customers and staff, but it seems that it is possible to give primary focus to your shareholders. Not so on projects; it is essential to take care of all the stakeholders. I return to the principal-agency relationship later in the chapter. For now I wish to discuss the governance structures and roles on projects.

Figure 15.1 illustrates the governance structures and roles on projects. The inner loop shows the three steps of governance introduced above:

1. Define the objectives.
2. Define the means achieving the objectives.
3. Define the means of monitoring progress.

As illustrated by Fig. 1.9, the means at one level of breakdown helps to define the objectives at the next level. So we start with the need for performance improvement, and to solve the problems (or exploit the opportunities) that are stopping us from (or would enable us to) achieving it. That is the client need. It is shown on the right-hand side of the benefits map (Fig. 2.5) and is the goal at the top level of Fig. 1.1. From the goal we define the desired outcome to be obtained from operating the change (the new asset) introduced by the project. These are new capabilities that will help us solve the problems or exploit the opportunities, either directly or indirectly as illustrated by the benefits map. The outcome is the means of achieving the goal but is the objective at the next level of the project, the middle level in Fig. 1.1. From the outcome we define the change that will deliver it, that is, the new asset, the extreme left-hand side of the benefits map (Fig. 2.5). This is the project output or

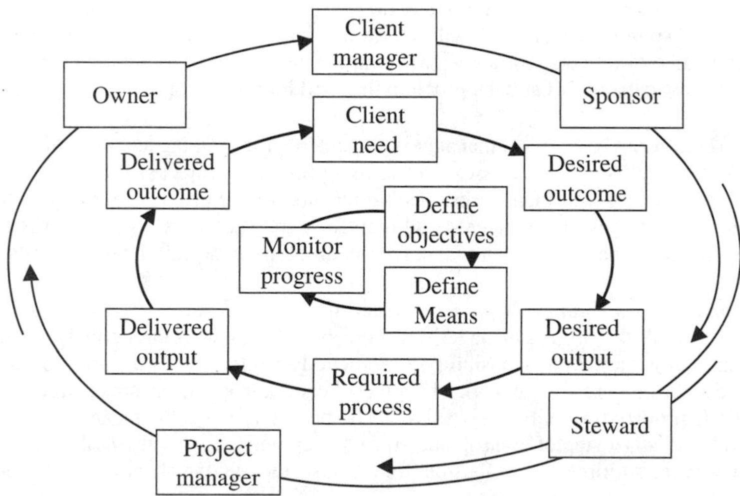

FIGURE 15.1 Governance structures and roles.

deliverables. The output is the means to achieving the outcome but is the objectives of the bottom level in Fig. 1.1. In Sec. 2.3, I suggested that when drawing the benefits map a certain amount of iteration may be required where you redefine the problems or opportunities in the light of the new capabilities that can be delivered by the asset, but you may also redefine the asset and the new capabilities to solve the problems you actually have. Don't deliver the industry-standard computer system if your needs are slightly different but also recognize that the more changes you make, the more likely it is to go wrong and it may be better to redefine your problems and final goal to achieve something that may work better.

Having defined the required project output, you can define the project process to deliver it using the approaches described in this book. As you undertake the project you want to monitor progress to ensure you deliver the desired output at a time, cost, and level of performance that enables you to repay your investment. As you deliver the output you check it will deliver the desired outcome, you use the benefits map to ensure that the outcome is used to solve the problem or exploit the opportunity to realize your goals.

Figure 15.1 also illustrates that there are at least four project governance roles involved. (I don't count the client manager as a project governance role.)

Sponsor. The sponsor is somebody from the client or user department who identifies the need for performance improvement, and there is a change that can be made that will deliver that performance improvement in a cost-effective way (the benefit will justify the cost). He or she begins to identify the possible new asset and capabilities (outputs and outcomes) that will solve the problem or exploit the opportunity, and begins to draw the benefits map. The sponsor approves the definition of the objectives (goals, outcomes, and outputs) on behalf of the user or client organization and approves the statement of requirements.

He or she also becomes the ambassador for the project, persuading the organization that the project is a good idea and trying to win resources in the form of money and people. He or she may be the holder of the budget that pays for the project or be a first report to that person. The ambassadorial role should continue throughout the project winning resources at the start and maintaining resources throughout the project.

Steward. However the sponsor will not be a technical expert and so will not be able to finalize the definition of the new asset and new capabilities on his or her own. He or she will need to involve a senior manager from the technical department to help design the new asset and define the capabilities it can deliver through its operation. I call this role the *steward.* The sponsor and steward will work together to finalize the benefits map and go through the iteration I spoke of above, revising the benefits map to obtain the best definition of the new capabilities and the problem that will be solved (or opportunity exploited).

Project Manager. The project manager is responsible for defining and managing the project process to deliver the new asset, and defining how the project will be monitored and controlled. He or she will then be responsible for monitoring progress during project execution to ensure the asset is delivered and is fit for purpose, that is, to ensure it is capable of delivering new capabilities that will solve the problem or exploit the opportunity.

Owner or Business Change Manager. I said in Sec. 2.3, when discussing the benefits map, that I think the project manager's responsibility ends with the delivery of the new capabilities. It is then the responsibility of somebody from the user department to ensure the new capabilities are used to work through the benefits map and ensure change is embedded and the problems are ultimately solved. The benefits map is the means of monitoring progress for this last step of control, and so each step needs to be measurable. The owner of the new asset is ultimately responsible for this last step of control, but he or she may delegate it to a business change manager.

Filling the Roles. The sponsor is essentially a preproject role defining the objectives (goals, outcomes, and outputs). This role also defines, through the benefits map, how the change will be embedded and performance monitored postproject. The owner and business change manager are postproject roles, responsible for embedding the change and achieving the performance improvement postproject. The sponsor and owner may be the same person, but do not need to be. PRNCE2 suggests you need two roles,[3] and calls the sponsor the project executive and the business change manager the senior user. Likewise the steward is a preproject role defining the objectives (outputs and outcomes), and the project manager is an intraproject role defining the means of obtaining the objectives and monitoring progress (through the project plan). The project manager and steward may be the same person, but don't need to be. On large, stand-alone projects, they are more likely to be the same person, but the steward may just be a senior manager from the technical or projects department. On small projects the steward tends to be the program or portfolio manager (Chap. 16). PRINCE2 calls the steward the senior user and the project manager just that.

I believe these four roles are essential. Particularly you need to know who is responsible for championing the project preproject and who is responsible for embedding the change and obtaining the desired benefit postproject (and that *is not* the project manager). Further, on anything but the simplest, smallest projects, there must be at least two people. The sponsor and owner should be from the user department and the project manager and steward from the technical or projects department. The sponsor must be an optimist and shoot for the moon; the steward must be a pessimist and bring the sponsor down to earth. If they are both optimists they will strive for the impossible; if they are both pessimists they won't strive to achieve what can be achieved. Tension between them is a good thing as long as it doesn't spill over into conflict.

15.3 THE PRINCIPAL-AGENT RELATIONSHIP

Ralf Müller and I[2] demonstrated three necessary conditions for project success (Sec. 3.3):

1. The project participants, especially the project manager and sponsor, should work together in partnership to achieve mutually consistent objectives.
2. The project manager should be empowered but not given total licence. The sponsor should set parameters within which the project manager should work, but the project manager must be given flexibility to enable him or her to respond to risk. If the sponsor imposes too much structure, the project manager has no flexibility to deal with the unknown; if the sponsor imposes too little, the manager has no guidance.
3. The sponsor should take an interest in progress.

The first two of these are illustrated in Fig. 15.2. Ralf Müller and I found in our sample that successful projects were clustered in the area of high cooperation and medium structure. However to operate there requires significant trust between the project manager and sponsor. We found the lowest success in the area of low cooperation and low structure. That was the worst position to operate the project. We found the lowest predictability in the area of low cooperation and high structure. In that quadrant, the project outcome was quite variable but never as successful as the area of high cooperation and medium structure. But it is in the area of low cooperation and high structure that many projects take place. The sponsor doesn't trust the project manager so adopts confrontational behaviours and imposes strict rules on the project manager's behaviour.

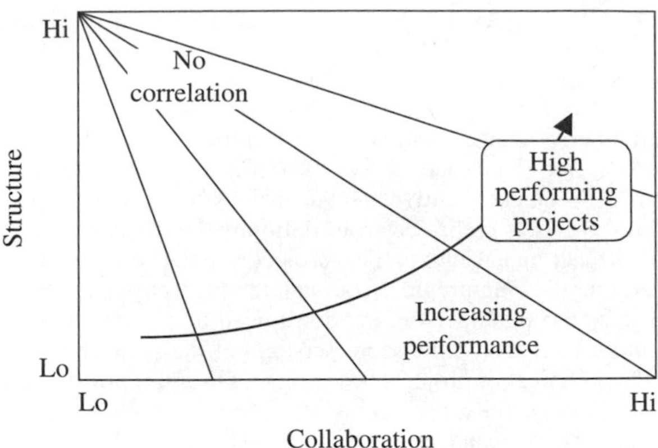

FIGURE 15.2 Cooperation and structure on projects.

Most people, when they spend any time thinking about it, recognize that it is good practice to treat the project as a partnership, and to work to mutually consistent goals. They recognize that the best way of achieving a successful outcome is for all the project participants to gain from that outcome. If the project outcome is going to be detrimental to a given party, then you can expect that either they will be working for project failure, or they will be trying to change the project outcomes to be more beneficial for them and so may push the project in unintended directions. If you think about it, that is obvious, and so why is it that people adopt uncooperative behaviours on projects? The principal-agent relationship provides an answer.

Michael Jensen[4] says that a principal-agent relationship exists if one party (the principal) depends on another (the agent) to undertake an action on their behalf. This is clearly the relationship between the sponsor (principal) and project manager (agent). Associated with this relationship are two problems called the *adverse selection problem* and the *moral hazard problem*, which lead to the lack of trust between the project manager and sponsor.

The Adverse Selection Problem. This problem receives its name from the fact that the principal (sponsor) has to choose the agent (project manager) to act on his or her behalf and has to do that on inadequate information. But then having appointed the agent, the principal cannot know for certain why they are taking the decisions they do, and whether the agent is acting in their (the principal's) best interest.

The Moral Hazard Problem. This problem arises because economic theory assumes that the rational human being will act rationally in any situation to maximize his or her beneficial outcome from the situation. That means the project manager will be taking decisions on the project to maximize his or her outcome and will only maximize the sponsor's outcomes en passant if their two sets of outcomes happen to be aligned. This is what I said above and is the need for the partnership. So if you are a sponsor on a project, don't expect the project manager to be working in your best interest if you have imposed a contract that will cause him or her to make a loss. He or she will cut corners or manufacture variations to turn his or her loss into a profit. Choose a form of contract that motivates the project manager to achieve your objectives.

In extremis, the moral hazard problem becomes opportunistic or even unethical behaviour. The project manager tries to make a profit wholly at the sponsors' expense. Usually, a desire for a long-term relationship with the customer or reputation within the industry stops the project manager doing this. His or her profit is maximized not just from this project, but over several projects, and so he or she behaves with propriety. But very occasionally it is truly just a once off relationship and the project manager may work to make maximum profit from this one job. It is a fear of that behaviour which causes the sponsor to adopt low cooperation and impose high structure. In the next section, I show how communication from the project manager can reduce that problem.

Bounded Rationaility. Often the project manager would like to work in the sponsor's best interests, but in fact it is human frailty that prevents him or her from doing so. Economic theory labels this "bounded rationality"[5] and it is caused by three elements of human frailty:

1. Inability to gather all the information relevant to the decision
2. Inability to fully process that information which is gathered
3. Inability to foretell the future and so flawlessly predict all the risks

The project manager ends up doing the best they can with the information he or she has, which is known as *satisficing*.

Agency or Transaction Costs. Against this background, the sponsor starts to impose structures on the project which create additional costs, which are over and above the cost of doing the work of the project. These are known as *agency costs*[4] or *transaction costs*.[6] Michael Jensen identifies four agency costs:

1. The *cost of forming and managing the contract* and the contractual relationship between the principal and the agent.
2. The *cost of communication* between the principal and the agent, and of reporting progress and controlling the work.
3. *Bonding costs* are things the agent does to win the principal's trust and support.
4. *Residual losses* arise because the project's outcomes are not exactly what the principal needs.

An example of a bonding cost is the agent's membership of professional bodies. Such membership gives the principal trust in both their competence and ethical behaviour. The agent had to prove their competencc to get professional membership and needs to behave ethically to maintain it. This is why project managers want membership of organizations such as Project Management Institute (PMI). Other bonding costs are gifts and invitations to sporting events. Because the agent has to make a profit, the principal has to ultimately pay for all of these things through increased project costs. Residual loss occurs either because the project manager is acting in his or her own best interest and not the principal's or because of bounded rationality, or both. The new asset does not work exactly as the principal requires and they therefore fail to get the full benefit from the project either because the project manager cut corners to maximize their profit or because the project manager didn't fully and perfectly understand what was required. This leads on to the need for good communication.

15.4 COMMUNICATION BETWEEN THE PROJECT MANAGER AND SPONSOR

Good communication is needed both to build trust between the owner and sponsor and reduce bounded rationality. I consider first what communication the project manager should give to the sponsor to help build his or her comfort and trust, and then what communication the sponsor should give the project manager to help him or her better understand the requirements and reduce bounded rationality.

Project Manager to Sponsor

I consider the content of communication, its frequency and form.

Content. The sponsor's lack of comfort is caused by the adverse selection problem and the moral hazard problem; he or she doesn't know for certain what decisions the project manager is taking or why, nor whether the project manager is taking the best decisions to maximize his or her (the sponsor's) outcomes. To help build the comfort of the sponsor, the project manager needs to give him or her information that will convince him or her that:

1. The new asset will function and be fit for purpose, that is, it will perform the desired new capabilities.
2. The right process has been adopted to deliver them in the optimum way.
3. It will be delivered within time, cost, and performance targets.
4. The project manager is behaving in a professional and trustworthy manner.
5. That appropriate controls are in place to achieve all of the above.

What data or information should the project manager give the sponsor to achieve the above? Bob Graham[7] suggests that when designing an information system you shouldn't ask what data you need, but ask instead what questions you need answering. The questions the sponsor needs answering are

Questions of product: Will the new asset be fit for purpose and perform to deliver the new capabilities?

Questions of process: Has the right process been adopted to deliver it in the optimum way, within constraints of time, cost, and quality?

Questions of surprise avoidance: Is the project manager behaving in a professional and trustworthy way, and are there any nasty surprises in the form of risks or other issues that are going to prevent any of the above?

With this communication there are two types of sponsor: those who take an interest in progress and those who don't. Ralf Müller and I found that when the project sponsor takes an interest in progress the project usually turns out well, and when they don't the project turns out badly. This is the third necessary condition for success above: the sponsor should take an interest in progress. Further, when the sponsor takes an interest in progress, he or she thinks the project is performing less well than it actually is, whereas if he or she doesn't take an interest in progress, he or she thinks the project is performing better than it is. Finally when the sponsor takes an interest in progress he or she usually wants more information than the project manager is willing to give. There is a tension here and you need to achieve a balance between two agency or transaction costs. Providing too much information takes time and effort and so costs money, but not providing enough leads to the project failing to perform as well as it should, and so contributes to residual loss. So how much information is right and how should it be delivered?

Timing and Means of Communication

How often should the communication be given and by what means? As we have seen there is a tension between the sponsor wanting too much and the project manager wanting to give too little. There are two methods of determining the timing of communication: by calendar time or by project events or milestones. I used to say that for the project manager to report progress to the sponsor once every 6 weeks on a project lasting 9 to 18 months was enough. However, Ralph Müller and I found that sponsors who took an interest would ideally like written performance reports once a week, but recognized that was too frequent from the point of view of the cost of producing the reports, and so were willing to compromise on once every two weeks. Thus the main written progress report from the project manager to sponsor should be made once every two weeks. This is calendar driven. That is not to say that the project manager will not also make a formal written report on the completion of a project milestone or project stage. That is event driven. But the main written progress report should be calendar driven and made once every two weeks on a typical project.

Ralph Müller and I also found that clients or sponsors are fairly schizophrenic about the reports. They usually trusted the written progress reports to give an accurate and realistic representation of project progress, but they didn't trust the written reports to give a true picture of risks and issues. Sponsors wanted a face to face meeting with the project manager to get a feeling of risks and issues. But they didn't trust the face to face meetings to give a true picture of project progress; they wanted the written reports for that. So clients in fact want two forms of communication from their project managers:

1. Written reports, delivered once every two weeks, reporting project performance data: time, cost, and functionality
2. Verbal reports, delivered once a week, reporting on risks and issues

And project managers want feedback on those reports to show that the sponsor cares, trusts them, and approves the process being adopted.

Sponsor to Project Manager

The project manager's need for information changes throughout the life cycle.

Start-Up. At start-up project manager (PM) needs to know the vision, mission, and purpose of the project, and the requirements in terms of the definition of the new asset and capabilities. The benefits map shows how the capabilities will be used and the problems they are intended to solve. If the PM can understand those things, it will help the PM deliver a system that is fit for purpose. If, for instance, you tell the PM that you require a customer requirements management system and nothing more, don't blame the PM when he or she delivers an industry standard system that doesn't solve the actual problems you have. You need to show the PM the benefits map and explain the new capabilities actually required and how they will be used to solve the problems you actually have. It is not the PM's responsibility to define the problems and determine the new capabilities; it is the sponsor's and steward's. So it is the role of the sponsor, with the help of the steward, to properly inform the PM, so the system as delivered is fit for purpose.

Implementation. During implementation the PM's information needs become more prosaic. The PM wants approval from the sponsor for the project process adopted to deliver the project's objectives. The sponsor is overall responsible for project governance, and so must approve the adopted means of delivering the objectives and of monitoring progress, and let the PM be aware of that approval. The PM also wants to know that senior management cares, and perhaps the knowledge that senior management cares helps achieve a successful outcome. The PM also wants to know that he or she actually has the trust of senior management, and what their flexibility is. As I said above, it is important to give the PM flexibility so the PM can respond to risk, but the PM needs to know what that flexibility is.

SUMMARY

1. Governance defines:
   - The objectives of the organization
   - The means of obtaining the objectives
   - The means of monitoring progress
2. There are two schools of governance:
   - The shareholder school, which says that the directors' sole responsibility is to maximize returns to shareholders.
   - The stakeholder school, which says they have responsibilities to other stakeholders as well including staff, customers, and the local community
3. In the project-oriented organization, there are three levels of governance:
   - Where the board's responsibility impacts on projects (Chap. 18).
   - Creating a context in which projects can thrive, including program and portfolio management (Chap. 16) and creating enterprise-wide project management capability (Chap. 17).
   - At the level of the individual project (this chapter).
4. The project is a temporary organization and so needs governance
5. There are four roles of governance of the project:
   - The sponsor who identifies the need for the project, defines the objectives and the means of embedding the change through the benefits map, and is ambassador for the project, winning resources for the project.
   - The steward who makes a technical input to the definition of the objectives bringing a pragmatic view.
   - The project manager who defines the project progress and is responsible for implementing the project and monitoring progress to delivery of the project's outputs.
   - The owner or business change manger who is responsible for embedding the change and ensuring the project's outputs and outcomes are used to achieve the desired performance improvement.
6. The sponsor and project manager are in a principal-agent relationship, which means their relationship is subject to:
   - The adverse selection problem
   - The moral hazard problem
7. The project manager may suffer bounded rationality which means he or she would like to precisely meet the client's requirements but can't because he or she:
   - Does not have all the information required
   - Cannot fully process that information he or she does have
   - Cannot foretell the future and so cannot predict all risks
8. The principal-agent relationship creates costs over and above the cost of works including:
   - The cost of forming and managing the contractual relationship
   - The cost of communication between project manager and sponsor, and vice versa
   - Bonding costs where the project manager binds the sponsor into the relationship
   - Residual loss because the new asset does not perform precisely as required
9. To be comfortable with project progress the sponsor wants answers to questions of:
   - Product: will the new asset be fit for purpose?
   - Process: has the optimum process been adopted to deliver it?
   - Surprise avoidance: are there any lurking issues and is the project manager professional and trustworthy?

10. The project manager needs to know:
- As much information as possible to avoid bounded rationality
- The client has approved the process and progress
- The client cares

11. The project manager needs to make two types of reports, both calendar driven:
- Written reports reporting project progress and performance data once every two weeks
- Verbal reports on risks and issues once a week

REFERENCES

1. Clarke, T. (ed), 2004, *Theories of Corporate Governance: The Philosophical Foundations of Corporate Governance*, London: Routledge, 2004.
2. Turner, J.R. and Müller, R., "Communication and cooperation on projects between the project owner as principal and the project manager as agent," *The European Management Journal*, 22(3), 327–336, 2004.
3. Office of Government Commerce, *Managing Successful Projects with PRINCE2*, 4th ed., London: The Stationery Office, 2005.
4. Jensen, M.C., *A Theory of the Firm: Governance, Residual Claims, and Organizational Forms*, Boston, M.A.: Harvard University Press, 2000.
5. Simon. H., *Models of Man*, New York: John Wiley & Sons, 1957.
6. Williamson, O.E., *The Mechanisms of Governance*, New York: Oxford University Press, 1996.
7. Graham, R.G., "Managing conflict, persuasion and negotiation," in Turner, J.R. (ed), *People in Project Management*, Aldershot, U.K.: Gower, 2003.

CHAPTER 16
PROGRAM AND PORTFOLIO MANAGEMENT

Up to now I have spoken very much about the project in isolation. However, the reality is that the vast majority of projects take place as one of a group of projects, either a program or a portfolio of projects. The traditional project management assumption is of the large, isolated project with a dedicated team in which:

- They deliver well-defined, independent objectives, which provide the full benefit on their own.
- They are relatively independent of other projects and operations with few minor interfaces.
- They have a dedicated team, wholly within the control of the project manager; the manager may desire a larger team, but he or she sets the priorities for the team's work day-by-day.

In the construction of a building, a fence is put around the construction site. The project will not be dependent on other projects, the only interface with the outside world being the connection of services across the boundary. People working on the construction site will be managed by the project manager, wholly within his or her control.

All the early books on project management werc written about this type of project but in my view it represents less than 10 percent of all project activity. The majority of projects take place as part of a program or portfolio of small- to medium-sized projects (SMPs), in which:

- They deliver mutually interdependent objectives where the full benefit is obtained only when several projects have been completed (Examples 1.3 and 16.1).
- They are dependent on other projects or operations for elements essential to their completion, such as data, new technologies, or raw materials.
- They borrow resources from a central pool and the resources remain within the control of the resource managers; the manager must negotiate release of the resources to the project and may loose them at little or no notice as the organization's overall priorities change.

Example 16.1 Related projects

A borough council I worked with was building a new shopping centre, sports complex, and car park together with new road access and new services. This was broken into five projects, which now could not be totally ring fenced. The road had a separate interface with the car park, the shopping centre, and the sports complex and those were linked to the services. Furthermore, the full benefit would not be obtained from the shopping centre and the sports complex until the link road and car park were completed. The sports centre and shopping complex would also be used by the same people so would generate benefit for each other.

I have used the terms small- to medium-sized, large, and major projects. It is common to categorise projects in this way. However, there is little agreement about what these mean and there is a wide difference between industries. What constitutes a large information systems project may be considered small in the construction industry. I saw an advertisement for a course which claimed to be about managing "mega" projects and went on to classify that as projects over $2 million!!! In the process plant industry that is miniscule. For a given organization, I define a large project as one which costs around 10 percent of company's turnover, a medium-sized one is one-tenth of that, and a small one is one-tenth of that. A large project usually justifies a dedicated team; small- to medium-sized projects share resources with other projects. A major project is 10 times bigger than a large project and the risk is such that no one organization can undertake the project on its own.

In this chapter I describe the management of projects as part of a group of projects. There are two types of project grouping, those that share common resources called a *portfolio* and those which contribute to a common or shared objective called a *program.* In the next section I define these more carefully and discuss their governance role. In managing a portfolio of projects, which share common resources, there are three issues: prioritizing projects to receive resources, sharing resources between those projects you have decided to do, and managing interfaces between the projects. In managing a program of projects which contribute to a common objective, the two issues are deciding which projects to do, and linking the program and project governance structures. Finally in this chapter, I describe different types of project office and their different roles.

This is an area that has changed considerably since I wrote the first edition of this book. As little as 10 years ago, people did not distinguish between programs and portfolios.

16.1 DEFINITIONS

I try to keep the definitions simple. Some people seem to engage in quite tortuous definitions. I don't think that is necessary. Simple definitions capture the key feature of what we are discussing.

Programs

A program of projects is a group of projects which contribute to a common, higher order objective. The parent organization has a change objective which may require contributions from several different areas, or several different types of project for its achievement. For instance in Example 1.3 to develop a palm nut oil industry, the Malaysian government needs to develop new plantations, transport infrastructure, factories, and sales outlets. In Example 16.1, the borough council is developing a shopping and leisure facility with shops, sports facilities, car parking, and road access.

Somewhat against the trend, I personally think there is little difference between program and project management. Yes there is a big difference between managing a $50,000 project and a $50 million program, but not any greater difference in my view than between managing a $50,000 project and a $50 million project. The problem is the word "project" is used to describe a wide range of endeavours, from $50,000 to $5 billion. I coined the word "projette" to describe anything under about half a million dollars. So when people say project management and program management are different they are comparing the $50 million program with the $50,000 project because that is the context they are talking about. They are not comparing the $50 million program with the $50 million project.

However, there are some subtle differences between projects and programs. The main one is with the nature of the objectives. Projects tend to have what are labelled SMART objectives: specific, measurable, achievable, realistic, and timelined. The most famous example is when in 1963 President Kennedy set the target of sending a man to the moon and bringing him back safely by the end of the decade. It was specific, measurable, and timelined (respectively). Whether in 1963 it was realistic and achievable is a bit open to question, though it turned out to be both of those things (as long as you don't believe the conspiracy theorists). The objectives of a program tend to be just smARt; they are less specific and so as a consequence less strongly measurable and timelined. For instance the palm nut oil industry is less specific than a palm nut plantation. It is not so specific at the start about how many plantations, factories, roads, and the like. In fact, as we shall see, you may start out with a target, but change your view as the program progresses and you get feedback on what you are achieving. With the actual shopping and sporting complex I describe in Example 16.1, there was only limited space in which to build it and so the objectives were quite specific, measurable, and timelined. But with a larger shopping, sporting, and leisure complex, it may be more like the palm nut plantation with the exact scope of the program less well defined at the start. In discussing program management, I will describe how to deal with situations where the objectives are just smARt.

Also with a program, the governance structure of the individual projects needs to be nested within the governance structure of the program.

Portfolios

A portfolio of projects is a group of projects which share common resources. (The program has common outputs; the portfolio has common inputs.) The resources may be money or people, but can also be data or technology. Thus with a portfolio there are several management issues:

1. You need to prioritize projects within the pool of resources you have. You only have a limited number of resources, and so only can do a limited number of projects.
2. Having chosen which project you want to do, you need to share the resources between them. The average number of resources should be in balance, but you may find that resources demands for different projects peak together and there may be unexpected events which also cause resource clashes.
3. Where projects are sharing data and technology, the projects become coupled, especially where one project is producing something another needs to use to make progress. The project plans need to be linked to manage that interface. Project interfaces are also risks which need to be managed as such.

Sometimes the projects in a portfolio may be totally independent other than they are sharing resources. Sometimes they may be contributing to a common outcome. The projects in Example16.1 are also sharing resources and so also constitute a portfolio, whereas the projects in the palm nut oil industry are all quite independent of each other. The plantations may be at opposite ends of the country and are well away from the factories. The U.K.'s Office of Government Commerce, in their standard *Managing Successful Programs*,[1] do refer to the projects comprising a program as a portfolio of projects, but the assumption is that they are more like Example 16.1, the palm nut plantation.

Investment Portfolio

The word *portfolio* was borrowed from the finance industry, where people have a portfolio of investments, and indeed the projects an organization is doing is part of its portfolio of investments. The consequence is that in the project management field the word portfolio can be used in two ways. It can refer to a collection of projects sharing common resources, or it can refer to the sum total of an organization's investments, including all the projects and programs it is doing. I use the term *investment portfolio* for the latter.

Sometimes, especially for smaller organizations, the firm's investment portfolio will consist of just one portfolio of projects. Sometimes it will consist of several programs, several large projects and several portfolios of miscellaneous small- to medium-sized projects. Certainly where an organization is doing several small, medium, and large projects, it needs to create separate portfolios for the small- and medium-sized projects. As we shall see in the next section, the small- and medium-sized projects cannot compete alongside the large projects for resources; they get lost in the noise. The small- and medium-sized projects each have to be treated as separate portfolios. When prioritizing resources at the organizational level, you need to prioritize resources between the large projects and programs, and the portfolios of small- and medium-sized projects and programs. Then you need to prioritize the small- and medium-sized projects and programs within those portfolios.

In Sec. 2.2, I suggested an organization needs to achieve a balance of products within its product portfolio, and introduced the Boston consulting grid (Fig. 2.3) as one way of illustrating that. Likewise an organization needs to balance the projects within its investment portfolio. It needs projects for:

- Innovation, including developing new products, services, or production processes
- Operational improvements
- Marketing, including launching new products or opening new market segments
- Strategic realignment

Within the last category there can be:

- Mandatory projects such as responding to new legislation
- Repositioning projects to reposition the organization up the performance curve (Fig. 2.2)
- Renewal projects to renew products or facilities

Governance

In the last chapter, I said there are three levels of governance in the project-based organization:

1. At the level of the board
2. Within the context of the organization linking the board to individual projects
3. At the level of the individual project

Programs and portfolios are part of the middle level. Both help define the objectives of projects, and ensure the right projects are done to achieve the organization's desired development objectives. Programs do this by identifying the higher order development objectives required and then define the program of projects required to deliver individual components of that. Portfolio management does it by deciding which projects best deliver the organization's development objectives and assigning resources to them. Both

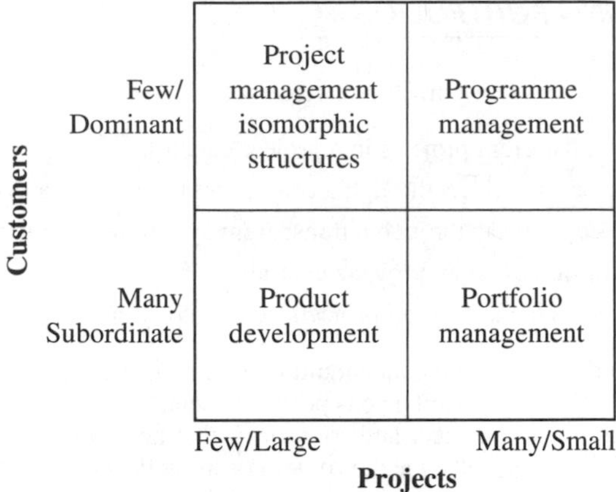

FIGURE 16.1 Four governance structures.

help define the method of achieving the objectives. Programs do that by identifying individual projects to deliver components of the objectives. Portfolios do that by sharing resources between projects and managing interfaces. Both also provide a means of monitoring progress.

Anne Keegan and I identified four different governance structures depending on the size of an organizations projects and customers[2] (Fig. 16.1). In this figure, the size of projects and customers is relative. A project is large if it represents around 10 percent of a company's turnover or more. It is small if it is around 1 percent or less. If projects are large, there are necessarily few of them to make up the company's turnover, and if they are small there are necessarily many of them. Likewise, a customer is large if it represents around 10 percent of turnover or more, and small if it is around 1 percent or less. Again if customers are large there are necessarily few of them. But they are also dominant; losing one such customer can have a huge impact on the company's turnover. If the customers are small, there are necessarily many of them, and losing one is less significant. This matrix defines four governance structures:

1. *Large projects for large customers:* These are traditional projects, which most of the early books are written for. The governance and organization structure of these projects tends to change as you move through the project life cycle, as illustrated in Table 6.5.
2. *Small projects for large customers:* This is program management. Concept and feasibility are done at the program level, so the project manager just applies the management cycle: plan and design, organize, implement, control progress, and link up. The projects themselves are homogeneous, with the team composition and size not changing during the project. Governance on the project is nested within that for the program. The program manager and sponsor interface with the customer, and the project manager responds to them.
3. *Small projects for small customers:* This is portfolio management. Again the projects tend to be fairly homogeneous, because they are small. The project life cycle and management life cycle tend to merge, and the project team is fairly homogeneous throughout its life.
4. *Large projects for small customers:* This tends to be product development, where you are developing a single product to be used by several customers. You develop portfolios of customers, that is, segment the market to understand their different requirements.

16.2 MANAGING PORTFOLIOS

I suggest a five-step process for portfolio management:

1. Maintain a list of all current projects in a project database.
2. Report the status of all projects through a central project-reporting system.
3. Prioritize and select projects through a transparent system maintained centrally.
4. Plan and assign resources on all projects centrally.
5. Evaluate the business benefits of all projects postcompletion.

In the investment portfolio; the data should be maintained centrally and the decisions taken at board level. Within a miscellaneous portfolio, centrally means within that portfolio with decisions taken by the manager and sponsor of that portfolio. When I use the word "project" in this section, it can also mean program or miscellaneous portfolio.

I also call these "steps," but the process will take place in a different sequence depending on the circumstances. An organization like that in Example 16.2 that needs to get to grips with portfolio management will work through the steps in the order listed above. The company in Example 16.2 needed to know what projects it was doing and the status of all projects. It needed to get to grips with the assignment of resources, and once it had done that it would need to prioritize the acceptance of new projects. But once an organization has got to grips with portfolio management, all five steps will be going on all the time within the portfolio. But within the life of a single project, they will take place in the order 3, 1, 4, 2, 5. I am going to talk through the steps roughly in the order they are encountered by the project.

Example 16.2 Failure in portfolio management

I worked with a well-known food manufacturing company at their London factory. They were having a problem completing key strategic projects. The company had had more than 50 percent market share for canned food in the United Kingdom, but had recently lost that position, but still had the largest market share. However, they risked losing that position in the next few years.

For the last four-and-a-half years they had been trying to build a new canning line. This was going to halve the cost per can leaving the factory gate. The project had originally been planned for two years, had already taken four-and-a-half years and had at least six months to run. It was at least 100 percent behind schedule. My client was the manager of a computer project to implement a material-monitoring system, which was also going to reduce the cost per can leaving the factory gate. He couldn't get factory managers to commit to attending project meetings. Production needs always took precedence over the project. Without user input, his project could not progress. The factory had also just started a Six Sigma implementation.

I started by planning the resource requirements of those three projects. It turned out that over the next six months they would require 30 percent of the factory managers' time. So if they were working Monday to Friday on production needs, they had to work Saturday and Sunday morning to fulfil their input to just those three projects. I next drew up a list of all the ongoing projects. There were 100. The three largest required 30 percent of the factory managers' time. The company had no idea.

I also asked for a plan and progress report on the new canning line. The company had recently spent £1 million buying a licence for mainframe project management software, and the only plans they were able to give me were schematic plans in a well-known PC-based system—for this project that was 100 percent behind schedule. The engineering director was sacked shortly afterwards.

Prioritizing Projects

The first step is to decide whether to add a project to the portfolio or not. To do that projects have to be rank-ordered according to a set of criteria. The two key criteria are the benefit of the project and its risk. These are plotted in Fig. 16.2. However, other criteria may be included in the ranking, including:

- Strategic importance
- Opportunity for learning
- Stakeholder acceptance

You may calculate a weighted average of all the criteria to determine the actual ranking of projects. In Fig. 16.2, there is a hurdle rate for acceptance of projects, the upper line. Projects in the top left-hand corner are acceptable. The line slopes up because higher risk projects are required to have higher returns. There is a lower hurdle rate, and projects below that, the bottom right-hand corner, are usually rejected. Projects in the middle band are borderline and other criteria may be considered.

The decision to accept or reject projects will be taken at a portfolio prioritization meeting. For the investment portfolio, this typically takes place once every three months. Within a miscellaneous portfolio it may take place more frequently. At the meeting you look at all the projects currently on the list of proposed projects. This will include all new projects proposed since the last meeting, and any brought forward from the last meeting. The prioritization is quite brutal. You know how many resources (money and people) you have available to work on new projects. You transfer projects from the proposed list to the list of new projects, working down in rank order, until you have exceeded the resource availability.

With one of my clients in the financial services industry it didn't quite end there. They would then look at the next project on the proposed list and ask if it was higher priority than one already in progress. If the answer was yes, the one in progress was cancelled and the proposed one added on top of the list of new projects. This would carry on until ones on the proposed list were not higher priority than ones in progress.

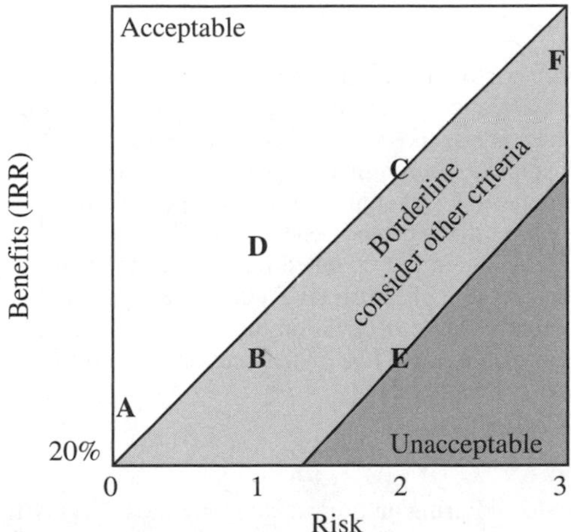

FIGURE 16.2 Benefit-risk diagram.

TABLE 16.1 Six Large Projects and Programs

Project	Cost (million)	IRR	Risk	Project type
Program A	$30	21%	0	Operational improvement
Program B	$30	22%	1	Operational improvement
Program C	$30	26%	2	Capital expansion
Project D	$30	24%	1	New product
Project E	$30	22%	2	New product
Project F	$30	26%	3	Strategic repositioning

It is still not the end for projects on the proposed list. For some you might decide their time is not now, but in the future their priority might increase, and so they may be carried to the next or a subsequent meeting. All other projects will be rejected and deleted from the list.

As a simple example, consider a firm with proposals for six large projects and programs each costing $30 million, 60 medium-sized projects each costing $3 million, and 600 small projects each costing $0.3 million. It has proposals for $540 million worth of projects but only has $300 million to spend. For the large projects, the benefit (measured by internal rate of return), risk (measured by where the project sits in Fig. 10.1), and project type are shown in Table 16.1. They are also plotted in Fig. 16.2. We have to decide how many of the large projects (and programs) we are going to do, and then that says how much money is left over for the medium- and small-sized ones. You can see from this simple example how the small- and medium-sized projects cannot compete alongside the large ones. They just cannot be seen in the noise. Programs A and C and Project D are clearly acceptable and Project E unacceptable. If you do A, C, and D you will have $220 million left over for the medium and small projects, perhaps $110 million to each portfolio. But you may decide you want to do one or both of B and F. They actually score equally on the benefit-risk plot, being on a line parallel to the two hurdle lines. You may prefer B because it is lower risk. You may prefer F because it offers opportunities for strategic repositioning and therefore potentially unknown returns over and above the direct ones. If you do one you have $180 million left over for the small and medium projects and if you do both $150 million left over. Your call.

Postproject Evaluation

I want to deal next with postproject evaluation. There is a fundamental flaw in the previous step: project sponsors are encouraged to inflate the potential benefit of their project and deflate the cost and the risk. If there is no check that projects deliver what their sponsors promise, then expect sponsors to continue to make their projects seem better than they are. Thus postproject evaluation is essential, and project sponsors must be held accountable if their projects fail to deliver the expected benefits (see Example 16.3). It is acceptable if the project delivers a benefit within the expected risk. For instance if Project F in Table 16.1 were to deliver an internal rate of return (IRR) of 20 percent, the hurdle rate at zero risk, that would be acceptable because it is within the range expected given the risk. But if it delivered an internal rate of return of just 15 percent, the sponsor should be made to explain why the benefit was less than predicted.

Example 16.3 Postproject evaluation

I worked with a bank, advising on project categorization. On my first visit they wanted to categorize projects as part of the project prioritization process. They were applying a process similar to that described above. But projects were failing to deliver their promised benefit and so the bank was failing to deliver its growth targets. At the time of my second visit nine months later, the CEO had changed. The new CEO took a different approach. He said they did not need to prioritize projects; they were a bank, access to money was not a problem. Any project that delivered the hurdle rate of internal rate of return, given its level of risk (Fig. 16.2) would be funded. But, a postproject evaluation would be done on every project and sponsors would be answerable if projects failed to deliver the promised benefits.

The Project List and Status Reports

I discuss these two together because the project status report in effect encompasses the project list. This needs to be kept simple. You need a single page showing all the projects in the portfolio and their current status and then a single page for each project. Figure 16.3 is a traffic light report showing the status for all the projects in a portfolio. For the chosen key performance indicators (Sec. 3.3), here cost, time, risk, and forecast first year revenue, each performance is shown as being on one of three status:

Green (light grey in Fig. 16.3): at or ahead of plan (say no worse than 5 percent over estimate)

Amber (grey in Fig. 16.3): just behind plan but not causing concern (say between 5 and 10 percent over estimate)

Red (Black in Fig. 16.3): causing concern (say more than 10 percent over estimate)

What you choose as the limits for the three status levels would depend on the levels of contingency and tolerances you set. For instance with Table 8.5 the raw cost estimate is 1000, and status green would be anything up to a forecast cost at completion of 1050; that is within the contingency. Status amber would be anything up to 1100; that is within the tolerance, and below the project's budget. Status red would be a forecast cost at completion over 1100; that is over budget. Perhaps black would be anything over 1200. Similar levels could be set for the estimated completion date, forecast risk, and predicted first-year revenue.

For each project, a single-page report is also produced to support the portfolio traffic light report. This can be in the form of a project dashboard (Fig. 3.1) or a single-page

Project Name	Budget	Cost	Time	Risk	Benefit	Status
Project 1	100	As planned	Problems	3	As planned	As planned
Project 2	200	As planned	As planned	3	As planned	As planned
Project 3	300	As planned	Problems	4	As planned	As planned
Project 4	900	Crisis	Problems	1	As planned	Crisis
Project 5	450	As planned	As planned	3	As planned	As planned
Project 6	600	Problems	As planned	2	Problems	Problems
Project 7	750	As planned	Problems	4	As planned	As planned
Project 8	800	As planned	As planned	3	As planned	As planned

As planned · Problems · Crisis

FIGURE 16.3 Traffic light report.

report such as that shown in Fig. 3.2. The advantage of the latter, incorporating the milestone report (Fig. 13.6), milestone tracker diagram (Fig. 9.16), and the earned value chart (Fig. 8.8) is that you can see the progress since the last report. With the project dashboard you just get a snapshot of the project as of today; it might have slipped further since the last report and you can't tell. Example 16.4 gives an example of the use of this in practice.

Example 16.4 Portfolio reports

I did a series of courses for project sponsors with an electronics company. On one of the courses, a senior manager was briefing the delegates and he said the managing director reviewed progress on the top 30 projects once a month. The delegates said that was impossible. It would take a day to review progress on each project so he would spend the whole month reviewing the projects. The answer was he received a traffic light report and a single page report for every project. Two-thirds of the projects (say 20) would be at status green. He would spend 5 minutes on each of those, the first two hours in the morning. Two-thirds of the rest (say 6 to 7) would be at status amber. He would spend 15 minutes on each of those, the rest of the morning. So that would leave three to four projects at status red. He would spend an hour on each of those, do a more thorough investigation, telephone the project manager and sponsor. The ones at status red would take all afternoon.

Sharing Resources

We now need to share the resources between those projects we have chosen to do. The resource requirements should on average balance across all the projects we have chosen to do, but there is a job to do here, for at least two reasons:

1. The resource that will dominate the prioritization of projects is money and so the demands for people may not be in balance and particularly not across all different resource types—some may be underutilized and others overutilized.
2. The resource demands may be in balance with the timescale we did the prioritization over (three months) but not day by day.

First, I will describe what people used to do to share resources and why it doesn't work, and then I will describe the technique I developed which is now accepted as best practice.

In the bad old days of project management, the common approach was to develop a plan for each project, with its resource requirements, and then combine all the individual project plans into one gigantic portfolio plan. That gave the total requirements for each resource. The computer was then asked to schedule all the projects so that the resource requirements did not exceed availability. Now the computer needs to be given a rule to prioritize one project or activity over another when there is a resource clash, and computers are dumb things, so once given a rule, they will apply it blindly and unquestioningly. One possible rule is to make Project A priority1, Project B priority 2, and so on. What happens? Project A gets what it needs. Project B gets what it needs from what is left. And Project C is stop-start-stop-start, never finishing, and wasting loads of money. Another rule is to give priority to activities by size of float. What happens? Every activity is scheduled by late start. You cannot abdicate management responsibility to the computer. You must retain management control. You do plan each project, but you must make decisions at a strategic level, and then plan each project within that framework. Thus I propose a six-step process for managing the prioritisation of resources across projects in a program:

1. Develop individual project plans, at the strategic (or milestone) level.
2. Determine the resource requirements and duration of the individual projects, at that level.

3. Incorporate each individual project into a rough-cut capacity plan (or master project schedule, MPS) as a single element of work, with an idealized resource profile based on the profile and duration calculated at Step 2. Time units here will be weeks or even months.
4. Schedule each project in the master project schedule, according to its priority, to achieve an overall resource balance, and assign it a time and resource window.
5. Manage each project to deliver its objectives within the time and resource window.

There are several provisos to this process:

1. If you are doing internal development projects, you can move projects around, perhaps delay the start, or extend the duration, to achieve a resource balance. If you are a contracting firm doing projects for clients, you don't have that luxury. You cannot tell a client you will start a project a month late because you don't have the resources. The only way contracting companies find to deal with this is by hiring subcontract staff. Typically, contracting companies have between 20 and 40 percent peripheral workers. It might be the forward plan shows the resource requirements dropping, but that is as an assignment comes to an end. The company has five bids out, expecting to win one on average. If it doesn't win any, it will need to lay contract staff off (much cheaper than making core workers redundant). If it wins one, it will retain all the existing workers. If it wins two it will need to find new workers. If it wins three, it will not be able to cope. How can you manage that uncertainty other than by using subcontract staff?
2. I first met an organization doing this in the ship repair industry. But the resource profile of all ships is the same. If you delay one project to solve a problem with shipwrights, you solve a similar problem with fitters, plumbers, riggers, and all the resources; not necessarily at the same time, but in phase. That does not work with projects involving professional people which tend to be much less homogeneous. As you solve a problem with hardware engineers, you create a bigger problem with systems analysts or marketing people. So even for internal projects people find they need to use peripheral workers.

This process involves three groups of people: portfolio managers, project managers, and resource managers (Fig. 16.4).

Portfolio Managers. Demands for new projects in the portfolio come to the portfolio managers. They ask project managers to plan the projects, defining the duration and summary resource requirements. The portfolio managers add the project to the rough-cut capacity plan. Resource managers give the portfolio managers the resource availability, and the portfolio managers give the individual projects a start and finish date and resource availability to balance the overall requirements.

Project Managers. Project managers then schedule the individual projects using the tools described in Part 2. In particular, they break the projects into work packages with individual start and finish dates. They then make resource demands on the resource managers in one of two ways as described in Sec. 6.2. Either they ask resource managers to second people onto the project (they have to do this for work packages involving resources from several departments), or they ask the resource manager to take responsibility for delivering individual packages of work (they can do this for single resource packages of work).

Resource Managers. Resource managers then need to resource projects in accordance with those demands. The demands should balance. But if the resource levelling in the rough-cut capacity plan is accurate to ±5 percent and there are 10 projects in the portfolio and each project has been broken into 10 work packages, the demands made on the resource

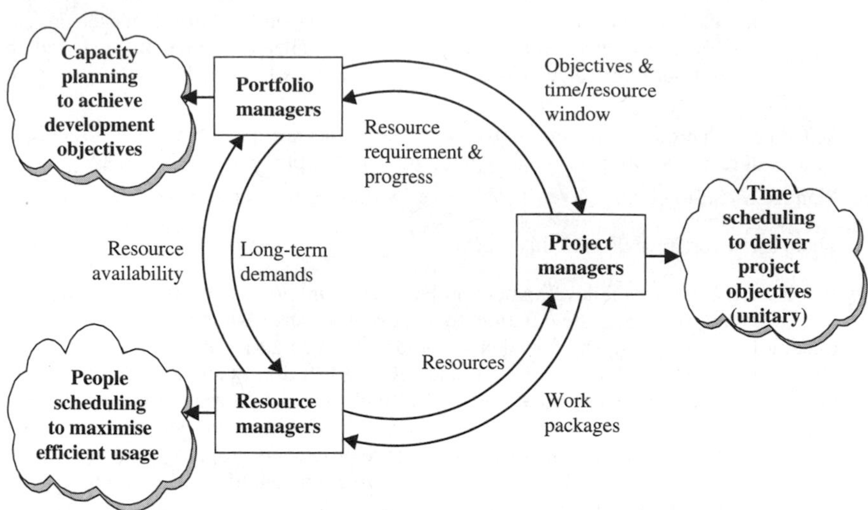

FIGURE 16.4 Portfolio management.

managers will only be accurate to ±50 percent (the error is proportional to $\sqrt{N}$, Sec. 5.1). Thus the resource managers still have a job to do smoothing the demands day by day.

Caveat. Maintaining the portfolio resource plan is not easy. As we have seen, there are problems with different profiles for different resources. Further unexpected events happen which can throw the plan out. But if you have the plan, you can cope with the unexpected events. You can work out their impacts and try to deal with them. If you don't have the plan you are completely out of control. It is not easy, it is not perfect, but it helps considerably.

Impact Matrix

I said above that resources are people and money, and in discussing prioritization I was dealing with money, and in discussing sharing of resources I was dealing with people. But resources can also be technology, data, plant, equipment, and the like. These create linkages between projects. Linkages are risks and so need to be managed. The risk management process is to identify the risk, assess the risk, and take action to reduce it, and this suggests how the linkages can be dealt with.

Identify. The *impact matrix* is a tool to identify the linkages. It is very simple. You put the projects down the rows and across columns and mark in the body of the matrix where there is a link. You are usually not concerned which direction the impact is so you only need a half matrix.

Assess. Rank the impacts as major, medium, or minor.

Reduce. Take action to reduce the major impacts. One way is to group the projects into subportfolios, where the major impacts are between the projects in the subportfolio and not with others. Then the control of those impacts can be the responsibility of one person.

16.3 MANAGING PROGRAMS

Programs follow a life cycle very similar to a project. The difference is that during the early stages—concept, feasibility, and design—you identify that to achieve your change objectives you need to do a number of unrelated things. Thus, rather than breaking the program into work packages, you break it into several subsidiary projects, which will be managed and delivered independently. There are four fairly significant differences between projects and programs. These differences are all related to each other, and in fact one follows the other.

smARt Objectives. The first, as I discussed above, is that for most programs the objectives are not as specific, measurable, and timelined as you would expect for a project. For the program in Example 16.1, the objectives were SMART. The program consisted of four projects, the shop, the sports centre, the car park, and the road access, and they were all clearly defined. However, on many programs this is not the case, and indeed is one of the great advantages of program management, you can start working on something when you don't know precisely what the program will encompass. You can't precisely define the change objectives you want to achieve, and so you can't precisely define all the outputs the program will deliver, but you can define enough to identify early projects with SMART objectives. The development of the palm nut industry in Example 1.3 was like this. In fact the early stages of the program may help you define what the later stages will deliver. I suggest the idea of a fish-tail program (Fig. 16.5). You have a vision of the end state you would like to achieve, perhaps two years away, but this is not precisely defined. So you start working on early projects (or projettes) which can be precisely defined, and link those into the growing system. That helps you clarify the end objective. Perhaps when you are 6 months into a 24-month program the end objective has been clarified and is now 2 months away, not 18. We will also see shortly, that you may never achieve the end objective, deciding at some point that you have done enough.

Early Benefits Realization. A major difference between the $50 million program and project is the opportunity for early benefits realization. With the project, you do all the work of the project, and only after you commission the project deliverable at the end can you begin to get any revenue returns. Thus a large amount of money is tied up before you can get any returns. This would happen if you were building a new power station, airport, or bridge; the whole thing must be finished before you can use any of it. But with the palm oil industry to the shopping and sporting complex that is not the case. With the former you can build one plantation and one factory, and use the income from that to fund the development of further plantations and factories. With Example 16.1, you could complete the road access, car park, and shop, and receive benefit from those while building the sports hall.

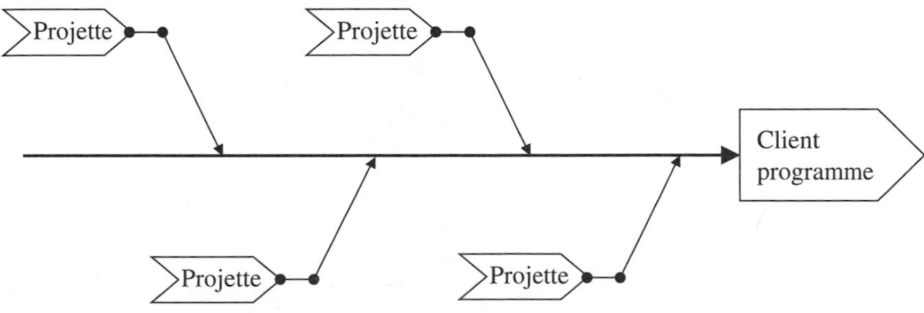

FIGURE 16.5 Fish-tail program.

With a larger out of town shopping centre you might do it in several phases, and use the income from the early phases to fund the later phases.

Cyclic Delivery. This then clearly lends itself to undertaking the program in phases. You undertake the projects in several phases (Fig. 16.6) gaining benefits realization from the early phases to pay for the later ones. But you also decide at the completion of each phase whether you want to continue with the program. The steps of the cycle are as follows:

Initiation: The first step, done once only, is initiation. During the concept, feasibility, and systems design stages, you plan what projects you think the whole program will entail and divide them into several stages. You then define precisely the projects for the first cycle.

Project execution: You execute the projects for the current cycle.

Project delivery and linkup: As the projects from the current stage are delivered, you measure what the program is achieving and what benefits are being realized. You then decide whether you want to continue with the program. There are several possible outcomes here:

- You may continue as first envisaged.
- You may terminate the program early because you are not achieving the benefits expected.
- You may decide to terminate the program early because the early stages delivered more benefits than expected, and so there is little left for later stages of the program to achieve.
- The program may be 95 percent complete, but it is not cost effective to do the last 5 percent.
- The program may have achieved all the objectives you originally envisaged, so you stop.
- The program has been so successful you decide to do additional work.

Renewal: If you decide to move into the next cycle, you need to define precisely the projects for this cycle, and revise the schematic definition of projects for planned future cycles. You then return to project execution.

Dissolution: If you decide not to continue, you dissolve the program. As part of this, you need to decide how any outstanding work will be dealt with.

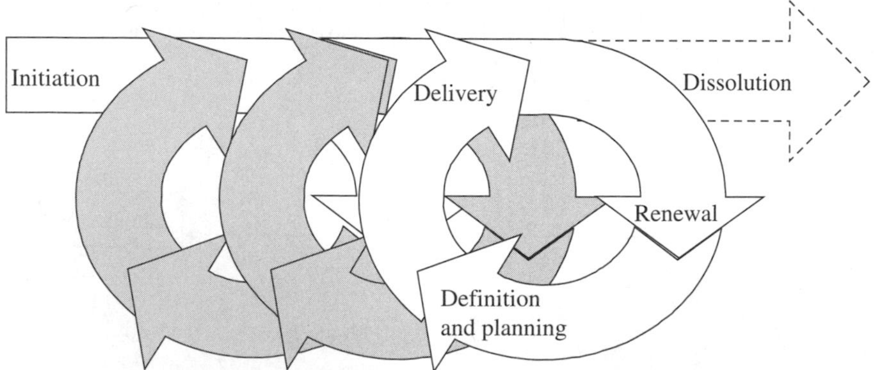

FIGURE 16.6 Cyclic delivery of programs.

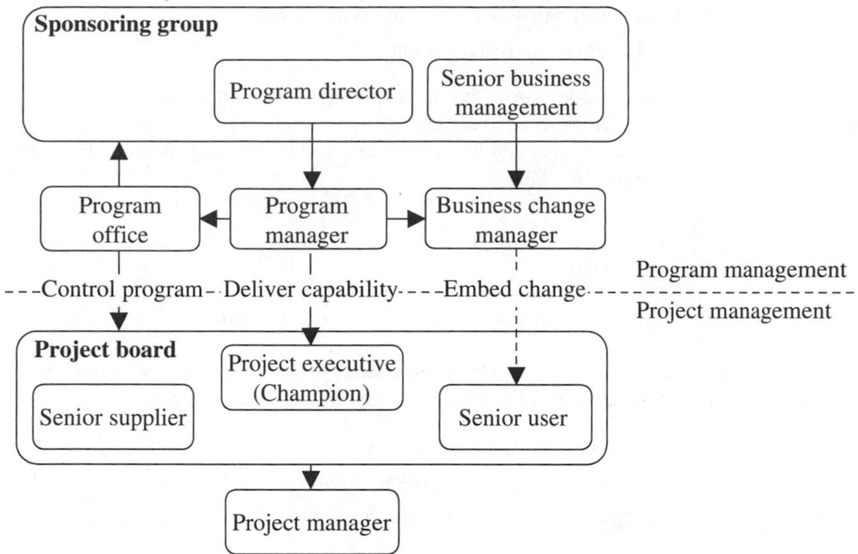

FIGURE 16.7 Governance roles on projects and programs.

Governance. All the roles identified for a project in Sec. 15.2 also exist for the program: the sponsor, the steward, the manager, and the owner or business change manager, and people fulfilling those roles on the project report to the person fulfilling the same role on the program (Fig. 16.7). In fact, the role of the business change manager, responsible for embedding the change and achieving the benefit was first identified in program management. The project manager cannot be responsible for embedding the change because he or she will be moving on to manage projects in the next cycle of the program. The project manager has to hand over the project outputs and the new capabilities and leave it to somebody from the business to embed the change and achieve the early benefits realization. On programs, the responsibility for delivering the new capabilities and the responsibility for embedding the change are clearly distinct, the former held by people from project and program management and the latter held by people from the business or user departments.

16.4 THE PROJECT OFFICE

Our understanding of the project office has also changed considerably in the past 10 years. It started life as something supporting a large or major project, where all the project planning and control support was located. Through a process which Monique Aubrey and Brian Hobbs[3] call "Balkanization" the role of the project office has divided and evolved. Jack Duggal[4] identified five roles for the project office:

1. Supporting planning and control on a large project or program
2. Decision support for portfolio management, as described in Sec. 16.2
3. Governance, including the development of policies, procedures, and systems and the production of compliance reports

4. Data and knowledge management, training, and consultancy
5. Communication and relationship management

The first of these is a temporary role that lasts as long as the project or program. All the rest are permanent. Thus the project office has evolved from a temporary governance role, associated just with a large project and ceasing to exist when the project ends, into a permanent feature of the organization.

Duties of the Project Office

To fulfil these roles, the project support office can undertake a number of duties:

Maintaining the Master Project and Program Plans. The project office maintains the master project, program, and portfolio capacity plans on a central (computer) system:

- For a large project, that will be a stand-alone plan.
- For a major project or program, it may be broken down into subproject plans.
- For a portfolio, the project office will maintain the capacity plan and individual project plans.

In all cases there must be clearly defined levels of access for different managers. All managers will need to interrogate the plans at all levels. However, they will only be able to make changes at their level of responsibility. Changes must be within the tolerance set at the higher level. If that is not possible, the approval of the higher level manager must be sought. Sometimes, the ability to make changes is limited to the project office staff. Managers can only recommend. In this way the integrity of the system is maintained.

Maintaining the Company-Wide Resource Plan. The resource aggregation at the project level provides the company-wide resource plan. The project office can take a company-wide view of the resource availability and assign resources to individual projects (within the constraints set by investment portfolio). Individual projects are not in a position to do this, unless they have a dedicated resource pool.

Providing Resource Data to the Project Initiation Process. When the organization is considering whether to initiate a new project, the project office can compare the resource requirements to projected availability. The project office does not have the power to veto a project, it is up to senior management to accept or reject it. However, if there are insufficient resources, senior management must decide whether to stop another project, or buy in resources from outside. That is extremely valuable information. Better not to start a project, than stop it half finished.

Issuing Work-to Lists and Kit-Marshalling Lists. At regular intervals, as agreed with project managers or as set by the company's procedures, the project office will issue work-to lists and kit-marshalling lists (Chap. 13). Giving this work to the project office ensures it is done regularly, and that it is done to a consistent style, in a way which people from across the organization can readily understand.

Facilitating the Control Process. The project office can manage the control process, relieving project staff of some of the bureaucratic processes, allowing the latter to concentrate on their project work. Figure 6.5 is a responsibility chart showing a procedure for this control cycle. The project office will of course facilitate the control of time, cost, quality, scope, resource usage, (organization), and risk. This activity requires the project office to:

- Progress, receive, and process the turnaround documents
- Analyze the consequences of the progress information
- Perform the what-if analysis
- Revise the plan with the appropriate manager
- Reissue work-to lists for the next period

Issuing Progress Reports. Following on from the control process, the project office can issue progress reports. These may go to:

- Project managers
- Program and portfolio managers
- Other senior managers
- Client

The reports issued will be defined by a procedures manual. The data gathered in turnaround documents may be used for other purposes, such as:

- Payroll
- Recording of holidays and flexitime
- Raising of invoices
- Recording project costs for the company's accounting systems

For the last, it is vital that costs are recorded by the project and sent to the accounts system, and not vice versa. With the latter, information can be received several months after costs are incurred, which is far too late for control. The data can be recorded separately for each system, but then it almost never agrees. The dispatch of this data, which may be electronic, will be done by the project office as part of the reporting process. It is important to review the data before dispatch, rather than allowing it to go automatically to ensure its integrity. However, this can be simplified by building in automatic checks.

Operating Document Control and Configuration Management. Projects can involve the transmittal of a large amount of information. The project office can coordinate that transmittal. This may include

1. Keep a library of progress reports for ready access by any (authorized) personnel.
2. Record all correspondence to and from clients and subcontractors. As part of this process, the project office may include acknowledgement slips, and monitor their return to ensure receipt of the correspondence. Technical personnel can be lax in the recording of correspondence, which can cause problems if there is a claim. To avoid this, some organizations insist that all outward correspondence goes via the project office, and a copy of all inward correspondence is logged there. Since all correspondence becomes part of the contract, the need to log it cannot be stressed enough.
3. Monitor all correspondence between project personnel. On a large project, this can drastically reduce the channels of communication. However, it is more efficient to have a central clearing point for communication on projects with as few as four people. This can be essential if people have not worked together before, on projects with tight timescales, and on projects involving research scientists, who do not tend to be very communicative.

4. Maintain the records for quality control and configuration management, to ensure they are properly completed, before work commences on the next stage.
5. Monitor the despatch of design information to site or subcontractors to ensure it is received and the latest information used. I have known of cases where drawings are lost in the post, and of course the intended recipients have no way of knowing they should be using new data. Acknowledgement slips solve this problem.
6. Manage issues. Issues can arise on a project, which may or may not lead to a change or a claim. The project office can manage the decision-making process.

Producing Exception Lists. As part of the control process, the project office may produce exception reports. They will produce variance reports at each reporting period, but exception lists will highlight items which have become critical.

Purchasing and Administration of Subcontracts. Where there is not already a purchasing department within the parent organization, the project can take over the procurement function. There is a view that in some project-based organizations a very high proportion of total expenditure on projects is through purchased materials or subcontract labour, and so this function should be within the control of project or program management.

Maintaining the Client Interface. The project office may manage the relationship with the client. This includes the issuing of progress reports, the control of communications, and the dispatch of invoices. It also involves producing reports against agreed milestones, and the maintenance of links with opposite numbers in the client organization so that any threats to the contract can be worked through together. Contacts with the sponsor and other decision makers can help to ensure continued support for the current contract, which will ease its delivery, and help to win new work.

Acting as a Conscience. Effective project management requires that all the control procedures described are well maintained. Some can become bureaucratic, and distracting for the technical staff. While the project is running smoothly, they can seem unnecessary, and not receive adequate attention. However, if the project does go wrong, the data and plans are required to plan recovery or defend a claim. It is then too late to start recording the data and maintaining the plans. It must be done from the start. The project office can relieve project staff of the bureaucratic burden. Because they maintain the plans as their day-to-day duties, they become efficient at it, so the cost of the administrative overhead is less than if project personnel do it. Indeed, the service and support they give can speed up the work of the project. In fulfilling this role, the PSO act as a conscience, because they ensure the regular reports are filed, and they will not let certain major milestones be met until appropriate documentation is completed.

SUMMARY

1. A program is a set of projects contributing to a common goal.
2. A portfolio is a set of projects sharing common resources.
3. There is a five-step process for managing portfolios:
   - Maintain a list of all current projects in a project database.
   - Report the status of all projects through a central project reporting system.
   - Plan and assign resources on all projects centrally.

- Prioritize and select projects through a transparent system maintained centrally.
- Evaluate the business benefits of all projects postcompletion.

4. To prioritize projects you need some way of rank-ordering the projects, and then choose projects up to the limit of resources. The two main criteria for rank-ordering projects are:
- Benefit
- Risk

5. There is a five-step process for sharing resources between projects:
- Plan each project individually.
- Determine its resource requirements.
- Add to the master project schedule or rough-cut capacity plan.
- Assign each project a time and resource window to balance resources.
- Manage individual projects within their time and resource window.

6. Programs differ from large projects in four ways:
- The objectives are less specific, measurable, and timelined, smARt, not SMART.
- You aim for early benefits realization, using revenue from early projects to fund later ones.
- The program is delivered through several cycles of projects.
- The governance structures are more focused.

7. The project office has up to five functions:
- Supporting planning and control on a large project or program
- Decision support for portfolio management, as described in Sec. 16.3
- Governance, including the development of policies, procedures, and systems, and the production of compliance reports
- Data and knowledge management, training and consultancy
- Communication and relationship management

8. The role of the project support office is to:
- Maintain the master project and program plans
- Maintain the company-wide resource plan
- Provide resource data to the project initiation process
- Issue work-to lists and kit-marshalling lists
- Facilitate the control process
- Issue progress reports
- Operate document control and configuration management
- Produce exception lists
- Purchase and administration of subcontracts
- Maintain the client interface
- Act as a conscience

REFERENCES

1. OGC, *Managing Successful Programs,* London: The Stationery Office, 2003.
2. Turner, J.R. and Keegan, A.E., "Mechanisms of governance in the project-based organization: The role of the broker and steward," *European Management Journal,* **19**(3), 254–267, 2001.
3. Hobbs, J.B. and Aubry, M., "A multi-phase research program investigating project management offices (PMOs): The results of phase 1," *Project Management Journal*, **38**(1), 2007.
4. Duggal, J.S., "The project, program or portfolio office," in J.R. Turner (ed.), *Gower Handbook of Project Management*, 4th ed., Aldershot, U.K.: Gower Publishing, 2007.

CHAPTER 17
DEVELOPING ORGANIZATIONAL CAPABILITY

In Chap. 16, we looked at program and portfolio management, a key component of governance of the organizational context, linking the objectives of projects to the strategic objectives of the organization, and defining which projects will be done to achieve those objectives. In this chapter, we look at another key component, developing organizational project management capability, which helps define the means by which the organization achieves its project objectives, and hence its strategic objectives. First I consider what we mean by organizational project management capability. Then I discuss the development of individual competence, a key component of organizational capability. But the organization can have capability over and above the competence of the individuals; the whole is much greater than the sum of the parts. So I consider four practices and four processes for developing capability, and four areas for managing the knowledge which underpins that capability. Finally, there are things that can stop an organization from improving its capability. These are known as competency traps, Sec. 17.6.

17.1 DEFINING CAPABILITY

Organizational project management capability comprises three components:

The Project Management Body of Knowledge

First, the organization needs to know what components of project management are relevant to the delivery of its projects, and how it will operationalize each one. It can seek guidance from a number of sources, the Project Management Institute (PMI) PMBoK,[1] the IPMA ICB,[2] or the Association for Project Management (APM) BoK.[3] Or else, you can adopt the components of project-based management as outlined in Parts 1, 2, and 3 of this book. The organization's definition of its body of knowledge will include the following:

Project Life Cycle. The organization needs to understand the life cycle appropriate for its type of projects, as outlined in Chaps. 1 and 11. There may be different life cycles appropriate for different types and sizes of project.

Management Cycle. The organization also needs to understand how it will operationalize the management cycle. It may adopt the very simple cycle suggested in Chaps. 1 and 11:

- Plan the work.
- Organize the resources.

- Implement by assigning work to resources.
- Control progress.

Or it may adopt a more extensive cycle as suggested by PMI comprising:

- Starting processes
- Planning processes
- Organizing processes
- Controlling processes
- Closing processes

The organization also needs to understand the difference between the management cycle and the project life cycle, and whether the difference matters for its size of projects. For instance, if the organization is involved in programs of small projects, it will apply the life cycle at the program level and the management cycle at the project level. If on the other hand it is involved in large projects, it will apply the life cycle to projects and the management cycle at each stage of the project as illustrated in Fig. 1.8.

The Functions of Project Management. The third component of the body of knowledge is the functions of management. In Part 2, I suggested six functions, managing scope, organization, quality, cost, time, and risk. I also suggested stakeholder management in Chap. 4. PMI suggests nine functions:

- Five of my six from Part 2.
- What it calls managing integration rather than organization.
- What it calls managing communication, covered under stakeholder management (Chap. 4) and communication between the sponsor and manager (Chap. 15).
- Managing human resources and procurement—I do not suggest that last two are unimportant, I have just made them beyond the scope of this book, each deserving a book in their own right.[4,5]

Project Management Methodology

Not only does the organization need to know how it is going to operationalize the individual components of the body of knowledge, it needs to put all of those together into an integrated project management methodology to deliver individual projects from concept to realizing benefit from the strategic change introduced (operation of the new asset). None of the professional associations' bodies of knowledge[1,2,3] actually recommends a methodology, though APM's[3] comes the closest. (I was part of the committee that developed the initial draft and we were told by APM to take it out.) The PRINCE2[6] process is a methodology but I would offer the methodology recommended in this book as illustrated by Fig. 1.11.

Technical and Craft Skills

This last component of organizational capability does tend to be underemphasized by the project management community. Yes, project management is a generic skill, and so the techniques can be applied to all types of project. But the techniques do need to be packaged in different ways depending on the type of project[7,8] and the organization does need to be expert in the basic technical and craft skills that are used in the projects it undertakes. If it has the necessary technical and craft skills, it will know how to tailor its project management methodology and the project management functions to its type of projects.

17.2 DEVELOPING INDIVIDUAL COMPETENCE

A necessary component of organizational capability is the competence of individuals. Without competent individuals the organization can have no capability. In this section, I consider how we define the competence of individuals, the components of competence, how it varies at different stages of an individual's career, and how to assess and develop it.

Defining Competence

There are two main ways of defining the competence of individuals.

Competency Model or Attribute Approach. The competency model or attribute-based approach, popular in the United States,[9,10] defines competence as the knowledge, skills, and personal characteristics required to deliver superior performance. A competency is an individual component of competence. There are surface competencies, knowledge, and skills, which can be easily measured and developed, and core competencies which are less easily measured. Inherent in this approach is the idea of threshold and differentiating competencies. Threshold competencies are the basic ones essential for doing the job, and differentiating ones are those leading to superior performance. The bodies of knowledge produced by the professional associations[1,2,3] define the threshold knowledge and skills required to manage projects. They therefore conform to this approach, but define the basic competencies rather than the differentiating ones.

Competency Standards or Performance-Based Approach. The other approach, rather than trying to measure the inputs to competence, tries to measure the outputs, and has been popular in United Kingdom and its former dominions (Australia and South Africa). Competence can be measured by performance in accordance with defined occupational, professional, or organizational standards.[11] Several performance-based standards for project management have been produced around the world.[12,13,14,15]

Combined Approach. Lynn Crawford[15] has developed a combined approach to defining competence. She suggests there are three components to competence:

Input competencies: The required knowledge and skills to do the job. Of these, knowledge tends to be a threshold competence, whereas skill tends to be a differentiating competencies.

Core competencies: The personal characteristics including motives, traits, and self-concept that improve performance. These are differentiating competencies.

Output competencies: The ability to perform required activities as defined by professional, occupational, or organizational standards.

Many years ago, I was introduced to a competency model by an Information Technology (IT) vendor. They defined three levels of competency, described by the words, I know, I can do, and I adapt and apply.

I know: This is the knowledge required to do the job.

I can do: This is the ability to apply the knowledge to perform routine tasks (skill).

I adapt and apply: This is the ability to apply your skill in unfamiliar situations to develop new methodologies to deal with those unfamiliar situations.

According to this model, an individual does not perform until they could adapt and apply. Further, it was not that a certain level of knowledge gave you an equivalent level of ability to do and that gave you an equivalent level of ability to adapt and apply. You needed quite a bit of knowledge before you could do, and substantially more knowledge giving additional ability to do before your could adapt and apply. But you would reach a threshold level of knowledge where more knowledge would not increase you ability to adapt and apply any more. Only other things, including experience, would do that.

They likened it to learning a language. You need a certain amount of vocabulary and grammar before you can compose simple sentences, but you need to be able to compose substantially more sentences before you can hold a conversation. I have a 2000-word vocabulary in French but have difficulty conducting conversations because I lack experience and self-confidence.

Explicit versus Tacit Knowledge. A related concept is that of explicit versus implicit or tacit knowledge.[16] Explicit knowledge is knowledge that can be codified and written down. It is mainly what we learn when we go on a training course or read a book. Implicit or tacit knowledge is inherent knowledge that we gain mainly through practice and experience. We know how to do things without thinking about them. You mainly drive a motor car using tacit knowledge; you can't think through the process of doing an emergency stop when one is called for. Similarly with a language, you start off holding a conversation using explicit knowledge, but it tends to be slow as you have to translate what you have heard and what you are going to say before replying. But with experience you conduct the conversation in the language and the language has meaning without translating.

In the models above, knowledge mainly means explicit knowledge, and tacit knowledge mainly falls under skills. Kolb[17] introduced a learning cycle for individuals (Table 17.1) showing them cycling between gaining tacit knowledge through experience and explicit knowledge through training and education. The two reinforce and support each other. Explicit knowledge gives you structures that explains why things work in the way you expect and having those structures makes you better able to test out concepts through your work experience and improve your tacit knowledge further, and develop your own new explicit knowledge, enabling you to adapt and apply.

Threshold Competence. I wish to explore a little bit further the concept of threshold competence. Lynn Crawford showed in her own research that project management competence as measured by the professional associations in their certification programs tends to be threshold competence.[15] A pass grade on the certification programs is necessary to be able

TABLE 17.1 Kolb's Learning Cycle for Individuals

		To	
		Tacit knowledge	Explicit knowledge
From	Tacit knowledge	*Concrete Experience* Creating tacit knowledge through on-the-job experience	*Observation and reflection* Understanding how and why things work
	Explicit knowledge	*Testing of concepts* Testing explicit knowledge in practice to see if it works	*Abstract concepts and generalizations* Codifying knowledge through education and training

to work effectively as a project manager, but a higher score does not make you a better project manager. It is other things that make you a better project manager (in your current job role), including years of experience.

I should qualify this. In the last chapter I said there was a substantial difference between the management of a $50,000 project and a $50 million project. The former tends to require primarily technical skills (as measured by the certification programs, especially PMI's), whereas the latter requires substantially more people and strategic management skills. Thus more knowledge won't make you better in your current role, but different knowledge is needed for different more advanced job roles.

Levels and Stages of Development

This leads to the concept of T-shaped managers (Fig. 17.1) illustrating different levels of management and different competence profiles required. The height of the upright shows the depth of professional involvement (primarily technical competence) and the width of the upright shows the breadth of that. The width of the cross-bar illustrates the range of professional contacts and the depth shows the amount of managerial competence required (mainly people management and strategic).

Team leader: A person's project management career starts as a team leader, managing a single discipline team, perhaps a $50,000 component of a larger project. Perhaps the team is just mechanical designers, and the team leader has to interface with other teams from the project, consisting of civil, electrical, and other engineers; or it is programmers, interfacing with testers and systems analysts. The team leader needs basic project management and people management skills. Their technical skills will probably be domain specific; that is, their own area of professional expertise will be the same as the team.

Junior project manager: The person will now be managing a small ($5 million) project, perhaps part of a larger program. There will be several teams on the project, several types of engineers, or teams of programmers, analysts, and testers. The manager is less involved professionally but has a wider range of skills to manage. The range of contacts has widened to other project managers in the program or portfolio, and line managers of the project team members. The manager now needs more people management skills, some other functional skills such as accounting to be responsible for project cost, and also needs to understand the project and program's contribution to corporate strategy. The manager's professional background is probably still domain specific: if it is an

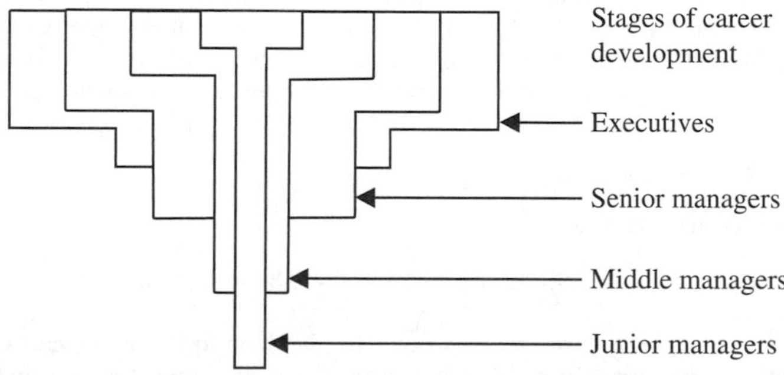

FIGURE 17.1 T-shaped managers.

engineering project they will be an engineer, and if it is an ICT project they will be from an ICT background. So, although there will be people on the project team from other professions they will be related professions.

Project or program manager: The person is now managing a larger, more complex project or program ($50 million) consisting of several unrelated professions, engineers, ICT people, sales and marketing personnel. Their background can no longer be domain specific, though it may be from the dominant profession. Their contacts go outside their parent organization to clients and suppliers. They need more strategic management and commercial skills.

Project or program director: The person is now managing the largest of projects or programs. Their contacts may now go outside the industry and international. They need directorial and governance skills.

Assessing Competence

Competence will be assessed against some measure of the desired competence for the job being performed. This will probably be a competence model for the job, and may specify the desired input competence (knowledge, skills, and experience), or the desired output competence (performance), or both. It may be based on one of the competence frameworks mentioned above: input competence for the bodies of knowledge of the professional associations[1,2,3] or output competence form one of the performance standards.[12,13,14] The competence model may be part of a job description.

At the individual's regular appraisal (once or twice a year usually), their current competence will be assessed against the model. Their competence will be assessed against the desired standards for the job they are currently doing to determine any shortfalls but may also be assessed against the standards for the next promotion or development they are looking for. In this way, the gap between their current and desired level of competence will be determined. This indicates the desired training or experience the individual requires to achieve the necessary development either to meet the requirements of his or her present job or progress to his or her next development.

Martina Huemann, Anne Keegan, and I suggested that this assessment should be conducted by the individual's line manager and not his or her project manager, because the timescale over which relevant decisions are taken is over several years, much longer than the project he or she may be currently working on.[18] However, the assessment should be based on appraisals conducted by the project manager. If the in-line appraisal is divorced from performance on the project then it is bad for the individual's motivation and for the cohesion of the project team he or she is currently working on. Some organizations have formalized in-project appraisals. One software consultancy we interviewed conducts an in-project appraisal at the end of every project, and once every three months if projects last longer, so the in-line appraisal, held once every six months, is based on at least two project appraisals.

Developing Competence

Competence needs to be developed over the short, medium, and long term.

Long Term. In the long term, the firm needs to plan how to fill its forecast competence requirements several years hence, developing project managers and other project professionals to work on the projects it expects to be doing. Anne Keegan and I found the engineering construction industry tends to take a very long-term view, taking 15 years to develop project managers capable of managing projects of $500 million or greater.[19] They try to identify people aged 25 who they can develop to manage projects of $500 million plus at the age of 40. We asked one interviewee how they identify potential candidates aged 25. He said, "People who are vocal with their ambition." In that industry they tend to adopt what Anne and I labelled the spiral staircase career. People are given a range of experiences throughout their career, moving between different types of job role. In the early stages of their career, the first two levels described above, they will tend to move between technical design and project roles. For the remainder of their career they may occupy technical roles, project roles, line management roles, and customer interfacing roles. There are at least two advantages of this:

- It develops more rounded individuals.
- It avoids the Peter principle.

The Peter principle occurs in functional, hierarchical line management, where people are promoted up the hierarchy. It states that people keep on being promoted until they reach a level at which they are incompetent, and there they stay for the rest of their careers—it is impossible to demote them back to the level where they were last competent. The result is organizations become staffed with incompetent people. In the project-based organization, with the spiral staircase career, people are promoted half or even quarter steps into different job roles. If they are uncomfortable in their new role, they can be moved sideways into one where they were last comfortable without any loss of face, and then progress up that ladder (see Example 17.1).

In the ICT industry, projects tend to be smaller and so people are developed for project management roles more quickly but they move from there to program management roles. The ICT industry tends to use pairing, where two people do a job which strictly one would do. By doing that it creates more innovative solutions and also develops twice as many people able to do the job.

One company that Martina Huemann, Anne Keegan, and I interviewed had what it calls development cells.[18] These are people responsible for identifying the future needs of the company and talent scouting for potential candidates who can be developed to fulfil them. By this method, they also avoid line managers Bogarting talent, holding on to good people both to the detriment of their career development, and the detriment of the firm and its needs for developing good people.

These long-term decisions need to be taken by line managers because the time horizon is very much longer than projects.

Example 17.1 Moving to a more suited position

One of the people Anne Keegan and I interviewed had been director of projects in his company on the company board and was moved to be project director on a $1.5 billion project for a major client that wanted the job done one-third faster than the previous record. This was a very high-risk project, and it was felt he had the skills to deliver it (which he did successfully). He was given a wage increase to go from being board director to project director because it was felt that this carried higher risk. But he was taken from a job where he was not comfortable (board director) to one where he was a perfect fit and could make a significant contribution to the company without any loss of face.

Medium Term. Medium-term development has a time horizon of one or two years, aiming to develop people to fill the gaps in their competence profile to make them more competent in their current job role, or to develop them for their next promotion. This will be in the form of training or perhaps education programs leading to master degrees. It will also be in the form of on-the-job experience to develop appropriate skills and tacit knowledge. The big issue is when at the routine appraisal in the line it is identified the individual needs to work on a certain type of project to get the developmental experience they need, and then shortly afterwards such an opportunity arises but their current project is only part way through. Do you move them to gain the experience or insist that they complete their current project, by which time the opportunity will have passed? Enlightened organizations move people for several reasons:

- It is beneficial to the organization to develop appropriate staff for future needs.
- It is beneficial to the individual.
- If you don't show commitment to people's development they may leave anyway.
- The current project is a development opportunity for somebody else.

These decisions must also be taken in the line because the time horizons are still longer than the duration of projects, but project managers need to be involved because they have to support the decisions.

Short Term. Individuals may also need to develop specific competencies to work on their current project. Project managers will now be very much more involved and may have to pay for the requisite training out of the project budget. Martina Huemann, Anne Keegan, and I came across two interesting examples of on-the-project training:

1. The first was a research and development organization, which would develop new technology early in the project and then test it under extreme conditions later in the project. They only had one opportunity to get the test right. So the project team members who had to do the testing went through expensive training in the use of the new technology and how to test it, and did several dummy runs under simulated conditions, so they could get it right when they had their one chance to do it.
2. The second was a software consultant that had to build up a project team from 32 to 96 people to do a development job for a client. The people joining the team had to be introduced to the legacy software and systems and strategy for the new system. Each person needed one week's training and only eight people could be trained at a time. So the training determined the rate at which people could join the team and it took eight weeks much to the annoyance of the customer.

17.3 DEVELOPING ORGANIZATIONAL CAPABILITY

The competence of individuals is a necessary component of organizational capability but not a sufficient component. Organizational capability is much more than the sum of the parts. Over the next two sections, I consider how to develop organizational capability describing four practices and four processes for developing it. There are four practices that can help organizations to develop organizational capability.

Procedures Manuals

A set of procedures manuals embodies the organization's project management capability. It sets out how the organization does the things I described in Sec. 17.1: the project life cycle, the management cycle, and the project management functions. Once captured, the procedures manual is the document that can be used to train and develop apprentice project managers.

Purpose of the Procedures Manual. There are several reasons why organizations need procedures manuals. They can provide

- A guide to the management processes
- A consistent approach and common vocabulary (see Example 17.2)
- A basis for company resource planning
- Training of new staff, especially apprentice project managers
- Demonstration of procedures to clients, perhaps as part of contractual conditions
- The basis for quality accreditation

Example 17.2 A common vocabulary

I worked in a company where the word "commissioning" was taken by the mechanical engineers to mean M&E trials, by the process engineers as the period following M&E trials during which the process was proved, by the plant operators as the period following process testing in which the first product was produced, and by the software engineers as all of those combined in which the computer control system is tested and proved.

A colleague reports working on a project to construct a petrochemical facility for which there were two project managers, one responsible for design, and one for construction. When asked what they understood by completion, one said completion of M&E trials, the other operating at 60 percent of design capacity. Both were working to the same day, even though the two dates are at least three months apart.

He reports another project to develop a computer system where when asked the same question people gave answers ranging from completion of beta test to the system has operated for 12 months without problem. Again they were working to the same date even though they were again at least 15 months apart.

On these projects, some people were going to judge them a success, and some a disaster.

However, some words of warning about the procedures manual:

Guidelines not rigid rules: I emphasized right back in Chap. 1 that the procedures should be guidelines not rigid rules. Every project is different and so the procedures need to be adapted to the needs of every project. The procedures represent organizational best practices, but the needs of the project and customer need to be accounted for (see Example 17.3). But I have said before, the more you change the procedures the more likely you are to make a mistake, so you want to use the standard as much as possible.

Different procedures for different types of project: It is also necessary to have different procedures for different types of project.[20] You need different procedures for different sizes of projects and for different technologies. Lynn Crawford, Brian Hobbs, and I gave guidance on how to categorize projects to choose appropriate procedures.[21]

Example 17.3 Adapting the procedures

I worked with a major design and construction contractor from the engineering industry. They would not let project managers manage projects until they knew how to adapt the procedures to the needs of the project; it was part of their tacit knowledge. Apprentice project managers were given the procedures and told to follow them to the letter as they shadowed their mentor, but they would not be let loose on their own until they knew how to adapt them.

Structure of the Procedures Manual. I suggest that the procedures manual should have the following parts:

Part 1—Introduction: This explains the structure and purpose of the procedures manual.

Part 2—Project strategy: This describes the approach to project management to be adopted by the organization, and the basic philosophy on which it is based. It will cover issues such as those described in Chap. 3 and Sec. 17.1. It describes the project model, introduces the stages of the life cycle to be followed (Part 3), and explains why they are adopted. It also explains the need to manage project management functions, and also the need to manage risk.

Part 3—Management processes: This describes the procedures to be followed at each stage of the life cycle. The inputs, outputs, and their components are listed, and the management processes required to convert the former to the latter are listed sequentially. Table 17.2 presents the contents page of a manual for an IT project, which shows that in some areas the breakdown was taken to between one and three levels below the project stage. It is adapted from manuals I have prepared for clients. In the procedures manuals of Table 17.2, I drew pictorial representations of the processes to achieve each stage or substage in the life cycle. Figures 17.2 and 17.3 are those for successive levels of breakdown. Where there is a lower level of definition, the process is shown as a fine box. Where the process is the lowest level, it is shown as a bold box. Against each bold box were listed the inputs, outputs, and steps required to achieve it.

Part 4—Supporting procedures: This part explains supporting procedures used throughout the project. It may describe the method of managing the project management functions in Part 2, scope, organization, quality, cost, time, and risk, or it may explain some administrative procedures, such as program and portfolio management, configuration management, or conducting audits and health checks (Sec. 18.3), or methods of data collection (including time sheets), or the role of the project support office. Only those important in the particular environment will be necessary.

Appendices—These may contain blank forms and samples.

Project Management Community of Practice

A project management community of practice provides a forum through which project managers can meet, exchange ideas, and form self-supporting networks. A community of practice can provide many benefits including:

- The support and mentoring of apprentice project managers
- A network through which project managers can meet and learn what other project managers are doing
- A development cell to identify good project managers worth developing further, for the benefit of the organization and their careers

I have worked with many organizations, including consultancies, software suppliers, and the military, who create communities of practice. A common feature is a regular meeting, usually about once every three months, in the form of a conference or seminar, where project managers can meet. A typical pattern is a seminar, lasting between two to four hours, with one or more internal speakers and external speakers, followed by a buffet supper. The attendees hear some new ideas from the external speaker(s), and something about project management within the organization from the internal speaker(s). During the buffet supper they can meet other project managers and talk about what they are doing.

TABLE 17.2 Contents Page for a Procedures Manual

		TriMagi ***Project success***
Contents		
		Introduction
		Program Management
PM		*Information Systems Project Management*
		Proposal and Initiation
P0		*Definition and Appraisal*
		Develop work breakdown
	P1	Develop milestone plan
		Work-package scope statements
	P2	Activity plans
		Develop project networks
	P21	Define specification and configuration
		Schedule resources and work
		Estimate resource and material requirement
	P213	Update project network
		Schedule project network
	P214	Produce resource and material schedules
		Schedule cost and expenditure
	P215	Assess risks
		Define controls
	P216	Appraise project viability and authorize
	P22	*Contract and Procurement*
		Develop contract and procurement plan
	P23	Make payments
		Execution and Control
		Finalize project model
	P233	Execute and monitor progress
		Control duration
	P234	Control resources and materials
		Control changes
	P235	Update project model
		Finalisation and Close-out
	P236	
	P24	
	P25	
	P256	
	P26	
P3		
	P31	
	P36	
P4		
	P41	
	P42	
	P43	
	P44	
	P45	
	P46	
P5		
Appendices		
A		Project planning and control forms
B		Supporting electronic databases
C		Sample reports
D		Staff abbreviations (OBS)
E		Resource and material codes (CBS)
F		Management codes (WBS)

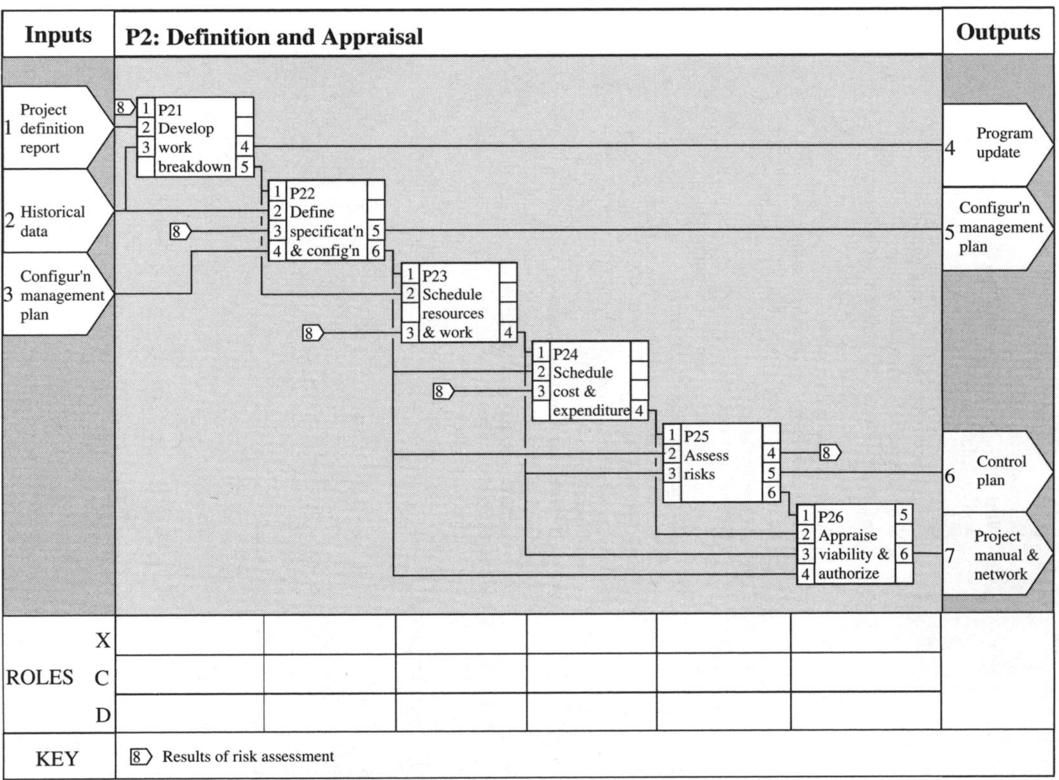

FIGURE 17.2 Pictorial representation of the stage, P2: *Project definition.*

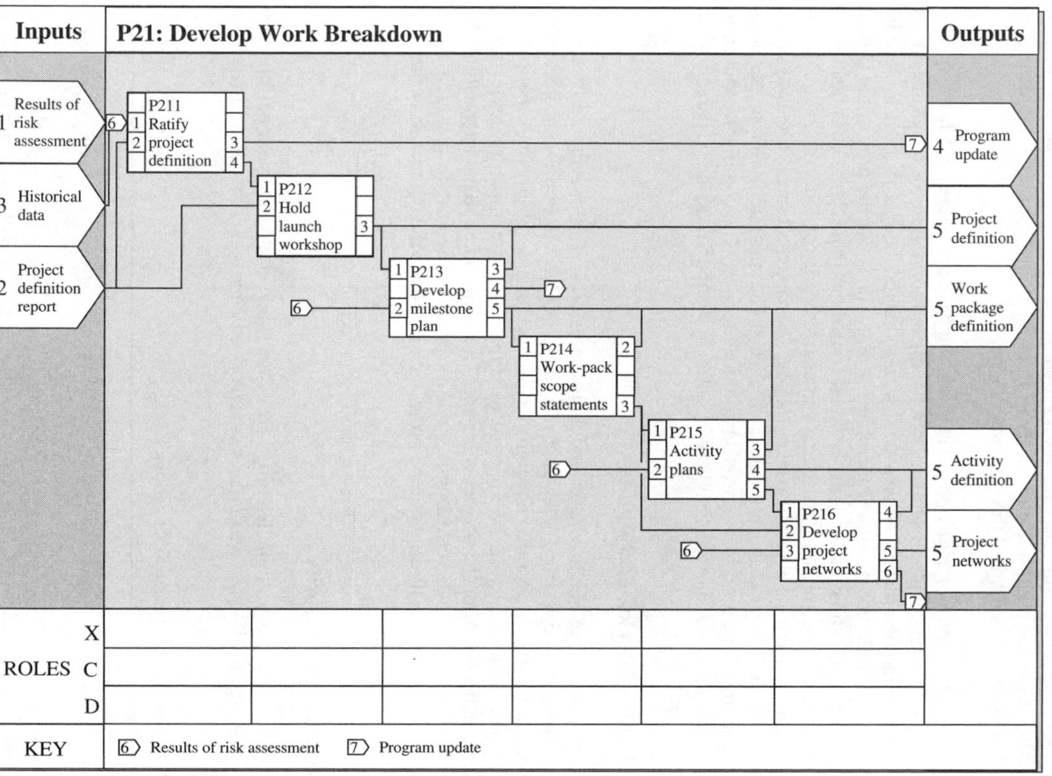

FIGURE 17.3 Pictorial representation of the step, P21: *Develop work breakdown.*

Perhaps then a month or two later they have a problem, and remember talking about something similar with someone else, and can contact them. I have been an external speaker at events lasting two hours to two days.

A Vehicle for Learning. The community of practice can provide a vehicle for learning within the organization. Table 17.3 is a model for organizational learning[16] in which the organization uses the community of practice to identify its tacit knowledge and make it explicit, and having made it explicit can codify it, improve it, incorporate it into its procedures, and them make it tacit again through practice.

Supporting the Community of Practice. The community usually does not form spontaneously, and cannot survive without support from the organization. It needs top management support to start it in the first place, and provide it with a strategic budget in order for it to be able to continue. It should not consume much money. But it does use the time of the person currently running it, and so he or she must be freed from other duties. In a consultancy for instance, the leader of the community must be given an overhead cost code to book his or her time to. The organization of the events also takes money. As an external speaker, I always have my expenses paid and am occasionally paid a small honorarium. The buffet supper costs money, and there may be a need for limited technological support. The community must also be provided with clear leadership to ensure it works effectively and to encourage participation.

Reviews, Health-Checks, and Audits

Project reviews, including health-checks and audits are a way of improving performance on the current project, and learning from past success and failures to improve project management in the organization overall. I am going to discuss these more fully in the next chapter. For now I just want to say that the emphasis used to be on *post*completion reviews, and so the learning happened after the project was over. This contributed to the attenuation discussed in the next section, where learning is always put off. The emphasis now is much more on conducting reviews throughout the project, especially at the completion of project stages or major milestones.

Benchmarking and Maturity

The final practice for organizational learning is benchmarking, comparing your performance with others, to try to identify your weaknesses, and work on improving those areas. In its basic

TABLE 17.3 Nonaka and Takeuchi's Learning Cycle

		To	
		Tacit knowledge	Explicit knowledge
From	Tacit knowledge	*Socialization* Sharing—creating tacit knowledge through experience	*Externalization* Articulating Tacit knowledge through reflection
	Explicit knowledge	*Internalization* Learning—acquiring new tacit knowledge in practice	*Combination* Systematizing explicit knowledge and information

form you compare your performance with people doing similar projects. This can be with projects you previously did, other projects in your department, other departments within your parent organization, directly with competitors, or with people from entirely different organizations. In reality it is usually difficult to compare directly with your competitors, since they don't want you looking at their projects, but I am aware of benchmarking networks where direct competitors feel the benefit of comparing with each other outweighs the risk involved. An alternative, rather than comparing directly with each other, is to compare with an industry-wide database. The European Construction Institute (*www.eci-online.org*) and the Construction Industry Institute (*www.construction-institute.org*) jointly maintain a database of 4000 projects for process plant construction. So you cannot compare your project directly with a competitor's project, but you can compare with industry averages.

An alternative to direct benchmarking is to use a maturity model to assess your performance, and indeed this is now more common. The first was the capability maturity model (CMM) developed by the Software Engineering Institute (SEI) of Carnegie Mellon University for software systems deveopment.[22] The original CMM model did not include project management, but the updated one, CMMI, did. PMI has now developed a maturity model for project management, the organizational project management maturity model (OPM3).[23] With a maturity model you answer a series of questions to determine your performance against a series of parameters, which you can then plot in a spider web model in Fig. 17.4. You can then compare your current performance to the desired scores for the next level of maturity you are aiming for. You identify areas where you fall short and can then work on improving your performance in those areas, hopefully while maintaining your performance in the areas where you already exceed the requirement. This is performing a gap analysis for the organization directly comparable to what we did for individuals above.

Maturity and the Four Practices. Table 17.4 contains a very simplified description of the five levels of maturity in SEI's CMM model. You will see at Levels 2 and 3 the organization works on improving its procedures and community of practice and at Levels 4 and 5 reviews and benchmarking. So the four practices I have described in this section directly contribute to maturity.

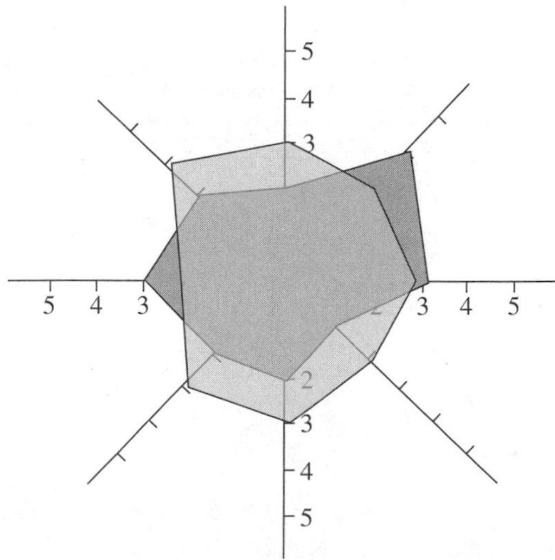

FIGURE 17.4 Spider web model.

TABLE 17.4 Levels of Maturity

Level	Requirements
1	Ad hoc processes, no guidance, no consistency
2	Procedures for individual processes, minimum guidance
3	Full procedures manual, group support
4	Processes measured, collection of experiences, and learning through reviews
5	Benchmarking, continuous improvement

Maturity and Organizational Performance. The concept of maturity does beg the question whether increasing maturity leads to increasing performance. Work in that area has been done at the University of California in Berkley.[24] They found that increasing maturity did lead to increasing performance, as illustrated in Fig. 17.5. However, their results were not statistically significant, so they identified a trend but could not confirm it. Their performance curve followed a learning curve, so the performance improvement going from one level to the next is only half what was obtained coming for the previous level to this. But the cost of achieving the performance improvement is twice what it was at the previous level. Bill Ibbs and Justin Reginato defined a *project management return on investment* (PM ROI) as the performance improvement obtained divided by the cost of obtaining increased maturity:

$$\text{PM ROI} = \frac{\text{efficiency gain \%} \times \text{annual project spend}}{\text{cost of achieving improvement}}$$

The efficiency gain is the combined improvement from cost and schedule performance improvement. Unfortunately, the ROI from one level to the next is only a quarter of that from the previous level to this. Many western companies find it is not worth progressing from Level 3 to Level 4, whereas firms in low-wage economies find the improvement worthwhile. In low-wage economies the cost of achieving improvement is less because wages are less, but "annual project spend" is higher because material costs are higher. In China and India, a large numbers of companies have achieved Level 5 on SEI's CMM model. In the West, only very large, project-oriented organizations can afford to go for maturity Level 5. Companies doing fewer projects need to stick at Level 3.

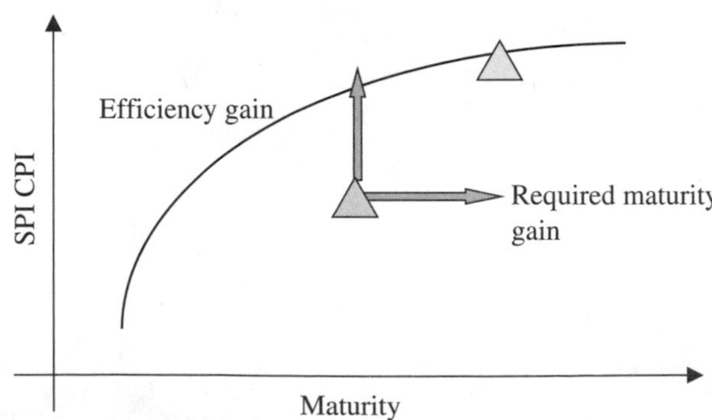

FIGURE 17.5 Increasing performance with maturity.

Bill Ibbs and Justin Reginato also found that the cost of project management also varied with maturity. Organizations with maturity Level 1 spent on average 3 percent of project costs on project management. This rose to 6 percent for companies with maturity level 3, and fell back to 3 percent for companies with Maturity Level 5. There are two issues here:

- If you need to stick at maturity Level 3, you need to work on reducing the cost of project management getting it back to 3 percent.
- This is a competency trap; organizations see the cost of project management increasing as they try to improve maturity, and do not want to spend more on project management, and so stay on low maturity.

17.4 IMPROVING ORGANIZATIONAL CAPABILITY

Anne Keegan and I found a four-step process for continuously improving organization project management capability[25] (Fig. 17.6):

Variation. Through trial and error and practical experience, you identify new ways of managing and delivering projects. The project management community and top management can help in this process by identifying weaknesses in the current approaches and encouraging people to try new ideas. I discuss competency traps below, but you need to avoid blaming culture and encourage people to try new ideas and accept the occasional mistake for the long-term benefit of finding a better way of working. Audits, reviews, and benchmarking can also help pinpoint weaknesses and identify a need of variation.

Selection. Through review processes, you determine those new practices which provide benefit through improved performance, and those which don't. The project management community, reviews, and benchmarking can also help in the selection process.

Retention. Through procedures manuals, you store the selected new ideas where they are accessible. The selected new ideas then need to be retained centrally, perhaps by the *project office* (Chap. 16), perhaps through knowledge management processes (Sec. 17.5), or

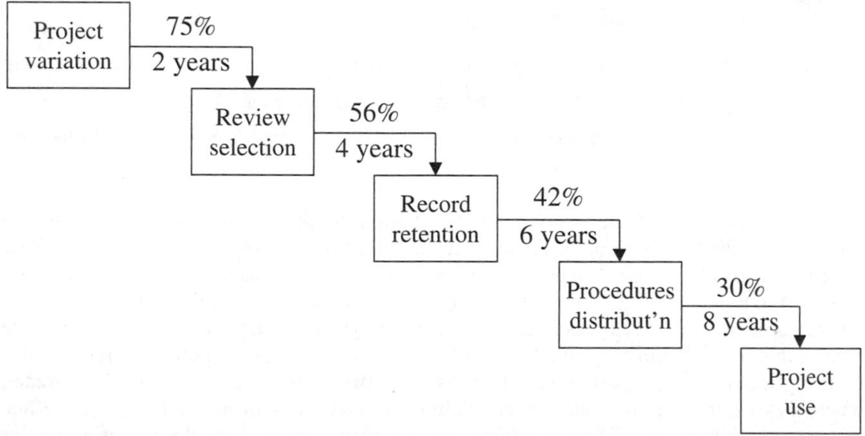

FIGURE 17.6 Four-step process for improving project management capability.

perhaps through story-telling in the project management community. They cannot be retained on a project, because the project is going to be disbanded. That is the problem of projects as temporary organizations; they cannot own and retain knowledge, it has to be owned and retained separately from the project, and thus the need for the fourth step.

Distribution. Through project management procedures and the project management community you distribute the selected new ideas to project managers on projects where they can be used to improve performance on future projects.

The first three steps of this cycle were first described by people working on the evolution of species, explaining how new genes arise by random mutation, are selected through survival of the fittest, and retained in the gene pool. The model was subsequently adopted by the management-learning literature. In a functional organization the new ideas are generated in the functional hierarchy, selected and retained there, where they are immediately available for use by managers. But Anne Keegan and I identified that in a project-based organization there is the essential fourth step, distribution. In a project-based organization, new ideas are generated on one project which is going to come to an end, and so need to be transferred to a central pool where they will be held and from there to managers working on new projects, where they can be used.

Attenuation and Delay. Anne Keegan and I also identified two further issues with the four learning practices (Fig. 17.6). First there is a loss of learning at each step of the process. Terry Cooke-Davies has identified that there is a 25 percent loss of information at each step meaning that less than one-third of the good new ideas that a project-based organization generates actually end up being used on new projects.[26] There are many ways suggested to overcome that problem:

- Make project reviews mandatory.
- Make the project office responsible for collating the results of reviews.
- Use the intranet to store and distribute the good ideas.
- Ensure the project management community is working effectively.

The first three all add up to the bureaucracy of the organization, and so you have to balance the benefit of the knowledge management and innovation, against the cost of processing the data.

There tends to be a delay between each step:

- With *post*completion reviews, there may be a delay before ideas are selected and retained.
- It may be two years before the next edition of the procedures are issued.
- The project manager at the coal face may be too busy to read the new procedures until the start of his or her next project.

There is a concept of the viscosity of information. Some information oozes through an organization like treacle taking years to go from new idea to use on a new project. But other information zips through like gas through a vacuum. This is especially true of information entered into the intranet. It is immediately available for use, and so if there is no control on what information is entered, yesterday's hearsay can become today's received wisdom. The ideal is that there should be some delay with variation and selection, perhaps three months, allowing ideas to be properly tested and distilled, before being entered into the intranet, where they would be immediately available for reuse on future projects. The project office can manage this process. Other organizations are using the ideas of discussion rooms or "wiki" space. In a wiki space, individuals can enter ideas they have, describe how they solved a problem, or describe a new management approach they used. Other people can then comment on that idea, say whether they tried it, what experience they had, and how valuable they found it. The new ideas are then tested and selected through trial and discussion. New ideas can also be tested through the project management community in the same way.

17.5 KNOWLEDGE MANAGEMENT

An essential part of the above process is retention and distribution, and this falls within the wider area of knowledge management.

The Donald Rumsfeld Problem

There are four types of knowledge that need to be managed:

- That you know you have
- That you don't know you have
- That you know you don't have
- That you don't know you don't have

Conventionally, knowledge management has focused on just the first of these. But I suggest you should also think about how to discover the other three.

Know You Know. This is explicit knowledge you know you have. There are four questions which help you develop a knowledge management system:

1. *Where are we?* Establish an inventory of your current knowledge management practices.
2. *Where do we want to be?* Consider your knowledge management needs, what knowledge you need to improve performance. You need to identify your business context and drivers, the context within which your projects take place. Then you want to identify the success factors and key performance indicators relevant to your projects, which will tell you what you need to manage to improve performance, and so what knowledge you need. Then you need to identify the characteristics of the knowledge: what are the sources of knowledge, who are the users, and what are the enablers and inhibitors?
3. *What is the gap?* This will identify a gap between your current knowledge management practices and those you need.
4. *What is the migration path?* So you can then plan the project to develop the knowledge management systems you need.

This is gap analysis again (Fig. 2.1) sometimes known as the Y-model because the first two steps should be conducted independently, with the results coming together at the third to identify the gap. The first two steps should be independent because you don't want to influence the other; you don't want to define your requirements based on what you have got, or have your understanding of your knowledge inventory influenced by what you need.

The knowledge management system needs to consist of four processes:

1. Knowledge generation:
    - What is the knowledge, where does it come from, and how is it captured?
    - How is data converted to information, information to knowledge, and knowledge to wisdom?

2. Knowledge transfer:
   - How is knowledge distributed from where it is generated to where it is used?
3. Knowledge location and access:
   - Where is it stored in repositories of data?
   - How is it transferred to those who need it?
4. Knowledge maintenance and modification:
   - Who has the right to add to it?
   - Who has the right to change it?

Don't Know You Know. There are two types of knowledge that fall into this category: tacit knowledge and what I call "X-files" (Example 17.4 explains why). I described above how to use Nonaka and Takeuchi's cycle, coupled with the project management community to make tacit knowledge explicit and thereby make it known. This can also help identify knowledge through random connections, perhaps resulting in two seemingly unrelated ideas being put together. The X-files need to be found by data mining or careful archiving, and then they should be indexed so that files can still be found by the search engine on the intranet.

Example 17.4 X-files

In the first episode of the X Files, Mulder and Sculley think they have found new evidence of aliens. In particular they find a person with a device implanted in his ear. This device is conclusive proof, so they send it to FBI headquarters. In the last scene, a man is seen walking in the depths of the FBI basement past all the archive (X) files. He reaches a row, pulls out a box, and throws in the device, and there are already five or six in the box. The knowledge (truth) is not "out there," it is in the X-files in the basement of the FBI where nobody knows about it. Don't know they know.

Know You Don't Know. This is easier. Research can be done in the normal places: the internet (Wikipedia and Google), research journals, and books. Benchmarking and reviews can also help. Organizations also conduct research workshops to improve their own understanding of a particular situation. Project start workshops (Chap. 12) are in fact an example of such a workshop.

Don't Know You Know. You don't know to look for it and it is difficult to discover in structured normative ways. It requires random searches and random interconnections. Having people work together in cross-discipline teams can throw up new ideas. Encourage people to meet and talk at the water cooler.

17.6 COMPETENCY TRAPS

Competency traps are things that stop us from learning. There may be a better, more efficient, or more effective way of working, but competency traps either stop us finding it, or stop us trying it even if we know it exists. In project-based organizations competency traps include the desire for safety and reliability, blaming culture, contracting practice, and so on.

The Desire for Safety and Reliability. You only have one shot at a project, and so there is often a preference for safety and reliability than efficiency and effectiveness. For instance, you may have two ways of doing something, one guaranteed to work with 100 percent efficiency and the other with 80 percent chance of success but 300 percent efficiency. The second way is on average two-and-a-half times better. But managers in project-based organizations often prefer the way that offers guaranteed success for the one time they are going to do it. If in a functional organization you are going to do it 100 times. The first 10 times you get it right 8 times and wrong twice, but be two-and-a-half times better off. The next 10 times you will have learnt from your mistakes and get it right 9 times and wrong once. From then on you will get it right every time and be 3 times better off. But in a project you only do it once and want it right that one time.

Blame Culture. This is related to the previous trap, but now looks at the decision from the perspective of the person making the decision, not the organization. If they work in a blame culture they will have an overriding preference for the safe, reliable option. Their assessment of the situation is different. If they choose the first option they have a certain chance of a quiet life. Nobody will notice. If they choose the second option they have an 80 percent chance of a quiet life, nobody will notice if it works. They won't get praised for the extra efficiency. But if it goes wrong they will get blamed, so they have a 20 percent chance of being hounded. They will choose the safe option.

In the early 1980s, I worked for a chemical manufacturer. There the attitude was if you don't take risks, you don't make profits, but if you take risks you make the occasional mistake. They therefore liked people who made the very occasional mistakes, and didn't like people who never made mistakes. People who never made mistakes weren't making profits. It was the exact opposite of a blame culture. Of course if you always made mistakes you got put on "special duties" and eventually shuffled offstage.

Contracting Practice. Standard contacting can be a competency trap. Don't expect a contractor on a remeasurement contract without a bonus to suggest a process improvement. They are going to lose profit. If you want your contractors to suggest improved ways of working, you need to offer them a bonus to do it.

Fear of Competitors Stealing Your Innovations. Some organizations don't innovate for fear that their competitors will steal their new ideas. It is a similar reason to why some organizations don't train their staff, for fear that they will leave and their competitors will get the benefit of the training. People are actually more likely to stay if they are properly developed. Enlightened organizations train their staff; enlightened organizations find ways of improving their processes. Yes their competitors will eventually adopt the ideas as well, but it will take them about two years, so the organization that does the research and development will always have two years lead.

Nonlinearity and Coupling of Projects. Projects are nonlinear coupled systems. To make improvements requires not just one bit to be changed on its own but the whole project to be changed. That can sometimes create complexity as it is difficult to design a new, integrated solution. Or it can create competition where each stakeholder wants to optimize the project outcome for themselves, resulting in an inferior outcome for the whole project. I discussed this in Chap. 3, when I discussed the need to obtain a balance of all the stakeholders' different objectives.

Traditional Project Management Thinking. One of the worst competency traps of all is traditional project management thinking. It preaches rigid control and certainty of estimates.

Closing the estimates early can often lock you into high-cost solutions. Often, to find the best solution for the project requires you to keep options open for as long as possible, and that requires you to maintain uncertainty of the outcomes longer than you may be comfortable with. Rigid plans, with rigid control, can also lock you into high-cost solutions at an early stage.

SUMMARY

1. Organizational capability comprises
   - The project management body of knowledge
   - An understanding of how to manage individual projects to deliver their objectives
   - Technical and craft skills
2. The project management body of knowledge comprises
   - The project life cycle
   - The management cycle
   - The project management functions: scope, organization, quality, cost, time, and risk
3. Project management competency is
   - Knowledge
   - Skills
   - Personal characteristics
   - To perform in accordance with defined standards
4. Different levels of management require different profiles of competence. Lower levels require more technical competencies. Higher levels require more people and strategic competencies.
5. Competence is assessed by gap analysis, comparing current competence to desired levels for future development.
6. There are three development horizons for individuals and organizations:
   - Long-term career plans and succession strategies
   - Medium-term education and experiential development
   - Short-term specific competency training
7. The first two of these need to be aligned with the line having time horizons longer than projects and the third with the project.
8. There are four practices for developing organizational capability
   - Procedures manuals
   - A project management community of practice
   - Reviews
   - Benchmarking
9. There are four processes for developing organizational capability
   - Variation
   - Selection
   - Retention
   - Distribution
10. There are four steps for assessing knowledge management needs
    - Assess where you are.
    - Define where you want to be.
    - Determine the gap.
    - Develop a knowledge management plan to close the gap.
11. There are four steps of knowledge management
    - Generation
    - Transfer

- Location and access
- Maintenance and modification

12. There are six competence traps in project-based organizations:
- Desire for reliability
- Blame culture
- Contracting practice
- Fear of competitors stealing ideas
- Nonlinearity and coupling of projects
- Traditional project management systems thinking

REFERENCES

1. Project Management Institute, *A Guide to the Project Management Body of Knowledge,* 3d ed., Newtown Square, PA.: PMI, 2004.
2. International Project Management Association, *IPMA Competence Baseline: The Eye of Competence,* 3d ed., Zurich: International Project Management Association, 2005.
3. Association for Project Management, *The APM Body of Knowledge,* 5th ed., High Wycombe, 2006.
4. Turner, J.R (ed.), *People in Project Management,* Aldershot, U.K.: Gower, 2003.
5. Turner, J.R and Wright, D., *The Commercial Management of Projects,* Aldershot, Gower: 2009.
6. Office of Government Commerce, *Managing Successful Projects with PRINCE2,* 4th ed., London: The Stationery Office, 2005.
7. Payne, J.H. and Turner, J.R., "Company-wide project management: the planning and control of programmes of projects of different types," *International Journal of Project Management*, **17**(1), 55–59, 1999.
8. Turner, J.R. and Müller, R., *Choosing Appropriate Project Managers: Matching Their Leadership Style to the Type of Project,* Newtown Square, PA.: PMI, 2006.
9. Boyatzis, R.E., *The Competent Manager: A Model for Effective Performance,* New York: Wiley, 1982.
10. Spencer, L.M.J. and Spencer, S.M., *Competence at Work: Models for Superior Performance,* New York: Wiley, 1993.
11. Gonczi, A., Hager, P., and Athanasou, J., *The Development of Competency-Based Assessment Strategies for the Profession,* Canberra: Australian Government Publishing Service, 1993.
12. Engineering Construction Industry Training Board, *National Occupational Standards for Project Management,* Kings Langley, Herts, U.K.: Engineering Construction Industry Training Board, 2003.
13. IBSA, *Volume 4B: Project Management.* In: *BSB01 Business Services Training Package Version 4.00,* Hawthorn, Victoria, Australia: Innovation and Business Skills Australia, 2006.
14. PMSGB, *South African Qualifications Authority Project Management Competency Standards: Levels 3 and 4,* Pretoria, South Africa: South African Qualifications Authority, 2002.
15. Crawford, L.H., "Assessing and developing the project management competence of individuals," In: Turner, J.R. (ed.), *People in Project Management,* Aldershot, U.K.: Gower, 2003.
16. Nonaka, I. and Takeuchi, H., *The Knowledge-Creating Company,* New York: Oxford University Press, 1995.
17. Kolb, D. A., *Experiential Learning: Experience as the Source of Learning and Development.* Englewood Cliffs, NJ : Prentice-Hall, 1984.
18. Turner, J.R., Huemann, M., and Keegan, A.E., *Human Resource Management in the Project Oriented Organization,* Newtown Square, PA.: Project Management Institute, 2007.
19. Keegan, A.E. and Turner, J.R., "Managing human resources in the project-based organization," In: Turner, J.R. (ed.), *People in Project Management,* Aldershot, U.K.: Gower, 2003.
20. Payne, J.H. and Turner, J.R., "Company-wide project management: the planning and control of programmes of projects of different types," *International Journal of Project Management*, **17**(1), 55–59, 1999.

21. Crawford, L.H., Hobbs, J.B., and Turner, J.R., *Project Categorization Systems: Aligning Capability with Strategy for Better Results*, Newtown Square, PA.: Project Management Institute, 2005.

22. Paulk, M.C., Weber, C.V., Curtis, B., and Chrissis, M.-B., *The Capability Maturity Model: Guidelines for Improving the Software Process*, Boston: Addison-Wesley Longman Publishing, 1995.

23. Project Management Institute, *Organizational Project Management Maturity Model,* Newtown Square, PA.: Project Management Institute, 2003.

24. Ibbs, C.W and Reginato, J., *Quantifying the Value of Project Management,* Newtown Square, PA.: Project Management Institute, 2002.

25. Keegan, A.E. and Turner, J.R., "The management of innovation in project based firms," *Long Range Planning,* **35**, 367–388, 2002.

26. Cooke-Davies, T.J., *Towards Improved Project Management Practice: Uncovering the Evidence for Effective Practices through Empirical Research*, PhD Thesis, Leeds,: Leeds Metropolitan University, 2000. Available: *www.dissertation.com.*

CHAPTER 18

GOVERNANCE OF THE PROJECT-BASED ORGANIZATION

We turn our attention now to the highest level of governance, where the board of directors takes an interest in the (key, large) projects taking place within the organization, where they:

- Set objectives for those (key, large) projects.
- Ensure people are appropriately empowered and motivated to enact the projects.
- Ensure appropriate controls are in place to ensure the projects achieve their objectives, both for the delivery of benefit and consumption of resources—they need to do this both to ensure the projects are profitable and from a compliance perspective to meet their responsibilities to their shareholders.

Traditionally, boards of directors and senior managers have ignored projects, taking a greater interest in routine operations. Projects were something taking place in the skunk works, managed by geeks. But under modern compliance regimes boards are responsible for the performance of projects, and so they have to take an interest. The United Kingdom's Association for Project Management (APM) has a special interest group (SIG) looking at the governance of project management, with specific focus on the overlap between the board and project management.[1] In the next section I give an overview of key points of this guide. I then further expand on three practices it recommends: audits, health checks, and end-of-stage reviews.

18.1 GOVERNANCE OF PROJECT MANAGEMENT

The APM guide suggests the aims of good corporate governance are to ensure

A1: A clear link between corporate strategy and project objectives:
- In the definition of the project (Chaps. 2 and 5)
- In the benefits and project governance roles (Chap. 15)
- In portfolio and program management (Chap. 16)

A2: Clear ownership and leadership from senior management (Chap. 15)

A3: Engagement with stakeholders (Chap. 4)

A4: Organizational capability (Chap. 17)

A5: Understanding of and contact with the supply industry at a senior level

A6: Evaluation of project proposals based on their value to the organization not capital cost

A7: A focus on breaking down development and implementation into manageable (Part 3)

Principles of Good Governance

In order to achieve these objectives, the guide suggests eleven principles of good governance of project management:

P1: The board of directors must assume overall responsibility for the governance of projects. They have a duty under modern compliance regimes to be able to predict future cash flows of the business, and this requires them to be able to predict outturn cost and future returns for all large projects, programs, and portfolios.

P2: Roles, responsibilities, and performance criteria for the governance of projects (and programs and portfolios) must be clearly defined (Chaps. 15 and 16).

P3: Defined governance arrangements, supported by appropriate methods and controls, must be applied throughout the project life cycle.

P4: Members of delegated authorization bodies have sufficient representation, authority, competence, and resources to take the decisions for which they are responsible. Such authorization bodies include

- Project or program steering committees, including sponsor, owner, steward, and project manager (Chap. 15 and 16)
- The portfolio selection committee (Chap. 16)

P5: There must be a coherent and supportive relationship between the overall corporate strategy and the project portfolio (Chap. 16).

P6: The project business case must be supported by sound and realistic data so decisions can be based on the knowledge that predictions are valid and the board can meet its duties under the compliance regimes

P7: All projects must have an approved plan with defined authorization points where the business case will be reviewed and approved. Decisions made at the authorization points must be clearly recorded. In Chaps. 1 and 15 I have shown that during the project the project governance structure must be aligned with the project; that is, the project manager must be put in control and empowered to take decisions. Senior and functional management do not like this because they are ceding control to the project manager. But by having clear authorization points, they only have to cede control between authorization points. End-of-stage reviews can be used as authorization points.

P8: There are clearly defined key performance indicators for reporting project status and for escalating risks and issues to appropriate levels (Secs. 3.2 and 16.2).

P9: The board and its delegated agents decide when independent audits of projects, programs, and management systems is required and implement such audits as required (Sec. 18.2)

P10: Project stakeholders are engaged at a level that is appropriate for their importance and in a way that fosters trust and cooperation (Chap. 4).

P11: The organization fosters a culture of continuous improvement and frank discussion and project reporting. The organization aims to be a learning organization (Secs. 17.3 and 17.4) and avoid competency traps (Sec. 17.6) especially those associated with a blame culture.

Components of the Governance of Projects

The APM guide identifies four components of the management of project management:

- Portfolio direction (PD)
- Project sponsorship (PS)
- Project management (PM)
- Disclosure and reporting (DR)

Tables 18.1, 18.2, 18.3, and 18.4 show key questions under each component and the governance principles to which they are related.

Compliance

As I have said, because of the modern compliance regime, boards of directors and senior managers have to take a much greater interest in projects and project management than they have traditionally taken.[2] Many of the principles of governance listed above are consistent with and supportive of the Sarbanes-Oxley Act 2002, as shown in Table 18.5.

TABLE 18.1 Key Questions for Portfolio Direction (PD)

	Issue	Related principles	Related aims
PD1	Are the organization's financial controls, financial planning, and expenditure review processes applied to both individual projects and the portfolio as a whole?	P2, P3, P5, P8	A6
PD2	Does the organization discriminate correctly between activities that should be managed as projects and other activities that should be managed as nonproject operations?	P3	
PD3	Is the organization's project portfolio aligned with its key business objectives, including those of profitability, customer service, reputation, sustainability, and growth?	P5	A1, A6
PD4	Is the project portfolio prioritised, refreshed, maintained, and pruned in such a way that the mix of projects continues to support strategy and take account of external factors?	P5, P7	
PD5	Has the organization assessed the risks associated with the project portfolio, including the risk of corporate failure?	P5, P6, P7, P8	
PD6	Is the project portfolio consistent with the organization's capacity?	P5	A4
PD7	Does the organization's engagement with project suppliers encourage a sustainable portfolio by ensuring their early involvement and by a shared understanding of the risks and rewards?	P10, P8	A5
PD8	Does the organization's engagement with its customers encourage a sustainable portfolio?	P10	A3
PD9	Does the organization's engagement with the sources of finance for its projects encourage a sustainable portfolio?	P10, P5	A3
PD10	Is organization assured that the impact of implementing its project portfolio is acceptable to its ongoing operations?	P10, P11, P5	A3, A4

TABLE 18.2 Key Questions for Project Sponsorship (PS)

	Issue	Related principles	Related aims
PS1	Do all major projects have competent sponsors at all times?	P1, P2, P3	A2
PS2	Do project sponsors provide clear and timely?	P3, P4	A2
PS3	Do sponsors devote enough time to the project?	P4	A2
PS4	Do project sponsors ensure that project managers have access to sufficient resources with the right skills to deliver projects?	P4	A4
PS5	Do sponsors own and maintain the business case, and are they accountable for realization of benefits?	P5, P6	A1
PS6	Do project sponsors hold regular meetings with project managers and are they sufficiently aware of project status?	P7, P8	A2
PS7	Are projects closed at the appropriate time?	P7, P8	A1
PS8	Is independent advice used for appraisal of projects?	P9	
PS9	Do sponsors adequately represent the project throughout the organization?	P10	A3
PS10	Are the interests of key stakeholders, including suppliers, regulators and financiers, aligned with project success?	P10	A3

TABLE 18.3 Key Questions for Project Management (PM)

	Issue	Related principles	Related aims
PM1	Is the board assured that the organization's project management processes are appropriate for the projects that it sponsors?	P1, P3, P7, P8	A2, A4
PM2	Do all projects have clear critical success criteria and are they used to inform decision-making?	P2, P5, P6. P7	A1
PM3	Are key success factors identified for all projects and are they used to inform decision-making?	P2, P3, P8	A1
PM4	Is key governance of project management roles and responsibilities clear and in place?	P2. P3	A2
PM5	Is the board assured people responsible for project delivery, especially project managers, are clearly mandated, competent, and have the capacity to achieve satisfactory project outcomes?	P2, P4	
PM6	Is authority delegated to the right levels, balancing efficiency and control?	P2, P3, P4	
PM7	Are project contingencies estimated and controlled in accordance with delegated powers?	P7, P8	
PM8	Are appropriate issue, change, and risk management practices implemented in line with adopted policies?	P8	A7
PM9	Are service departments and suppliers able and willing to provide key resources tailored to the needs of different projects and to provide an efficient and responsive service?	P10	A3
PM10	Are project managers encouraged to develop opportunities for improving project outcomes?	P11	A4

TABLE 18.4 Key Questions for Direction and Reporting (DR)

	Issue	Related principles	Related aims
DR1	Where responsibility for disclosure and reporting is delegated or duplicated, does the board ensure that the quality of information that it receives is not compromised?	P4	A2
DR2	Does the organization use measures for both key success drivers and key success indicators?	P5, P6, P7	A1
DR3	Does the board receive timely, relevant, and reliable information of project forecasts, including those produced for the business case at project authorization points?	P6	
DR4	Can organization distinguish between project forecasts based on targets, commitments, and expected outcomes?	P6	
DR5	Does the board receive timely, relevant, and reliable information of project progress, including the identification of risks and their management?	P7, P8	
DR6	Do project processes reduce reporting requirements to the minimum necessary?	P7, P11	
DR7	Are there threshold criteria that are used to escalate significant issues, risks, and opportunities through the organization to the board?	P8	
DR8	Does the board seek independent verification of reported project and portfolio information as appropriate?	P9	
DR9	Does the board reflect the project portfolio status in communications with key stakeholders?	P10	A5
DR10	Does the business culture encourage open and honest reporting, including being supportive of whistle blowers?	P11	

18.2 CONDUCTING AUDITS

Principle P9 states that the board of directors should decide when and if independent scrutiny of a project may be required. An independent review is called an audit, and is described in this section. However, project teams should also be encouraged to conduct internal reviews of themselves; they are called health checks and are described in the next section.

Purpose of Project Audits

Audits may be conducted at several points throughout a project for the following reasons:

Check that the Design is Correct. One of the primary contributing factors to the success of a project is to ensure it is correctly established and designed in the first place. This means that

- The purpose of the project has been correctly identified.
- The objectives set will deliver that purpose.
- The new asset will achieve those objectives.

TABLE 18.5 The Principles and the Sarbanes-Oxley Act (2002)

Section of the Act	Particular sections	Relevant principles
106 Foreign public accounting firms	(a) Applicability to such firms	P2, P9
108 Accounting standards	(b) Recognition of standards	P3, P11
201 Services outside the scope of practice of auditors	(a) Prohibited activities	P9
202 Preapproval requirements	(a) Audit committee action	P9, P11
204 Auditors report to audit committee	(k) Such reports	P3, P9, P11
302 Corporate responsibility for financial reports	(a) Regulations required	P2, P3, P5, P6, P8, P9, P11
401 Disclosure in periodic reports	(j) Off balance sheet transactions	P6, P7, P8, P9, P10, P11
404 Management assessment of internal controls	(a) Rules required	P2, P3, P4, P5, P6, P7, P8, P9, P11
406 Code of ethics for senior financial officers	(a) Code of ethics disclosure	P2, P3, P11
407 Disclosure of audit committee financial experts	(a) Rules defining "financial expert"	P2
407 Real-time issuer disclosures	(1) Real Real-time issuer disclosures	P8, P11
906 Corporate responsibility for financial reports	Failure of corporate officers to certify financial reports	P2, P3, P4, P6, P8, P11
1102 Tampering with record or otherwise hampering an official proceeding	©Corrupt behaviour	P3, P11

- The new asset is designed in accordance with the inherent assumptions.
- The design information used, including any research data, is valid.

A check of the design may be conducted by a red team as described in Sec. 7.2.

Ensure the Quality of the Management Processes. A second major contributor to success is the use of qualified management processes. An audit can be conducted at any time during a project to determine whether it is being managed in accordance with best practice, and that usually means in accordance with defined procedures, perhaps as set out in a manual. Such an audit is most effective when conducted about one quarter of the way into a stage, as the pattern of management has been set by that time, but work is not so far advanced that mistakes cannot be recovered.

Learn from Past Success. If a project has gone particularly well, then a review can help to identify what contributed to success. These reviews are usually best conducted at the end of a project, although it can then be difficult to gain people's commitment as they are keen to move on. However, it is usually easier to get people to review their successes than their failures.

Avoid Past Mistakes. Likewise, if a project has gone particularly badly, then it can be instructive to determine what mistakes were made, so they can be avoided in the future. However, people can be very defensive in these circumstances (unsurprisingly).

Types of Project Audit

In order to achieve these objectives, three types of project audit may be conducted.

Project Evaluation Audit. A project evaluation audit is an independent check of the feasibility or design studies. It is an enforced review of the investment appraisal as it currently stands, and the assumptions on which it is based. The auditors check the validity of the data used in the feasibility and/or design studies, and the conclusions drawn from it. Often the original design team may have been overoptimistic because they have a subjective commitment to the project. It is important that the auditors are truly independent, and that they do not share the same commitment or they may merely repeat the mistakes.

Internal Audit. An internal audit, or health check, is a quality control check of the management processes, conducted either by independent auditors or by the project team to ensure best practice is being followed, and hence that the project as defined will be delivered to quality, cost, and time. (Usually only the design or execution stages will be audited.) An audit will be conducted about one-quarter into the stage and will cover everything from progress of the work itself, to the procurement and marshalling of materials. The auditors will check

- The validity of the data being gathered.
- How it is being used to generate management reports.
- How those reports are being used to take timely and effective action, to ensure that the project meets its quality, cost, and time targets.

Postcompletion Audit. The successes and failures of a project are reviewed in a postcompletion audit. The scope of a postcompletion audit may be very similar to an internal audit, but now the auditors are checking past practice with the knowledge of how the project actually turned out. A postcompletion audit may be conducted:

- As an informal review by the project manager and his or her team
- At a formal debriefing meeting
- At the same time as a end-of-project party (Sec. 14.4 and Example 14.4)
- As a detailed review by external (independent) consultants

Conducting Audits

There is a seven-step process to conducting an internal or postcompletion audit:

Conduct Interviews. How you conduct interviews is a matter of style. You should always have some agenda of topics you wish to cover. Some people prefer to use a questionnaire, working through the questions in methodical order. My own preference is to have a list of broad topics I wish to cover. I explain them to the interviewees at the start, but then allow them free rein. Before closing the interview I ensure all topics have been covered. I find I learn more this way. Like Agatha Christie's detective, Hercule Poirot, I find nobody can spin a consistent web of deceit, so if you let them talk, they must eventually tell you the truth. If you ask a set of closed questions, it is very easy for them to be economical with the truth. The topics covered should address the standards of good practice which you are using as your basis, as described below.

Analyse Data. You should check the data being used, to determine its validity. The data gathered must be relevant, give a true representation of progress, and be processed in such a way that errors are not introduced. For data handled manually, there can be errors of transcription. These are usually unwitting, but they can be deliberate. It is the norm to find that when data is entered manually into several computer systems it does not tally. I once spoke to a project manager in a firm of engineering contractors who said it was common for project accounts and company accounts to differ by up to 5 percent, which he thought acceptable. To avoid errors of transcription, electronic means of data entry are used now.

Sample Management Reports. Reports used to monitor progress are checked to ensure they are relevant and representative of progress, and they enable the manager to spot divergences from plan easily, so that they can take quick, effective action. The reports may be used by the project manager, work package managers, or senior managers including the sponsor, champion, or steering committee.

Compare against a Standard of Best Practice. The information gathered about how the project is being managed is compared to a model or standard of best practice. Clearly, while you are conducting the early steps, you bear your model in mind. However, I find it is better to gather the information freely, because you then actually find out what is going on. If you merely ask whether the standard is being followed it is very easy to miss the gaps and it is very easy for people to mislead you. The standard of best practice may be a procedures manual used by the organization (Table 17.2 and Figs. 17.2 and 17.3), or a diagnostic procedure prepared by a firm of consultants. The standard will be hierarchical, presenting a series of important issues and questions at each stage throughout the life cycle of a project, or against each element of work in a standard work breakdown. This enables the auditor to focus on those areas which are important to the project at hand, rather than wading through a list of irrelevant questions. Each stage of this life cycle is supported by a series of questions against each parameter.

Repeat Steps 1 to 4 as Necessary. The comparison may raise further questions about the data, or the management processes used. Alternatively, you may realize there are things which were not adequately covered during the initial interviews. You may need to return to one or more of Steps 1 to 4 until you are satisfied everything has been adequately covered. My style is to conduct a preliminary set of interviews with senior managers to try to establish their views of the problems. As a result of that initial set of interviews, and my experience of similar organizations, I draw up a more detailed audit plan covering selected topics from the audit procedure. I then work through Steps 1 to 4 according to that plan. After that first full time through, I typically have 80 percent of the information I require. One or two more selected interviews may then give me all the information I can reasonably expect to get.

Identify Strengths and Weaknesses of the Management Approach. Through comparison of the information gathered with the audit procedure, you can identify strengths and weaknesses of the management approach used on projects in the organization, either on the project being audited or in general. I always believe it is important to identify both strengths and weaknesses for two reasons:

- You learn as much by reinforcing strengths as you do by eliminating weaknesses.
- People are more receptive to bad news if you start by giving them good news; even when reviewing an utter disaster it can make people feel that not everything they did was wrong.

Define Opportunities for Improvement. From the strengths and weaknesses you can identify areas where improvements can be made. Clearly you should aim to eliminate weaknesses. However, the application of the good points may be patchy, and so you can look to widen their scope, or you can find ways of improving their efficiency, and thereby make their application stronger still.

Emotions

Emotions can run high during an audit by external assessors. If the purpose of an audit is to work out why a failed project went badly wrong that is almost impossible to avoid. With an internal audit, the external assessors should try to present themselves not as policemen, there to check up on the project team and find fault, but consultants there to help the team achieve a successful project. An internal audit may be conducted for several reasons including:

(a) The project is a key strategic project which must not go wrong. There is no indication that anything is wrong, but the board would like the project checked because it is better to pick up any potential problems early. The audit is preventative medicine.

(b) There is an indication that all is not well with the project, so the board of directors would like it recovered before it goes more seriously wrong, while it can still be put back on track with little additional cost.

(c) The project is in serious difficulties and the board feels it needs significant external help to recover.

Interestingly it is in the second case where the team may be the most uncooperative. In the third case they may feel guilty and actually seek help. In the first case if they understand the true purpose of the audit they may welcome it. It is in the second case where they may be most trying to hide what is going on because they feel their reputations are at stake. But that is the key. In all three cases the auditors must say they are there to help the project team, not check up on them. Everybody wants a successful project and the auditors are there to help the project team achieve that, and their reputations are best served by supporting the auditors.

18.3 CONDUCTING HEALTH CHECKS

An audit is a check conducted by an external group of people. The board may also require the project team to conduct a check on themselves. But doing a check on one's own performance is something which should be encouraged. It is a common syndrome to make a mistake and realize if you had spent five minutes thinking about what you were doing you may have foreseen the problem. There is a saying that when you have been chased up a tree by the alligators it is too late to drain the swamp. It is good practice to develop the habit of taking a step back from the coal face every now and again to do a self-review. This is part of the quality assurance process (Sec. 7.2). I suggest two types of health check:

The projectivity diagnostic: This reviews not an individual project but the project management capability of the organization. It is a simplified maturity model.

The project success diagnostic: This is a check on an individual project.

Both techniques are primarily qualitative. The idea is to identify areas of weakness, but also, and more importantly, to identify differences of opinion within the project team. The diagnostic questionnaires ask people to rank their views about various issues on a scale of 1 to 6. We then use simple arithmetic calculations, spreads, variances, means, and differences to highlight where differences of opinion lie, and where weaknesses in the approach to project working within the organization lie. However, these calculations are designed to focus attention, not calculate some answer, like the number 42, which will determine whether or not your project will be successful. Having undertaken the diagnostic exercise, you will want to spend as much time working on determining why differences of opinion exist and then to eliminate them, as you will spend trying to reduce the impact of areas of weakness.

The Projectivity Diagnostic

The projectivity diagnostic (Table 18.6) can be conducted at any time to assess the health of project working in the organization, or in the start-up stages of an individual project to induct people into project-based ways of working. The concept of projectivity is used to represent an organization's ability to achieve its development objectives through project-based working. Organizations with low projectivity are unable to deliver projects effectively, and therefore consistently fail to achieve their strategic objectives. The projectivity diagnostic is designed to help you identify how well projects are established, planned, organized, executed, and controlled in your organization. There are no right or wrong answers to the questions. For some it will be a worry if the responses are not what you expect. For instance, if the majority of people say they cannot clearly see the link between organizational strategy and projects or if they think there are no established, clear principles and guidelines for project work, then that will be a cause for concern. However, this diagnostic is primarily designed to help you identify areas of agreement and disagreement in your project team (in its widest sense).

Using the Questionnaire. There are 106 questions grouped into five main problem areas. These are areas identified by Kris Grude[3] as those where projects consistently fail (Sec. 3.3):

- Foundation and infrastructure for project work
- Planning and estimating
- Organizing and cooperating
- Controlling and leading
- Executing and obtaining results.

The questionnaire asks people to rate each question on a scale of 1 to 6, where 1 equals false and 6 equals true. The questions are designed so that sometimes 1 indicates poor performance, and sometimes 6, so that people do not get into the habit of ticking every answer 4 to 5, but actually have to think about what the question is asking them. You should give the questionnaire to a wide variety of people within the organization:

- Senior managers representing sponsors, champions, and customers
- Peer groups representing professional colleagues, resource providers, users, and other stakeholders
- Project workers, representing designers and implementers
- Project managers

TABLE 18.6 Projectivity Diagnostic

No	Statement	Score	X	S	V	P	D
Problem area 1: Foundation and infrastructure for project work							
1.1	It is easy to see the relation between our project and overall business plans	1 2 3 4 5 6				6	
1.2	We have established sufficiently clear principles and guidelines for project work	1 2 3 4 5 6				6	
1.3	Our principles and guidelines for project work are understood by all involved parties	1 2 3 4 5 6				6	
1.4	Our principles and guidelines for project work are accepted by all involved parties	1 2 3 4 5 6				6	
1.5	The client/user roles and responsibilities are defined before start-up	1 2 3 4 5 6				6	
1.6	In our projects, the project team's roles and responsibilities are defined before start-up	1 2 3 4 5 6				6	
1.7	In our projects, the client/user keeps to agreed prioritizations (tasks/time/resources)	1 2 3 4 5 6				6	
1.8	Our project management is not very good at keeping to agree prioritizations	1 2 3 4 5 6				6	
1.9	In our projects, line managers contribute loyally to decision processes according to their responsibility	1 2 3 4 5 6				6	
1.10	In our projects, line management keep to agreed time limits for decisions	1 2 3 4 5 6				6	
1.11	In our projects, line management often reverse decisions that have been taken	1 2 3 4 5 6				1	
1.12	In our projects actual resources are committed as part of our planning process without line management being made aware	1 2 3 4 5 6				6	
1.13	Management makes sure that agreed resources for project work are made available at the right time	1 2 3 4 5 6				6	
1.14	Available resources for project work are taken into consideration in our business plans	1 2 3 4 5 6				6	
1.15	Our management plan so that development personnel do not get tied up in maintenance	1 2 3 4 5 6				6	
1.16	Our management plan so personnel are relieved of operational tasks when given project tasks	1 2 3 4 5 6				6	
1.17	We have sufficient and adequate tools and methods for planning projects	1 2 3 4 5 6				6	
1.18	We have sufficient and adequate tools and methods for organizing projects	1 2 3 4 5 6				6	
1.19	We have sufficient and adequate tools and methods for reporting and controlling progress	1 2 3 4 5 6				6	
1.20	We have sufficient/adequate tool and methods for reporting and controlling quality	1 2 3 4 5 6				6	
1.21	We have sufficient/adequate tools and methods for reporting and controlling time	1 2 3 4 5 6				6	

(*Continued*)

TABLE 18.6 Projectivity Diagnostic (*Continued*)

No	Statement	Score	*X*	*S*	*V*	*P*	*D*
Problem area 1: Foundation and infrastructure for project work							
1.22	We have sufficient/adequate tools and methods for reporting and controlling cost	1 2 3 4 5 6				6	
1.23	We have clear policies/procedures for prioritizing between projects	1 2 3 4 5 6				6	
1.24	We have clear policies for handling prioritization problems between operational tasks and project tasks	1 2 3 4 5 6				6	
1.25	It happens quite often in our projects that the project team and the client/user do not have a common understanding of the deliverables	1 2 3 4 5 6				1	
1.26	In our projects, everybody has the necessary knowledge of the procedures/methods/tools we use for project management	1 2 3 4 5 6				6	
1.27	I have the necessary skills to plan and organize projects	1 2 3 4 5 6				6	
1.28	I have the necessary skills to monitor and control projects	1 2 3 4 5 6				6	
1.29	I have the necessary skills to handle people's relationships and resolve conflicts	1 2 3 4 5 6				6	
1.30	Our project procedures/methods/tools/ are bureaucratic and tedious	1 2 3 4 5 6				1	
1.31	Our project procedures/methods/tools/ help us obtain commitment from all parties involved	1 2 3 4 5 6				6	
1.32	Our project procedures ensure goal direction and effective use of resources	1 2 3 4 5 6				6	
	Sum						
	Average						
Problem area 2: Planning and estimating							
2.1	Our overall project plans are understandable and give a good overview/description to all relevant parties, not just the specialists	1 2 3 4 5 6				6	
2.2	Our project plans are too generic	1 2 3 4 5 6				1	
2.3	We make project plans that are much too detailed and activity oriented	1 2 3 4 5 6				1	
2.4	Our plans are tailor-made for the task and focus on what is unique/important for progress	1 2 3 4 5 6				6	
2.5	Our project plans have imbedded quality control	1 2 3 4 5 6				6	
2.6	We have layered planning, where we focus on results and activities separately	1 2 3 4 5 6				6	
2.7	Our plans focus too much on completion date, too little on intermediate results/dates	1 2 3 4 5 6				1	
2.8	We often change our plans during the project	1 2 3 4 5 6				1	
2.9	Our plans always make it easy to control the achievement of intermediate and end results	1 2 3 4 5 6				6	

TABLE 18.6 Projectivity Diagnostic (*Continued*)

No	Statement	Score	*X*	*S*	*V*	*P*	*D*
Problem area 2: Planning and estimating							
2.10	Our project plans ensure that we do things in the right sequence, so that we do not have to do things over again	1 2 3 4 5 6				6	
2.11	Our project plans secure effective utilization of resources	1 2 3 4 5 6				6	
2.12	In our plans, we build quality assurance of the process as well as results	1 2 3 4 5 6				6	
2.13	We have a planning process that stimulates creativity and finding new solutions	1 2 3 4 5 6				6	
2.14	Our planning processes invite involved parties to participate and stimulate communication	1 2 3 4 5 6				6	
2.15	All involved parties are 100 per cent committed to our plans once agreed	1 2 3 4 5 6				6	
2.16	We have formalized estimating procedures to ensure maximum quality and commitment	1 2 3 4 5 6				6	
2.17	Our project plans always have a realistic completion date	1 2 3 4 5 6				6	
2.18	Our resource and cost estimates are unrealistic	1 2 3 4 5 6				1	
2.19	It sometimes happens we change our time and cost estimates because we don't "like" them	1 2 3 4 5 6				1	
2.20	We often set time and cost estimates too low for "selling" reasons	1 2 3 4 5 6				1	
2.21	In our projects, goals for individual's work are not precise	1 2 3 4 5 6				1	
2.22	In project planning, we often overestimate our own and other people's competence and skills	1 2 3 4 5 6				1	
2.23	In project planning, we often overestimate our and other people's available time and capacity	1 2 3 4 5 6				1	
2.24	With us, everybody can participate in estimating and planning their own work	1 2 3 4 5 6				6	
2.25	With us, everybody feels a personal responsibility for their own estimates	1 2 3 4 5 6				6	
2.26	In estimating we often do not account for non-productive time (illness, interruptions, etc.)	1 2 3 4 5 6				6	
2.27	In project planning, we often "forget" activities	1 2 3 4 5 6				1	
	Sum						
	Average						
Problem are a 3: Organizing and cooperating							
3.1	In our projects, the right people are always involved in the right activities	1 2 3 4 5 6				6	
3.2	Key people are often not available for the project at the time when planned	1 2 3 4 5 6				1	
3.3	People on the project are often not motivated	1 2 3 4 5 6				1	

(*Continued*)

TABLE 18.6 Projectivity Diagnostic (*Continued*)

No	Statement	Score	*X*	*S*	*V*	*P*	*D*
Problem are a 3: Organizing and cooperating							
3.4	We lack communication procedures and channels within our projects	1 2 3 4 5 6				1	
3.5	We lack communication procedures and channels between projects	1 2 3 4 5 6				1	
3.6	We have agreed and formalized the flow of information before start-up	1 2 3 4 5 6				6	
3.7	We organize our projects so we secure effective consulting processes	1 2 3 4 5 6				6	
3.8	We organize our projects so we secure effective decision-making processes	1 2 3 4 5 6				6	
3.9	Our way of organizing projects ensures maximum flexibility of people	1 2 3 4 5 6				6	
3.10	Nobody complains about lack of information in our projects	1 2 3 4 5 6				6	
3.11	In our projects, everybody knows and accepts their own role and responsibility	1 2 3 4 5 6				6	
3.12	Nobody knows what other people are doing on the project	1 2 3 4 5 6				1	
3.13	We seldom have conflicts within the team that result from bad cooperation	1 2 3 4 5 6				6	
3.14	We seldom have conflicts with clients or users that result from bad cooperation	1 2 3 4 5 6				6	
3.15	Our projects are ineffective because too many people/functions are involved	1 2 3 4 5 6				1	
3.16	In our projects, responsibility for tasks and decisions is connected directly to individuals, so there is no doubt	1 2 3 4 5 6				6	
3.17	We are organized to use the shortest possible route of communication between two persons	1 2 3 4 5 6				6	
3.18	In our projects, the project organization is more a formality than for real cooperation	1 2 3 4 5 6				1	
3.19	We are organized for resolving conflicts when they arise	1 2 3 4 5 6				1	
	Sum						
	Average						
Problem area 4: Controlling and leading							
4.1	In our projects, reporting has no purpose because it is never used for anything	1 2 3 4 5 6				1	
4.2	Reporting is used to watch team members	1 2 3 4 5 6				1	
4.3	Reporting is used in our projects to badger team members	1 2 3 4 5 6				1	
4.4	Reporting in our projects is used to discuss constructively necessary corrective action	1 2 3 4 5 6				6	
4.5	Our project plans are not arranged so that we can report against them for monitoring	1 2 3 4 5 6				1	
4.6	In our company, the project managers do not have the necessary authority	1 2 3 4 5 6				1	

TABLE 18.6 Projectivity Diagnostic (*Continued*)

No	Statement	Score	*X*	*S*	*V*	*P*	*D*
Problem area 4: Controlling and leading							
4.7	Project managers are too concerned with details of the technical content of the project	1 2 3 4 5 6				1	
4.8	The project managers are too pedantic	1 2 3 4 5 6				1	
4.9	Project managers will always try to cover up the problems to show a successful façade	1 2 3 4 5 6				1	
4.10	The project managers spend too little time managing the project	1 2 3 4 5 6				1	
4.11	The project managers cannot lead planning processes that result in realistic plans	1 2 3 4 5 6				1	
4.12	The project managers are unable to follow up methodically	1 2 3 4 5 6				1	
4.13	Project managers cannot inspire others	1 2 3 4 5 6				1	
4.14	In our projects, we have periodical meetings with fixed monitoring procedures that always result in concrete decisions on progress	1 2 3 4 5 6				6	
4.15	By monitoring our plans we are always able to see the need for corrective measures in time	1 2 3 4 5 6				6	
4.16	When we can't take corrective action it is always the clients/users' fault	1 2 3 4 5 6				1	
	Sum						
	Average						
Problem area 5: Project execution and delivering results							
5.1	Due to our way of working we are good at getting unfamiliar people working together	1 2 3 4 5 6				1	
5.2	In our projects, we use complicated methods too often	1 2 3 4 5 6				1	
5.3	In our organization, everybody has their own way of doing things	1 2 3 4 5 6				1	
5.4	Our projects are often subject to uncontrolled changes	1 2 3 4 5 6				1	
5.5	Our projects lack formal start-ups	1 2 3 4 5 6				1	
5.6	Our projects lack formal close-outs	1 2 3 4 5 6				1	
5.7	Lack of documentation is a frequent problem	1 2 3 4 5 6				1	
5.8	Insufficient quality control is a problem	1 2 3 4 5 6				1	
5.9	We often deliver inferior quality	1 2 3 4 5 6				1	
5.10	Our clients/users often report that they are pleased with the way we conduct our work	1 2 3 4 5 6				6	
5.11	We often deliver superior quality	1 2 3 4 5 6				6	
5.12	Our clients/users often report that they are pleased with the results we deliver	1 2 3 4 5 6				6	
	Sum						
	Average						

Analysing the Results. The results can be analysed in several ways:

Within groups: When analysing the results within groups, you will see whether the group:

- Agrees on the organization's performance in all areas
- Thinks that the organization's performance falls short in any areas.

These can be broken down, as follows:

(a) *Agreement:* In looking to see whether the group agrees on the answers to questions, you will look at the spread of answers. You can record two measures of spread:
- The spread S: The difference between the highest and lowest score for the group against that answer.
- The variance V: Calculated as

$$V = \sqrt{\Sigma(x - X)^2/N}$$

where x is the individual score
N is the number of people in the group
X is the mean score for that question, $X = \Sigma\, x/N$

Recording the answers in a spreadsheet enables you to calculate the mean, spread, and variance easily. I suggest you do not include the X, S, and V columns on the questionnaires you give to the people completing them; they are there to help you analyse the responses. Where there is a high spread, 3 or greater, at least some members of the team disagree about the response to that question. Where there is also a high variance, 2 or greater, there is fundamental disagreement among team members about the answer to the question. (A high spread but low variance indicates that only one or two members of the team disagree with the majority opinion.) The reason for any disagreement is worth exploring, and can be made part of the team-building process. I have kept the mathematics simple because we are interested in qualitative comparisons, not quantitative results or statistics. This is a qualitative exercise; the numbers are just a way of helping to focus attention. You do not need to worry about such things as confidence limits because they are not relevant here.

(b) *Performance:* You can analyse the results to see where they indicate poor performance. The polarity P of each question shows which end of the scale indicates good performance (1 or 6). (Again I suggest you do not include this column on the questionnaires for completion.) You can compare the average answer to each question, X to this polarity and calculate the difference D to determine where the team think the organization falls short in performance. A difference of 2 or 3 will indicate below average performance and 4 to 5 poor performance. The reason why the team think the performance is below average or poor will be more interesting than the fact that they do, and exploring the reason can again be part of the team-building process.

(c) *Problem areas:* By calculating the average of the differences D for all questions within each of the five problem areas, you can determine which problem areas the group considers are weaknesses of project management within the organization. Because you expect some questions to indicate acceptable performance, an average difference of 2 or 3 will indicate poor performance, and an average difference of 4 or 5 will indicate very poor performance.

Between groups: You can repeat the comparisons between groups. Primarily, you will inspect the mean answers X question by question to see whether one of the groups differs from other groups. Differences are quite likely between managers, team members, users, and so on. Exploring the reasons for differences is more important than the existence of the differences. Similarly, you can inspect the overall results on the problem areas as more of a threat than do the other groups. (Obviously, if all of the groups view one of the questions or one of the problem areas as a threat, then that will be addressed in the comparisons within groups. Here we are only looking for differences between groups.)

The Success or Failure Diagnostic

The second health check is based on the research by Wateridge[4] into the success or failure of projects described in Chap. 3. The health check is contained in Table 18.7. There are 85 questions in five parts:

Part 1 helps identify appropriate success criteria for your project (Sec. 3.1).

Part 2 helps identify what success factors you should focus on to achieve those criteria (Sec. 3.3).

Part 3 checks you are using appropriate tools and techniques for the management of your project (Part 2).

Part 4 checks you have an appropriate range of skills in the project team (Sec. 17.1).

Part 5 helps identify how well the project is being executed and managed.

The main emphasis again is on checking the consistency of view of all the members of the project team and stakeholders. Indeed there are no right and wrong answers to Part 1. The diagnostic can be given to a similar range of people as the projectivity diagnostic and the answers analysed in a similar way.

18.4 END-OF-STAGE REVIEWS

APM's principle P7 states that all projects must have an approved plan with authorization points and that decisions made at those authorization points must be clearly documented. I have said throughout this book that because projects are coupled systems, the governance structure of the project must be aligned with the project process. That is, the project governance team, particularly the project manager and project sponsor, must be empowered to manage the project process towards the project's outputs and outcomes. They must be empowered to take decisions in the best interests of the project and given the flexibility to deal with risk as they encounter it. Now many line managers in the functional hierarchy do not like this, and you find great resistance. But it has to be done in the best interest of the project. However, project authorization points are predefined points where corporate governance can take back control, to approve progress to date and forecasts for the forthcoming stages, before releasing power back to the project manager and project sponsor for the next stage. Many authorization points are aligned with major milestones, or end-of-stage transitions. They may be known as:

- Stage-gate reviews
- Tollgate reviews
- End-of-stage transition
- Gateway reviews

All the milestones in the P column in the milestone plan for the CRMO Rationalization Project (Fig. 5.2) are end-of-stage reviews. The PRINCE2 process has end-of-stage transitions built into it through the processes known as "Managing Stage Transition."[5] Under PRINCE2, the project manager is empowered to take the project to the next stage transition as long as the project performance remains within defined tolerances. But at stage transition he or she has to seek approval from the project steering committee to proceed to the next stage. Corporate or program governance take back control at those points. Figure 1.4 and Table 8.4 illustrate end-of-stage reviews. Table 18.8 gives a generic end-of-stage review processes.

TABLE 18.7 Success or Failure Diagnostic

No	Statement	Score	*X*	*S*	*V*	*P*	*D*
Part 1: Success criteria							
1.1	The success criteria are defined	1 2 3 4 5 6				6	
1.2	The success criteria are agreed	1 2 3 4 5 6				6	
1.3	I believe the success criteria are appropriate	1 2 3 4 5 6				6	
1.4	The project should achieve quality constraints	1 2 3 4 5 6				6	
1.5	The project should be a commercial success	1 2 3 4 5 6				6	
1.6	The users should be happy	1 2 3 4 5 6				6	
1.7	The sponsors should be happy	1 2 3 4 5 6				6	
1.8	The project team should be happy	1 2 3 4 5 6				6	
1.9	The project meets its stated objectives	1 2 3 4 5 6				6	
1.10	The system should achieve its purpose	1 2 3 4 5 6				6	
1.11	The project should be delivered on time	1 2 3 4 5 6				6	
1.12	The project should be delivered to cost	1 2 3 4 5 6				6	
1.13	The project should contribute to the organization's overall business strategy	1 2 3 4 5 6				6	
1.14	There is a clear relationship between the project and business plans and strategies	1 2 3 4 5 6				6	
1.15	The project team do not appreciate the important success criteria	1 2 3 4 5 6				1	
1.16	I am confident the project will be a success	1 2 3 4 5 6				6	
1.17	The project goals are clear to me	1 2 3 4 5 6				6	
1.18	The goals have been explained to the team	1 2 3 4 5 6				6	
1.19	I can explain the benefits of the project	1 2 3 4 5 6				6	
1.20	The project has an unrealistic completion date	1 2 3 4 5 6				1	
	Sum						
	Average						
Part 2: Success factors							
2.1	Estimates for the project are realistic	1 2 3 4 5 6				6	
2.2	Project estimates are optimistic	1 2 3 4 5 6				1	
2.3	Estimates were made in consultation with the person allocated to the task	1 2 3 4 5 6				6	
2.4	The project has been planned strategically	1 2 3 4 5 6				6	
2.5	Project plans are understandable to all	1 2 3 4 5 6				6	
2.6	The project plans are often changed	1 2 3 4 5 6				1	
2.7	Plans focus on completion date and not on intermediate results/dates	1 2 3 4 5 6				1	
2.8	The project plan effectively utilizes resources	1 2 3 4 5 6				6	
2.9	I am happy with the plans and estimates	1 2 3 4 5 6				6	
2.10	The project participants are motivated well to achieve the project objectives	1 2 3 4 5 6				6	
2.11	Responsibilities are not well delegated	1 2 3 4 5 6				1	
2.12	The clients and users know their roles and responsibilities	1 2 3 4 5 6				6	
2.13	I am happy with the leadership shown by senior management	1 2 3 4 5 6				6	
2.14	I am happy with the leadership shown by project management	1 2 3 4 5 6				6	

TABLE 18.7 Success or Failure Diagnostic (*Continued*)

No	Statement	Score	*X*	*S*	*V*	*P*	*D*
Part 2: Success factors							
2.15	Communication and consultation channels have been effectively set up	1 2 3 4 5 6				6	
2.16	There is poor communication between the project participants	1 2 3 4 5 6				1	
2.17	The users are involved effectively	1 2 3 4 5 6				6	
2.18	Communication channels are poor	1 2 3 4 5 6				1	
2.19	Project managers do not fully report project status to sponsors/users' teams	1 2 3 4 5 6				1	
2.20	Corrective measures are always taken in time when the project encounters problems	1 2 3 4 5 6				6	
2.21	All roles and responsibilities are well-defined	1 2 3 4 5 6				6	
2.22	All parties are committed to the plan	1 2 3 4 5 6				6	
2.23	Resources are available at the right time	1 2 3 4 5 6				6	
2.24	Procedures for handling priorities are adequate	1 2 3 4 5 6				6	
2.25	Quality assurance is not a major aspect of the projects	1 2 3 4 5 6				1	
	Sum						
	Average						
Part 3: Tools, techniques, and methodologies							
3.1	Tools, techniques, and methods for planning the project are adequate	1 2 3 4 5 6				6	
3.2	Tools, techniques, and methods for controlling the project are adequate	1 2 3 4 5 6				6	
3.3	Tools, techniques, and methods for organizing the project are adequate	1 2 3 4 5 6				6	
3.4	I agree that the tools, techniques, and methods used are appropriate	1 2 3 4 5 6				6	
3.5	The development tools and methods are sufficient for the project	1 2 3 4 5 6				6	
3.6	The management tools and methods are sufficient for the project	1 2 3 4 5 6				6	
3.7	The development tools and methods are poorly applied on the project	1 2 3 4 5 6				1	
3.8	The management tools and methods are poorly applied on the project	1 2 3 4 5 6				1	
3.9	The chosen methodologies stifle creativity during the project	1 2 3 4 5 6				1	
3.10	There are established methods which are to be used	1 2 3 4 5 6				6	
3.11	These established methods are being used on this project	1 2 3 4 5 6				6	
3.12	I believe these methods are appropriate for the project	1 2 3 4 5 6				6	

(*Continued*)

TABLE 18.7 Success or Failure Diagnostic (*Continued*)

No	Statement	Score	*X*	*S*	*V*	*P*	*D*
Part 3: Tools, techniques, and methodologies							
3.13	There are computer-based tools available for this project	1 2 3 4 5 6				6	
3.14	Computer-based tools are being used effectively	1 2 3 4 5 6				6	
3.15	The project uses methods for assessing and managing risks	1 2 3 4 5 6				6	
	Sum						
	Average						
Part 4: Skills							
4.1	There are the necessary skills available to plan the project	1 2 3 4 5 6				6	
4.2	There are the necessary skills available to organize the project	1 2 3 4 5 6				6	
4.3	There are the necessary skills available to control the project	1 2 3 4 5 6				6	
4.4	There are the necessary skills available to develop the system	1 2 3 4 5 6				6	
4.5	Project management are unable to handle fully the human relations aspects	1 2 3 4 5 6				1	
4.6	Conflicts are resolved satisfactorily	1 2 3 4 5 6				6	
4.7	The project plan overestimates the skills and competences of the team	1 2 3 4 5 6				1	
4.8	Project management is astute in dealing with the politics of the project	1 2 3 4 5 6				6	
4.9	Project management is unable to inspire others	1 2 3 4 5 6				1	
4.10	Project management is good at getting the project team working together	1 2 3 4 5 6				6	
	Sum						
	Average						
Part 5: Execution							
5.1	A life cycle approach is being applied	1 2 3 4 5 6				6	
5.2	I agree with the life cycle used	1 2 3 4 5 6				6	
5.3	An effective start-up meeting was held for this project	1 2 3 4 5 6				6	
5.4	The right people are allocated to the project	1 2 3 4 5 6				6	
5.5	Project team members are carrying out appropriate activities	1 2 3 4 5 6				6	
5.6	Project resources are selected well	1 2 3 4 5 6				6	
5.7	There are no problem areas during the project	1 2 3 4 5 6				6	
5.8	I do not foresee any problem areas on the project	1 2 3 4 5 6				6	
5.9	The management of the project is excellent	1 2 3 4 5 6				6	
5.10	The project team has appropriate members at appropriate times	1 2 3 4 5 6				6	
5.11	The project risks were assessed at the outset of the project	1 2 3 4 5 6				6	

TABLE 18.7 Success or Failure Diagnostic (*Continued*)

No	Statement	Score	*X*	*S*	*V*	*P*	*D*
Part 5: Execution							
5.12	I believe that the assessments of risks are appropriate	1 2 3 4 5 6				6	
5.13	Project risks are being managed well	1 2 3 4 5 6				6	
5.14	The deliverables are fully identified	1 2 3 4 5 6				6	
5.15	The deliverables are quality assured constantly	1 2 3 4 5 6				6	
	Sum						
	Average						

PRINCE2 is designed for medium-sized projects. The United Kingdom government, through the Office of Government Commerce, has developed a gateway review processes for larger projects.[6] There are six gateway reviews:

0. Strategic assessment
1. Business justification
2. Procurement strategy
3. Investment decision
4. Readiness to service
5. Benefits realization

TABLE 18.8 End-of-Stage Reviews

	End of stage		
Item for review	Concept	Feasibility	Design
Management	Need for performance improvement Change identified Appoint sponsor First draft of benefits map	Business plan Outputs and desired outcomes defined Appoint project manager Risks identified	Finalize business plan Appoint team
Design	High-level options	Identify and assess options Select preferred option	Complete design
Planning	High-level scheme	Milestone plan	Activity plans
Cost	Order of magnitude	±30%	±10% Risk analysis Review benefits
Procurement	Options considered	Contract strategy Invitation to tender (ITT) prepared	Issue ITT
Users	Early consultation	Review user requirements	Finalize user requirements
Compliance	Health, safety, and environmental (HSE) issues identified	HSE plan	Implement HSE plan

The first takes place at the program level, which is why it is labelled 0. Whereas for PRINCE2 it is assumed the project will be managed internally, for larger projects it is assumed they will be contracted out, and so the focus of this gateway review process is on tracking the contracting process.

SUMMARY

1. There are eleven principles of the good governance of project management:
   - The board of directors is overall responsible.
   - Roles and responsibilities for the governance of projects must be clearly defined.
   - Defined governance arrangements must be applied throughout the project life cycle.
   - Members of authorization bodies must be properly empowered.
   - The project portfolio must be linked to corporate strategy.
   - The project business case must be based on sound and realistic data.
   - All projects must have an approved plan, with defined end-of-stage review points; decisions made at end-of-stage reviews must be fully documented.
   - There must be defined criteria for reporting status and defined escalation criteria.
   - The board must decide when independent audit of projects is required.
   - Project stakeholders must be engaged at an appropriate level.
   - The organization must foster a culture of openness and continuous improvement (and avoid competency traps).
2. Poor governance results in
   - No link between corporate strategy and projects
   - Lack of ownership of projects and their results
   - Poor engagement with stakeholders
   - Poor enterprise project management capability
   - A lack of engagement with suppliers
   - Poor evaluation of project proposals
   - Lack of focus on breaking a project down into manageable steps
3. The APM model has four elements of governance
   - Portfolio direction (PD)
   - Project sponsorship (PS)
   - Project management (PM)
   - Disclosure and reporting (DR)
4. Project audits will be conducted to
   - Check the design.
   - Ensure appropriate management processes are being used.
   - Learn from previous successes and failures.
5. There are three types of audit:
   - Project evaluation audits to check the validity of the design
   - Internal audits to check that a project underway is sound
   - Postcompletion audits, usually to find why a project went wrong
6. There are seven steps in conducting an internal or postcompletion audit:
   - Conduct interviews.
   - Analyse data.
   - Sample management reports.
   - Compare against standard of best practice.
   - Repeat steps 1 to 4 as necessary.
   - Identify strengths and weaknesses.
   - Define opportunities for improvement.

7. Informal internal audits conducted by the project team on themselves are called health checks.

8. There are two types of health check suggested:
- The projectivity diagnostic to check the working environment supports project-based management
- The success or failure diagnostic to ensure the project has been established according to the principles of Chap. 4

9. End-of-stage reviews are an essential part of project governance, where the project manager and sponsor hand authority back to corporate governance to approve progression to the next stage.

REFERENCES

1. Association for Project Management, *Directing Change: A Guide to Governance of Project Management*, High Wycombe, U.K.: Association for Project Management, 2004.
2. Thomas, J., Delisle, C., and Jugdev, K., *Selling Project Management to Senior Executives: Framing the Moves that Matter,* Newtown Square, Pa.: Project Management Institute, 2002.
3. Andersen, E.S., Grude, K.V., Haug, T., Katagiri, M., and Turner, J.R., *Goal Directed Project Management*, 3rd ed., London: Kogan Page/Coopers & Lybrand, 2004.
4. Wateridge, J.H., "IT projects: a basis for success," *International Journal of Project Management,* 13(3), 169–172, 1995.
5. Office of Government Commerce, *Managing Successful Projects with PRINCE2,* 4th ed., London: The Stationery Office, 2005.
6. Office of Government Commerce, *The OGC Gateway™ Process: Gateway to Success,* London: Office of Government Commerce, 2004.

CHAPTER 19
INTERNATIONAL PROJECTS

In this chapter I look at international projects, that is, projects involving parties from two or more countries. This is not strictly a governance issue, but is a significant type of project which I think deserves space. These projects have very specific problems, particularly the problem of cultural fit. Indeed, the term international projects can involve a multitude of different types of projects with a range of features. I consider international projects and their management. I describe different types of international projects and their characteristics. I then list common problems in the management of international projects and describe how to overcome them. I describe the issue of cultural fit and the work done by Gred Hofstede and what it says about the approach of different nationalities to the management of projects. I then describe how to work with an international project team and international partners.

19.1 TYPES OF INTERNATIONAL PROJECT

International projects come in many forms.

Projects in Your Own Country for a Foreign Client

There are many reasons why a foreign client may want to make an inward investment into your country: to develop new markets, make use of local expertise, or gain access to raw materials. In this case, you will have the familiarity of working in your own environment, within your own legal system, and with familiar subcontractors. The main difficulty arises from working with a client of a different culture and with an unfamiliar way of doing business. It will be important to understand their different approaches and to try to accommodate them. You may expect that since they are working on your own home ground, the client should make an attempt to respect your local culture and ways of working. However, it can still be valuable to understand their culture, so that you can understand their ways of working, not unwittingly offend them, and also help them fit into your environment.

Projects in Your Own Country Using Foreign Contractors

You may use a foreign contractor because you need to buy expertise not available in your own country, or because they are cheaper than local alternatives, or because you are compelled to do by European or other international competition laws. As a contractor working for a foreign client you may be required to use a subcontractor nominated by the client. The problems are again mainly ones of different cultures and ways of working. Now you may expect even more that the supplier will respect your way of doing business, especially if they want to break into your local market. However, it can still be valuable to understand their approaches just to avoid any misunderstandings (see Example 19.1).

Example 19.1 Working as a contractor overseas

I once spoke to an American partner of Accenture, who told me that when they sent a consultant to an assignment in mainland Europe the consultant was sent on a two-week language and cultural awareness course. If they were being sent to Britain they were not but he felt they should be. English and American English are different, and sometimes significantly so. (You tell the English to walk on the pavement and they will walk on what the Americans call the sidewalk; you tell Americans to walk on the pavement and they will walk in the middle of the road on what the English call the tarmac.) On a visit to the United States, I found myself asking questions by stating what I thought to be the answer, but having it interpreted as a statement of my belief by my hosts. (I have the same problem in Holland and China, but not in France.) I commented on this to one of my hosts, and she said she found it arrogant to ask a question by stating what I thought to be the answer, yet it is normal behaviour in England and France.

Projects in a Foreign Country for Which You Are the Client

Now we are working in a different business environment and legal system to which we have to adapt. We may expect our suppliers to respect our cultural traditions, and as the client we may have some influence in that respect. However, it may still be necessary to understand local traditions, just because we are in the minority, and because we may unwittingly cause offence (Examples 19.2 and 19.3).

Example 19.2 Respecting local traditions versus expecting your traditions to be respected (1)

Under American federal law, American companies are forbidden from giving bribes anywhere in the world. Under German federal law, German companies are forbidden from giving bribes in Germany. But if they are operating in a part of the world where offering "commission" is part of standard business practice, then they are allowed to pay it, and it is treated as an allowable business expense for tax purposes in Germany. Employees of American companies have told me that when working overseas one of the reasons for forming joint ventures is to pay the "commissions" via the joint venture partner.

Example 19.3 Respecting local traditions versus expecting your traditions to be respected (2)

While running a course in Malta, I spoke to a Maltese who had just completed an assignment with the U.S. Navy in Naples. He had been employed as a consultant, but effectively worked as an employee of the U.S. Navy and so was bound by their ways of doing business. He said that he had been told that he would be sacked if he accepted so much as a cup of coffee from a contractor as that could be interpreted as a bribe and he certainly was forbidden to pay anything that could be interpreted as a bribe. He said it was virtually impossible to work in Naples without lubricating the wheels with commissions, and an Italian contractor would be deeply offended if you refused coffee they offered you.

Working as a Contractor for a Foreign Client in Their Country

You may have been employed for your expertise, you may have been used because your own government provided aid and required that a certain element of the contract should be procured in your country. The aid may even have been in the form of services rather than cash. As well as problems of cultural differences, risks you may encounter include:

- The financial risks and credibility stakes may be high.
- As the client is employing you for your expertise, they may not know very much about the project and the scope may not be well defined.
- Because of this lack of knowledge the client may not have full confidence in the project.
- Project management and interfaces with the client may be executed in a foreign language.
- The client may have a significantly different cultural background and not be confident in your project management techniques.
- With fewer shared cultural and commercial assumptions, the chances of a damaging misunderstanding arising are much greater.

A solution to many of these problems is to include local nationals in your project team. This has the benefit of enabling you to avoid many of the language, cultural, and social difficulties as well as opening doors for you in the country. Under normal circumstances you will be expected to work within local traditions (see Examples 19.1, 19.2, and 19.3). The exception is aid projects, where as a representative of the donor country you may have greater expectations of the locals conforming to your ways of working (except see Example 19.4).

Example 19.4 A foreign aid failure

The Jamaica Maritime Training Institute project lasted for more than 13 years with Norwegian aid money. The project was originally planned for three years. However, as the project neared its original end date, local job opportunities for the Jamaican staff were limited. Hence, they had no desire to complete the project. This was well understood by the local authorities. The prevailing prognosis, after 13 years of project work, was that at least another five to seven years of work are needed before the original goal, as it was formulated, could be reached. This did not even include the development of a counterpart staff competent enough to take over administrative and technical responsibilities!

Projects in a Country for Clients Also Alien to the Country

This is likely to be for a multinational company used to operating worldwide. Such a company is likely to be fully aware of most of the related problems. When it decides to proceed with an international project it is usually after stringent research and development studies and most of the main potential pitfalls have been addressed. A typical example of such a project might be the building of a refinery for an international oil company in the Middle East. Characteristics of such projects might be:

- Well-defined project scope.
- Stringent contractual and funding conditions.
- The client will closely monitor all aspects of the project in an extremely professional way.

- The client may well insist on various aspects of the project being carried out in a very prescribed way and may require you to utilise some of its existing facilities.
- Contract law may be that of the country where most of the work is to be executed.

The client will most likely have a much better appreciation of the overall context and factors affecting the potential success or failure of the project than you will. It is therefore essential that you talk to them at all times, maintain their confidence, and use their expertise.

Multinational Joint Ventures

This type of project is often the most difficult to execute, not from the technical viewpoint, but from the complexity of dealing with a number of different national bodies each with its own aims and priorities. Features of these projects include:

- Complex multinational contractual and funding arrangements.
- Multinational project teams.
- Relatively poor project definition at the outset.
- A requirement to observe and maintain national interests.
- The project may be spread out over a wide geographical area if each participating nation expects to execute its own share of the work.
- Good communications are of paramount importance.

With this type of project it is essential that the organization structure is set up correctly and implemented from the start. Lines of responsibility, authorities, and demarcations must be clearly understood at all levels. A good principle is to ensure truly multinational teams are established in each major work location. This provides an informal communication facility between nations and helps to avoid cultural and language problems.

19.2 THE PROBLEM OF INTERNATIONAL PROJECTS

Having considered some of the types of international projects, we can identify some of the problems that arise in their management. In Sec. 4.4, I defined a virtual team as one with a boundary within the team that increases the cost of communication across the boundary. Many of those boundaries create problems on international projects.

Culture

The main problem is one of culture. Our approach to personal relationships, doing business, and project management are determined by our basic mental programming. The lily pond model (Fig. 19.1) illustrates that our behaviour is the visual representation of our attitudes and beliefs, which is determined by our values and basic programming, which in turn is based on our unquestioned assumptions about what is right and wrong. Gerd Hofstede[1] identified that our assumptions are based on our family, education, linguistic, gender, social, regional, religious, and ethnic background and these influence our behaviour as individuals, in groups, and as professionals. When working on international projects we need to understand the approaches of different cultures, to be able to work with people and predict behaviours, and not to give and take offence. The next section deals with this in greater detail.

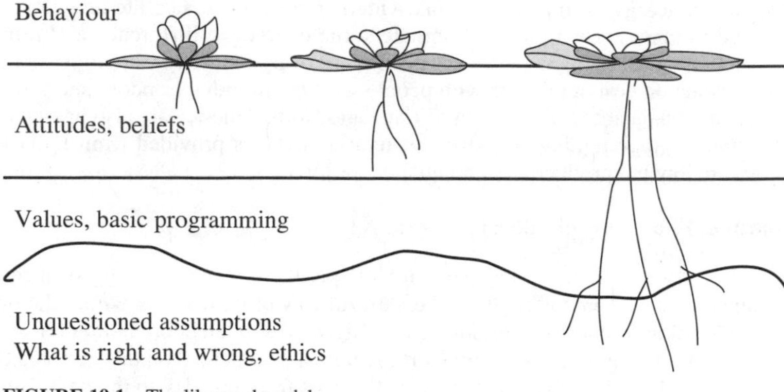

FIGURE 19.1 The lily pond model.

Distance

The second problem arises from the degrees of distance. The most obvious dimension of distance is geographic remoteness, but there can be other dimensions of distance as well:

- *Time zone:* It can be easier for someone in Britain to work with someone in South Africa, and someone in New York to work with someone in Argentina than people in New York and Britain to work together, because of overlapping working hours in the first two cases.
- *Organizational behaviour:* In organizations that encourage individualistic behaviour or strongly functional working, people working in adjacent offices can be remote from each other (see Example 19.5); new people joining must learn the language, jargon, and ways of working before they can work effectively (see Example 19.6).
- *Language and culture:* These cause degrees of distance as discussed above (Fig. 19.2).
- *Professions:* Each comes with its own jargon and mental models, which can cause as much remoteness as language and culture.

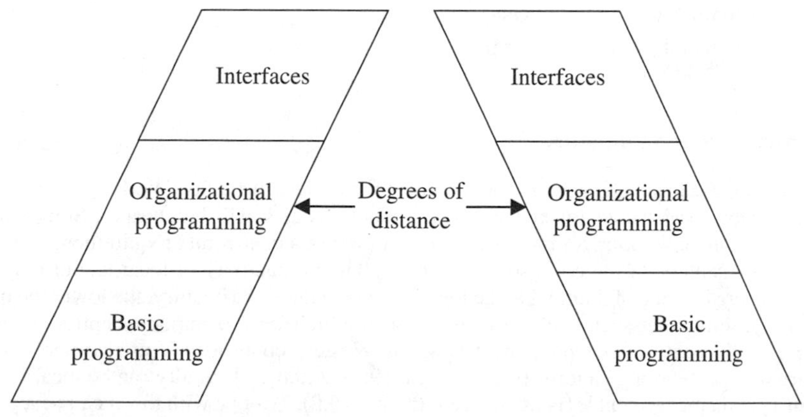

FIGURE 19.2 Degrees of distance.

Modern technology such as e-mail, fax, video conferencing, satellite telephones, and Internet and intranet are helping to eliminate some degrees of difference and reinforce others (see Example 19.5 again). Indeed, people are using modern technology to achieve 24-hour working on design projects with people working in India, London, and California sharing a common database. Firms from Europe and North America are also having design work, computer programming, and even secretarial services provided from India where wage rates are low but productivity and quality are high.

Example 19.5 Organizational remoteness (1)

When I worked at Henley Management College, there was something of a ritual of morning and afternoon coffee. It can be a useful way of networking within the organization and asking someone a question as an alternative to telephoning them or sending an e-mail. A visiting academic from North America commented on this and said that in his university he did not see most of his colleagues from one day to the next. He came in at 8.30 in the morning, went straight to his office, came out only to give lectures or go to the library, and went home at 5.00 in the evening or even later. Academic research assessment techniques which reward individual performance greater than team performance reinforce this behaviour.

Example 19.6 Organizational remoteness (2)

My sister-in-law on joining the consultants McKinsey was handed four typed pages of acronyms and told to learn them by the next day, or she would not be able to work effectively.

Organization, Management, and Communication

International projects often require more complicated organization structures in order to deal with a number of factors including:

- Collaboration with joint venture partners
- Special requirements from funding agencies
- National interests, and the requirement to use local labour and suppliers
- Local administrative requirements
- Providing facilities for ex-patriot personnel

Productivity and Logistics

The need to use local labour, transport, or storage can cause difficulty. Working abroad you may employ local labour for nonspecialised functions as it will be cheaper. Sometimes, especially where working for the government, it may be a contractual requirement to make a given percentage of the costs sourced locally. The productivity of local labour may be lower. There is a rule of thumb that the lower the wage rates of a country, the lower the productivity, sometimes such that unit labour costs are higher. (One major exception is India where productivity rates tend to be as high as Western countries.) You also need to be aware of local working patterns (see Example 19.7). Another difficulty can be local social security and employment legislation (see Example 19.8). To cope with this it can be a good idea to employ all local labour through a local joint venture partner.

Example 19.7 Local working hours

I ran a three-day course in Egypt. I asked the local organizers what working patterns I should have during the day, and they said whatever I normally did in the United Kingdom. On the first day we started at 09.00 and broke for lunch at 12.30, returning at 13.30. The audience were asleep for most of the afternoon. I asked what the problem was and they said that their normal day was to start at 07.00 and work for seven hours until 14.00. They would then have lunch followed by a siesta in the heat of the day. We did that on days two and three of the course.

Example 19.8 Local employment law

Someone on a course at Henley Management College had worked on a project to build a new airport in Nairobi. At the end of the four-year project they found that under Kenyan employment law it was almost impossible to sack someone if you had employed them for more than 12 months continuously—difficult if you are working on a transient project.

Local Legislation and Regulation

Finally, work in a country often has to be done under the law of that country, including:

- Contract law
- Business law
- Employment law
- Health and safety law
- Environmental protection and planning regulations
- Commissions

For this reason it can often be critical to employ a local agent who can ensure that you meet all the local requirements. In the Middle East having a local agent in itself is a necessity to guide you through local business practices and to pay appropriate commissions. In some countries it is a legal requirement for doing business in that country.

19.3 MANAGING CULTURE

Culture is the most significant problem on international projects.

Dimensions of Cultural Difference

There have been many studies into the nature of cultural difference.[1,2,3] Table 19.1 shows some of the dimensions identified by Gerd Hofstede[1] and Fons Trompenaars .[2]

Hofstede. Gerd Hofstede[1] identified four parameters of cultural difference:

1. *Power distance:* The extent to which the less powerful person in a society accepts inequality in power and considers it as normal.
2. *Individualism:* The extent to which individuals primarily look after their own interest and the interest of their immediate family (husband, wife, and children).

TABLE 19.1 Dimensions of Culture due to Hofstede[1] and Trompenaars[2]

Dimension	Measures
Power distance	Respect for authority
Individualism vs. collectivism	Personal ambition vs. group cooperation
Internal vs. external	Motivation for self or society
Achievement vs. ascription	Importance of status and performance
Uncertainty avoidance	Risk aversion
Masculinity	Differentiation between gender roles
Attitude to time	Attitude to deadlines
Short term vs. long term	Attitude to investment returns and results
Universalist vs. particularist	Ethics, principles of right and wrong
Specific vs. diffuse	Attitude to legal processes and personal trust
Neutral vs. emotional	Willingness to express feelings

3. *Uncertainty avoidance:* The extent to which people are nervous of situations they consider to be unstructured, unpredictable, or unclear, and the extent to which they try to avoid such situations by adopting strict codes of behaviour and a belief in absolute truths.
4. *Masculinity:* The extent to which the biological existence of two sexes is used to define different roles for men and women.

Some of Hofstede's findings are plotted in Figs. 19.3 and 19.4 (Regions and countries are represented by the codes shown in Table 19.2). In Fig. 19.3, developing countries and

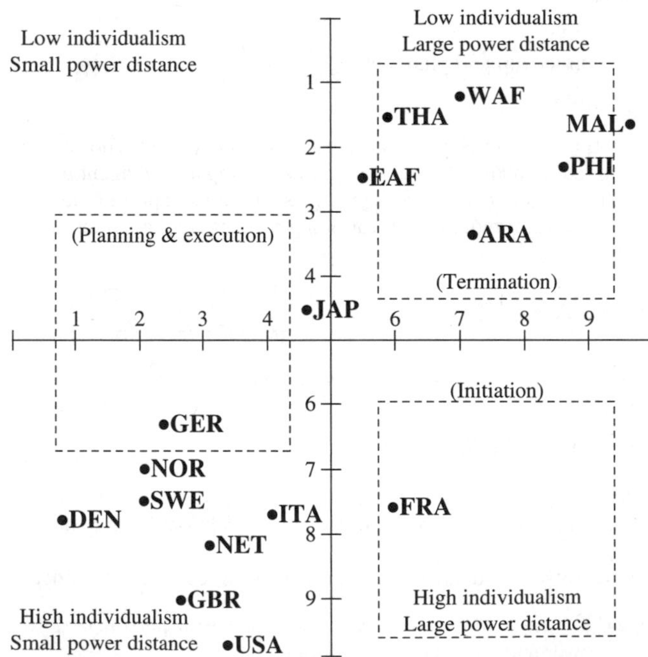

FIGURE 19.3 Country plot against Hofstede's cultural factors *power distance* and *individualism* against assumed preferred behavioural attitude scores within each project stage.

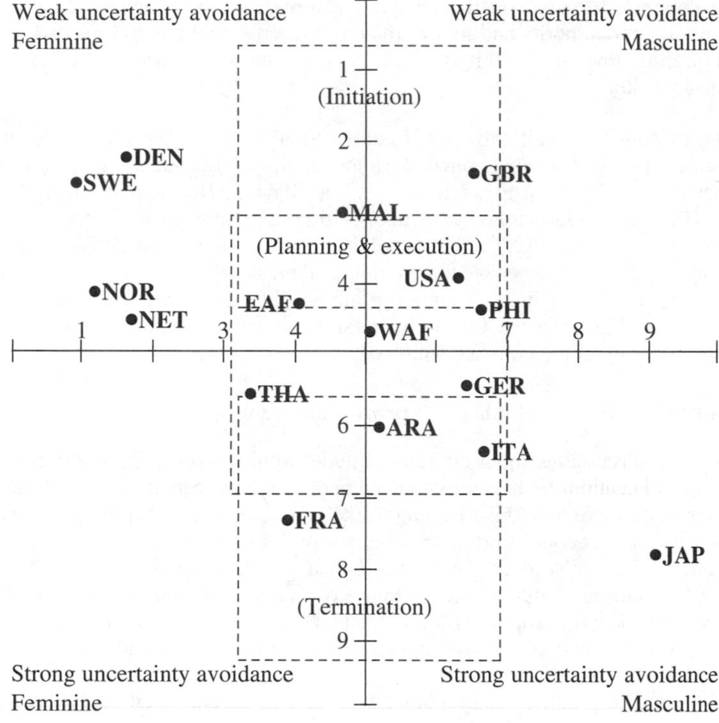

FIGURE 19.4 Country plot against Hofstede's cultural factors *uncertainty avoidance* and *masculinity* against assumed preferred behavioural attitude scores within each project stage.

TABLE 19.2 Country Ranking of Fitness for Project Management

No	Country	Code	Initiation Score	Planning Score	Execution Score	Closure Score	Total Score
1	Germany	GER	6.10	2.17	2.17	2.49	12.93
2	Italy	ITA	5.56	3.86	3.86	2.43	15.71
3	France	FRA	5.10	4.80	4.80	3.61	18.31
4	USA	USA	5.23	5.03	5.03	4.28	19.57
5	Netherlands	NET	5.89	4.66	4.66	4.44	19.65
6	Norway	NOR	6.70	4.28	4.28	5.09	20.35
7	Gt Britain	GBR	5.45	5.12	5.12	5.26	20.95
8	Arab Nations	AR	5.48	5.11	5.11	6.49	22.19
9	East Africa	EAF	6.07	4.57	4.57	7.13	22.38
10	Sweden	SWE	6.57	5.17	5.17	6.41	23.32
11	Denmark	DEN	7.18	5.15	5.15	6.44	23.92
12	Japan	JAP	8.10	5.48	5.48	5.62	24.68
13	Thailand	THA	7.43	5.16	5.16	7.43	25.18
14	West Africa	WAF	6.74	5.91	5.91	8.21	26.77
15	Philippines	PHI	5.59	6.80	6.80	8.73	27.92
16	Malaysia	MAL	6.07	7.93	7.93	10.04	31.97

Western countries form two district groups. The former are in the first quadrant implying a greater respect for authority and society than in Western countries. In Fig. 19.4, there is no pattern meaning that "masculinity/femininity" and "uncertainty avoidance" are unrelated to national wealth.

Attitudes to Time. We all know stories about Southern Europeans being free with time and Northern Europeans being punctual. (I am convinced it is due to the length of the day in winter; you can't afford to waste short winter days in Northern Europe, at least you couldn't if you were a farmer 500 years ago.) Attitudes to time reveal different cultural programming (see Example 19.9). Germans believe events are controlled by planning and respecting deadlines. Things have to be ordered. Time is limited and cannot be wasted and lost. Keeping people waiting is insulting; it implies they are not busy and therefore unimportant. A project leader meets few problems stressing the importance of missed deadlines. Plans are carefully thought out and followed.

Example 19.9 The Northern European dance card

I have had discussions about different attitudes to time with a Chinese friend who was born in China and lives in Beijing but studied for many years in England. She is aware of both cultures. When I visit Beijing, I like to arrange to meet up with my friends and colleagues. I start contacting them about a month in advance to arrange when to meet up. I have a limited number of evenings there and want to tightly schedule them all. But when I contact my Chinese friends they say, "Great that you are coming. When you arrive, give me a ring and we will arrange to meet." I get frustrated because I want it all arranged in advance, like ducks in a row. But the Chinese like to arrange their social lives at the last minute.

I liken it to a dance card. When I was 16 and 17 and went to the New Zealand equivalent of the high school prom, we were given a dance card. All the music to be played during the evening was prescribed, and we were given a card with it listed. We had to arrange in advance who we were going to dance each number with, and were only allowed three dances with the partner we had come with. It all had to be tightly scheduled; the ducks in a row. It is very Northern European. It would not appeal to the Chinese—and I don't think it appealed to 17-year-old New Zealanders either, even in 1970.

In other parts of the world, the Middle East and Japan, time is seen through much longer lenses. It flows organically and things come together at appropriate moments. This view does not discount persistence in effort and thriftiness with resources. Emphasis is placed on doing things at once, particularly getting relationships established. Doing things as they arise, means interruptions which derail forward plans. Deadlines are seen as movable because it is more important to ensure relevant issues are attended to when they occur so that continuity is maintained. Imagine, then, the confusion when a Japanese company attempting to establish a project with an American organization feels that a meeting to sort out how to proceed is urgent. They try to arrange it for two days time but are told that senior American executives do not have time in their busy schedules within the next two months! The Japanese believe the Americans are not taking them and the relationship seriously. The Americans think the Japanese do not realise how busy they are running other aspects of their business.

What Needs Defining. Some cultures like everything spelt out in detail (see Example 19.9 again), assuming that unless things are stated they become woolly and a source of disputes. This view is adversarial, concentrating on areas that *may* cause dispute. A different perspective is to focus on the common interest at the centre of most projects, the Chinese *guan* xi. Building sufficient contractual infrastructure to provide shape and a way of working is thought to be important. The conflicts of interest are expected to be worked out as they occur to meet the specific circumstances. The aim is to build and preserve a relationship that will realise the project's purpose. The English project manager brought up in the adversarial tradition can expect a pragmatic approach to contracting when working with French partners. It does not mean that they are commercially careless or not astute business people.

Cultural Profile of Project Managers

We often assume project management is a discipline with universal rules applied uniformly worldwide. This views project management as a systems science with mechanistic systems applied universally. Project management is a social science, some people even describe it as an art, which will be applied differently by different cultures. Svein-Arne Jessen[4] proposed that the requirements for power difference, individualism, and uncertainty avoidance varied throughout the life cycle. He deduced the level of each required at each of four stages (Table 19.3 and Figs. 19.3 and 19.4.) (Masculinity he assigned a median score throughout since it appeared to have no effect on performance.)

During initiation, *power distance* should be high, as this is when the manager must give priority to the requirements and direction of top management (or the client). *Individualism* should also be high, as there is a need for creativity and innovative thinking during this stage; *uncertainty avoidance* should be low, as feasibility demands the ability to think in new directions and uncover new solutions, which often means risk, change, and unpredictability. During planning and execution the picture changes. *Power distance* should be low, as people who do the work should also be responsible for planning and executing it (Chap. 5). The main purpose of planning and execution is to ensure the prescribed goals are achieved and the project team members are the best people to decide the method of achieving it. During closeout attitudes should change again. *Power distance* should be high, as evaluation of the work done and results obtained are the responsibility of top management, because they are able to evaluate the work objectively and also because they are able to view the project in its wider context. *Individualism* should be low for the same reason. *Uncertainty avoidance* should be high, as the termination needs to be a well-structured process (Chap. 14) ending with the achievement of the project's objectives and ensuing benefits. Furthermore, the project team may feel insecure about the future, and so the manager should aim to maximize their security.

Doing a least squares fit of a country's Hofstede scores to the requirements for effective project management derived from his survey, Svein-Arne Jessen deduced a country's performance at each stage of the project life cycle and overall (Table 19.1). The results show that project management is typically a Western approach to problem solving. It is probably also not surprising that Germany is top of the list; their very systematic industrial approach

TABLE 19.3 Preferred Cultural Approach at Each Stage of the Life Cycle

Trait	Feasibility	Design	Execution	Close-out
Power distance	High	Low	Low	High
Individualism	High	Medium	Medium	Low
Masculinity	Medium	Medium	Medium	Medium
Uncertainty avoidance	Low	Medium	Medium	Low

could well have been the model for the initial development of project management in the United States in the early 1950s. Arab countries and East Africa are also in the upper half of the list, showing either an in-built ability in these cultures to use the project approach, or a very strong and perhaps forced implementation of project management in these countries by Western cultures, which may have directly affected their behaviour.

(1) Most European countries fit into the accepted mould for project management having the right structural tools for systematic planning, organizing, and executing projects. They have self-confidence (high individualism) for taking on challenging tasks and doing them independently (low power distance), and accepting and fighting risks (low uncertainty avoidance). Their weakness occurs during start-up and termination, when it is necessary to ensure that the organization is doing the right projects and ensuring that the completion of the project results in the required benefits.

(2) Scandinavian countries, which often regard project management as typifying their cultures, score fairly low. They are well known for managing nearly everything through projects, with the result that organizations have far more projects than they have resources to handle, and large files of projects almost never terminated.

(3) The United States, which invented the concept of project management 60 years ago, scores well in both diagrams but their small power distance and their high acceptance of risk, expressed as weak uncertainty avoidance, could result in weak project termination, implying unnecessary time and cost overruns.

(4) Japan seems not to fit the project profile particularly well in any diagram having too strong uncertainty avoidance and lacking a profile that triggers project initiation, planning, and execution. As we know this has not prevented powerful industrial development in that country. It is probably also not so remarkable that the project approach is less used in Japan. Instead they prefer approaches such as production programming and quality circles which fit more with their cultural preferences and are the backbones of their success.

(5) Developing nations score fairly low on many of the described project management features. Indeed, these are the same factors for which they are often criticized by Western aid providers. However, they score fairly well on project initiation and they have a good balance between femininity and masculinity. Furthermore, their balanced uncertainty avoidance is a great advantage during project planning and execution. Here their fit is much better than, for instance, the Scandinavian countries.

We can match in project performance between pairs of countries. This provides interesting comparisons. For instance, the USA and Great Britain score highly in Africa, the Middle East, and South East Asia, and, surprisingly, the host country also acts as teaching agent with their greater ability at project initiation. It is also surprising that the Scandinavian countries, which in Gerd Hofstede's analysis came out with very much the same cultural profile behave quite differently when compared on the different project phases. Norway, for instance, performs well at planning and execution in East Africa, while Sweden is much better off in the Middle East. Hence, contrary to the common belief that the Western-oriented techniques of project management are just straightforward procedures anyone can learn and implement, there are considerable cross-cultural problems in using the approach in non-Western countries. Usually, insufficient focus is given to the fact that project management is not just a technique; it is an attitude of mind. Project management originated in Western countries and its popularity has been steadily growing but the outcomes have not always been in line with expectations particularly in developing countries. Traditionally this has been explained as weakness in the local human resources and the remedy being more training in the different mechanics of project execution, often in a

Western setting. However, the reason may well be a weak understanding of local needs by Western countries particularly needs beyond the project scope which are hard to articulate and define in Western terminology. Furthermore, many Western cultures are weak in both the initiation and the termination phases due to their individualistic attitudes towards authority, risk, and quality of life. In summary, in spite of its increasing popularity and widespread appeal, the many pros and cons of the project approach should be given serious consideration before implementing it.

The Project Leader's Role in Managing Differences

Faced with such a bewildering array of factors, all conspiring to reduce project performance, project managers might be forgiven for wanting to cut and run! However, there are emerging a range of strategies that companies and project managers can employ in order to realise the full synergistic potential of cross-boundary project teams.

Project Manager Selection. Too often we find wholly unsuitable project managers are dumped into complex situations. No one has taken care to think about which sort of person and experience is best suited to making a success of these complex roles. Ralf Müller and I discuss the importance of obtaining cultural fit when selecting the project manager.[5] Before managing such projects, project managers should have had experience working in different organizations, managing a range of disciplines and, ideally, having lived and worked in more than one country, preferably as a member of project teams (see Example 19.10).

Example 19.10 Project manager selection

I audited a company where the head of estimating wanted to try his hand at project management. The first and only project he managed was in Israel, and the project was 100 percent overspent.

Awareness of Own Programming. It is sometimes surprising for project leaders to realise that to work in a multicultural environment they need to be aware of their own mental programming. If the project leader is from the company and country owning the project, the automatic assumption is that things will be done "our" way. One challenge for a project leader is to balance and evolve the demands of the interface between their "home" organization, the client organization, and culture, and team members. Companies who work internationally find that to be successful they have to modify their own thinking and working practices. Cross-cultural working is a two-way street, not colonisation.

Awareness of Others' Cultural Programming. Working in multicultural environments requires the project leader to appreciate that things will be done, seen, and understood differently. Project leaders need to be curious, not shocked, and should demonstrate interest in finding out and understanding different people's world views. They need to respect values leading to behaviours alien to them, but important to the individuals and society to which they belong. Assuming things will be done "our way" only pushes differences underground so that they become embedded blockages. This easily creates an atmosphere of winners and losers which can prejudice effective delivery.

Leadership and Membership of Project Multicultural Teams: What to Do? For teams to work effectively, the roles and responsibilities of the leader and team members, both individually and collectively, must always be agreed. If the team is composed of people from different cultures, expectations of leadership and membership differ. Clarifying degrees of equality, responsibility, and accountability of the leader and members is fundamental. So team start-up and team building is vital for success. The activities well known in team-building events are just as important, but extra dimensions need to be added for international teams. There are three dimensions that have to be orchestrated to achieve high performance.

1. Ability to discuss and respect established ways of working
2. Awareness of own cultural programming
3. Awareness of others' cultural programming

Ability to Discuss and Respect Established Ways of Working. This means building a team culture where cross-cultural issues are openly discussed, so that appropriate ways are found to integrate all needs. In addition to formal team-building sessions, informal contacts between team members, suppliers, clients, and other stakeholders establish and nurture networks and create links that enhance mutual understanding, curiosity, and mutual respect. In low-definition cultures, informal relationships and getting to know individuals are considered more important than formal relationships (see Example 19.11).

Example 19.11 Respecting local ways of working

An English project manager commissioning a chemical plant in Latin America recognised he and his family would spend a lot of time meeting local dignitaries, suppliers, politicians, and government officials if he was to set up and hand the plant over. He explained that when at home he rarely saw anybody from work. He was keen on gardening and being with his family. However, he realised his new job would place a new set of responsibilities on him and his family in a new culture. Social activities connected with work had to be undertaken.

Accelerate Personal Network Development. The development of good personal relationships between people who have shared experiences is one of the most potent ways a project manager can influence project performance. However, companies in long-term joint ventures can also influence the wider networks through frequent job interchange, personal mobility, lateral career moves, interorganization conferences, meetings, and training courses. The more the webs of relationships between the organizations intertwine the better. This must be done not only at the top, but at all levels of the organizations concerned.

Language. Decide early on a common working language. Provide accelerated language training for all those whose first language is not the chosen language. Work hard on those for whom it is the first language to modify the way they speak. They must think as if it were a foreign language and should talk slowly, enunciate clearly, and avoid slang or jargon. Simultaneously, however, find ways to make it easier for those who are learning by translating key documents into several languages and by having a newsletter in more than one language.

Cross-Border Coaches. Identify people across the organization who have an awareness of the dimensions of difference and use them as coaches or mentors to the project team, either on training courses or available to advise less-experienced people about how to operate effectively in such environments. Such coaches can be supplemented by more formal cultural briefings about different countries that are increasingly available from specialist organizations.

Communications Infrastructure. New communications technologies are powerful tools for project managers, but they frequently fail to live up to their promise. The key lesson is not to fall into the trap of believing that e-mail, electronic and video conferencing, groupware, and other technologies get people communicating. The personal relationships and networks need to be built in part first and then the technologies can help dramatically to develop these networks further. Get the basics in place first; good telephones, several fax links, and a good directory of who is who, what they do, and how they can be contacted. Supplement these with project start-up workshops, where all the key players get to meet each other personally and work together, and you will have rapidly created the basic technical and interpersonal infrastructure you need. Out of this the need and scope for more sophisticated methods will emerge more clearly.

The Overseas Project Team

The character of any team is determined by the quality of its senior personnel. This is especially true of the international project where individuals are thrown together more closely and there is less scope for toleration of personalities who do not fit in. A good project team does not just happen. It is achieved by hard work, particularly by the project manager, and this work has to be done before the overseas team is mobilised. The selection of the project team itself also requires careful consideration. Factors which should be taken into account when selecting personnel are discussed in the following sections.

Ability to Work Well with Others. This is probably the most important characteristic. The turnover on international projects is caused more by poor interpersonal relationships than by deficiencies in technical skills. Character deficiencies are more serious in the close confines of the international project than they are in the home office (see Example 19.10).

Prior Experience on Overseas Appointments. This is always a good pointer but check with previous employers or managers. The person you are considering may be the one person who did not fit in with the rest of the team.

Stability Under Pressure and Ability to Cope. There are many pressures on the expatriate staff member and his other family whilst working overseas. These include

- Working and making decisions on the spot with limited support
- Coming to terms with different working patterns and practises of other foreign nationals
- Overcoming language barriers
- Domestic pressure from the family due either to working abroad on single status or to the family themselves trying to come to terms with the problems of living abroad

Versatility. The overseas team requires personnel who are able to cope with every situation which may confront them, be it the breakdown of a much-used computer system or the emergency repair of a broken down car under hazardous conditions. You have to have personnel skilled in their own technical disciplines but it pays to look for hidden talent as well!

Patience and Diplomacy. Overseas personnel at all levels have to have interpersonal skills and be able to relate to the nationals in their host country.

Professional Ex-Patriots. These are people who spend a high proportion of their lives working on major overseas contracts for a variety of clients. They can bring a vast range of experience to a project team not only in a technical sense but also in such important areas as knowledge of local customs, how to get things done and general environmental awareness.

The International Partner

Many problems of overseas working can be overcome by working with partners experienced in the countries concerned, especially choosing a local partner. In order to be confident of the relationship, you need to be careful in partner selection.

Nature of The Company. Since your fortunes are linked to how well partners do their job and respond to risk, a clear view is needed of their reliability in the face of risks. There should therefore be a thorough review of their financial strength, backing, track record in the technology and markets, and their strengths and weaknesses. This is especially important with new partners.

Relationship with Government. Since many overseas projects involve export credit guarantees, Third World aid, or other financial aspects impinging on government relationships, the effectiveness of a partner's relationships with its own government could be crucial. For particularly large or controversial projects, it may be necessary to create contacts between national governments, in which case it is vital to have good links established at both company and government level.

Attitude to Risk. Risks on international projects include not only normal contractual risks such as bid and performance bonds, penalties, damages, but also major additional risks such as climatic conditions, delays, and damage in port and freight handling, and security of storage. The most important interface with partners is their readiness to tolerate the extra cost of responding to these risks. An essential prerequisite to agreeing the scope of shared work is to define clearly roles and responsibilities, and the channels of communications on solving joint problems. The awkward issues lie in ensuring precise monitoring and identification of problems early enough for joint management decisions.

Market and Logistics Capability. In the context of the market, the partner's competence in handling an international project should complement and be integrated with one's own, so that actions to clients, authorities, local interests, and government agencies are consistent and tactfully effective.

SUMMARY

1. International projects may include
   - Projects in your own country for a foreign client
   - Projects in your own country using foreign contractors
   - Projects in a foreign country for which you are client
   - Working as a contractor for a client in their home country

    - Projects in a foreign country for clients also alien to that country
    - Multinational joint ventures
2. Problems on international projects are created by
    - Culture
    - Degrees of distance
    - Organization, management, and communication
    - Productivity and logistics
    - Local legislation and regulation
3. Dimensions of cultural difference include
    - Uncertainty avoidance
    - Power distance
    - Individualism
    - Masculinity
    - Role of time
    - Consideration of detail
4. In order to manage these differences, managers need to:
    - Select an appropriate project manager.
    - Be aware of the programming of themselves and others.
    - Use appropriate leadership styles.
    - Discuss and respect established ways of working.
    - Accelerate personal network development.
    - Use appropriate language.
    - Use cross-border coaches.
    - Develop a communications infrastructure.
5. In putting together a management approach for international projects, you need to:
    - Choose appropriate staff.
    - Choose an appropriate local partner.

REFERENCES

1. Hofstede, G., *Culture's Consequences: Comparing Values, Behaviors, Institutions and Organizations Across Nations*, 2d ed, Thousand Oaks, CA: Sage Publications, 2001.
2. Trompenaars, F. and Hampden-Turner, C., *Managing People Across Cultures* (*Culture for Business*), London: Capstone Publishing, 2004.
3. House, R.J., Hanges, P.J., Javidan, M., Dorfman, P.W., and Gupta, V., Culture, Leadership, and Organizations: The GLOBE Study of 62 Societies, Thousand Oaks, CA: Sage Publications, 2004.
4. Jessen, S-A, *The Nature of Project Leadership*, Oslo, Norway: Scandinavian University Press, 1993.
5. Turner, J.R. and Müller, R., *Choosing Appropriate Project Managers: Matching Their Leadership Style to the Type of Project*, Newtown Square, Pa.: Project Management Institute, 2006.

CHAPTER 20
EPILOGUE

In Chap. 3, I described key success factors in project management and from them derived five principles of good project management. I end by summarizing the five principles and the key success factors.

20.1 PRINCIPLES OF PROJECT MANAGEMENT

The five principles of good project management are

Manage through Structured Work or Product Breakdown. Use a breakdown structure:

- To delegate responsibility
- To define the scope
- To isolate risk
- To isolate changes

Focus on Results. Focus on what to achieve, not how to do it:

- To control scope
- To give a flexible but robust plan (using rolling-wave planning)

Balance Objectives through the Breakdown Structure. Achieve a balance:

- Between areas of technology
- Between technology and culture (people, systems, and organization)

Negotiate a Contract between the Parties Involved. All planning is a process of negotiation:

- Between the owner and contractor
- Between the project team members
- Through bipartite discussion
- By trading benefits for contributions

Adopt Clear and Simple Management Reporting Structures. Keep it simple. Use single-page reporting, nested through the breakdown structure, to give:

- Visibility
- Clarity
- Commitment

20.2 KEY SUCCESS FACTORS

The success factors were listed under four headings:

Foundations

- Align the project with the business.
- Gain the commitment of your boss and involved managers.
- Create shared vision, a sense of mission.

Planning

- Use multiple levels through a breakdown structure.
- Use simple friendly tools, one sheet per level.
- Encourage creativity, by delegating to experts through results.
- Estimate realistically.

Organizing and Implementing

- Negotiate resource availability.
- Agree cooperation.
- Define management responsibility.
- Gain commitment of resource providers through the shared mission.
- Define channels of communication.

Control

- Integrate plans and reports.
- Formalise the review process, through
  - Defined intervals
  - Defined agenda.
  - Defined criteria
  - Controlled attendance.
- Use your sources of authority as a project manager.

APPENDIXES

APPENDIX A
PROJECT DEFINITION REPORT FOR THE CRMO RATIONALIZATION PROJECT

TriMagi **Project Definition Report**		
Project: **Project Sponsor:** **Project Manager:**	CRMO Rationalization Project Steve Kenny Rodney Turner	
CONTENTS		Page
1. Project Context	Background Benefits map	
2. Project Definition	Purpose/scope/ objectives/work areas	
3. Project Strategy	Project success	
4. Work Definition	Milestone plan Milestones Milestone scope statements	
5. Organization	Responsibility chart Stakeholder register	
6. Project Plans	Activity plans Risk register	
7. Project Control	Project planning cycle Project control cycle	
8. Project Appraisal	Project cost estimate Milestone cost estimates	

Author:	JRT	Date:	1 March	Issue:	A

TriMagi
Background

Project:	CRMO Rationalization Project
Project Sponsor:	Steve Kenny
Project Manager:	Rodney Turner

TriMagi Communications is in business to supply visual, voice, and data communication networks based on its leading edge in glass fibre and laser technology. It will supply two-way cable television services to domestic and educational customers and data communication networks to these and commercial customers. It supplies telecommunication services through its cable and data networks. It will be the first-choice provider in the European countries within which it operates. It currently operates in its home base of the Benelux countries, (Belgium, the Netherlands, and Luxembourg) but plans to expand into other European countries.

With its expansion in Europe, TriMagi Communications intends to rationalise its Customer Repair and Maintenance Offices (CRMOs) in the Benelux countries, starting in its home base in Holland. There are currently 18 CRMOs in the region. Each office is dedicated to an area within the region. An area office receives all calls from customers within the area reporting faults. The fault is diagnosed either electronically from within the office or by sending an engineer to the customer's premises. Once diagnosed, the fault is logged with the field staff within the office and repaired in rotation. Each area office must cope with its own peaks and troughs in demand. This means that the incoming telephone lines may be engaged when a customer first calls and it can take up to two days to diagnose the fault.

To improve customer services, the company plans to rationalise the CRMO organization within the region with three objectives:

- Never have engaged call receipt lines within office hours.
- Achieve an average time of two hours from call receipt to arrival of the engineer at the customer's premises.
- Create a more flexible structure able to cope with future growth both in the region and throughout Europe and move to "enquiry desk," dealing with all customer contacts.

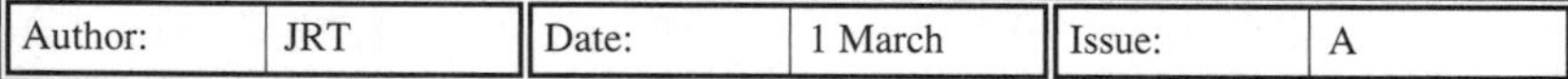

Author:	JRT	Date:	1 March	Issue:	A

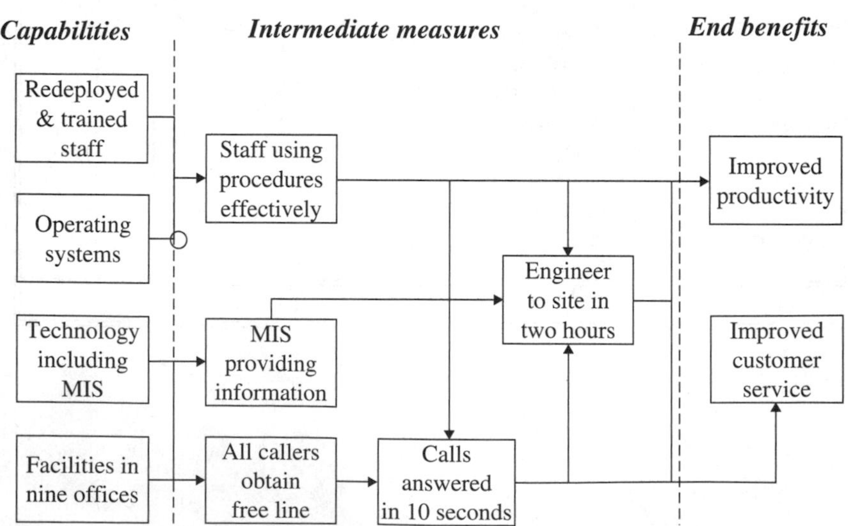

TriMagi
Project Definition

Project: **Project Sponsor:** **Project Manager:**	CRMO Rationalization Project Steve Kenny Rodney Turner
Purpose	The purpose of the project is to rationalize the CRMO organization: 1. To improve customer service so that: – All customers calling the receipt offices obtain a free line. – All calls are answered within 10 seconds. – The maximum time from call receipt to arrival of an engineer on site is two hours. 2. To improve productivity and flexibility so that: – The costs are justified through productivity improvements. – The call receipt offices made part of a unified enquiry desk, but there are no redundancies so all productivity improvements are achieved through natural wastage, redeployment, or growth.
Scope	The work of the project includes: 1. Changing from the existing structure of 18 area offices to 3 call receipt offices, 2 diagnostic offices, and 4 field offices 2. Investigating which of the two new CRMO networking technologies is appropriate for the new structure and to implement the chosen ones 3. Refurbishing the nine new offices to current standards 4. Training and redeploying staff to meet needs of operation of new CRMOs 5. Installing hardware to connect the CRMOs to the Customer Information System and to implement a statistical package to analyse fault data It is expected that the first call receipt and diagnostic offices will be available in five months time and the project will be complete in nine months. The work of the project excludes the retrenchment of any staff who are surplus to requirements within the CRMO structure; they will be passed to central personnel for redeployment on other expansion projects; with the implementation of the new Customer Information System, the call receipt offices may within the next two years be incorporated into unified "enquiry desks" dealing with all customer contacts. However, it will not be the project team's responsibility to achieve that integration.
Outputs	The outputs of the CRMO Rationalization Project are: 1. CRMO facilities have been installed in nine offices, (three call receipt offices, two diagnostic offices, and four field offices) within nine months. 2. Appropriate networking technology has been selected and implemented together with statistical MIS to achieve the required customer service levels. 3. Appropriate operating systems have been designed and implemented together with procedures to achieve the required customer service levels and productivity improvements. 4. Staff has been trained and redeployed to fill new positions, and vacate old positions. 5. First offices should be operational within five months and the work complete within nine.
Areas of work	To achieve the project's objectives, following areas of work are required: A *Accommodation*: Refurbish new offices and install hardware and furniture. (There is only one floor area available in the region large enough to take the first call receipt and fault diagnosis offices. The remaining eight offices must be housed in existing CRMO space.) T *Technology:* Decide on networking technology to be used; implement statistical MIS, and networking technology in new offices. O *Organization*: Communicate all changes to the staff involved, define the operation of the new CRMOs, and train and redeploy staff to fill new positions. T *Project:* Plan the project, organize the resources, and obtain financial approval.

Author:	JRT	Date:	1 March	Issue:	A

TriMagi **Project Success**	
Project: **Project Sponsor:** **Project Manager:**	CRMO Rationalization Project Steve Kenny Rodney Turner
Deliverables	The project will deliver to the parent organization: – Three call receipt offices, two diagnostic offices, and four filed offices – The technology to support the operation of the new system – Operational procedures for operation of the new system – Working methods to support the new system – Adequate numbers of competent people to support the new system
Success criteria	The project will be judged successful if: – There are never any engaged telephones in call receipt. – An engineer always arrives on site within two hours of a call being logged. – There are improvements in flexible working and productivity. – There are fewer customer complaints. – The new structure supports the company's expansion plans.
Stakeholders	Relevant stakeholders include: – The board of the parent company – Managers in the CRMO organization – Staff in the CRMOs – Customers – Managers of the new regions being established etc.

Author:	JRT	Date:	1 March	Issue:	A

TriMagi Milestone Plan

Project:	Rationalization of the Customer Repair and Maintenance Organization
Project Sponsor:	Steve Kenny
Project Manager:	Rodney Turner

Date	P	O	A	T	Milestone Name	Short Name	End Date
5-Mar	P1				When the project definition is complete including benefits map, milestone plan, and responsibility chart	Project definition	
30-Apr				T1	When the technical solution including appropriate networking and switching technology has been designed and agreed	Technology design	
22-Mar		O1			When a plan for communicating the changes to the CRM Orgaization has been agreed	Communicaton plan	
15-May		O2			When the operational procedures in the CRM Offices has been agreed	Operational procedures	
31-May		O3			When the job design and management design is complete and agreed	Job and management design	
31-May				T2	When the functional specification for the supportiong management information system (MIS) has been agreed	MIS funcational spec	
15-Jun		O4			When the allocation of staff to the new offices, and recruitment and redeployment requirements have been designed and agreed	Staff allocation	
15-Jun				T3	When the technical roll-out stratgey has been defined and agreed	Technical roll-out plan	
15-Jun			A1		When the estates roll-out stratgey has been designed and agreed	Estates roll-out plan	
30-Jun	P2				When the budget for implementation has been determined and provisional fianancial authority obtained	Financial approval	
15-Jul			A2		When sites 1 and 2 are available	Sites 1 and 2 available	
15-Jul		O5			When the management changes for sites 1 and 2 are in place (first call receipt and first diagnostic offices)	Management changes	
31-Aug				T4	When the system is ready for service in sites 1 and 2	Systems in sites 1 and 2	
31-Aug		O6			When a minumum number of staff have been recruited and redeployed and their training is complete	Redeployment and training	
15-Sep			A3		When sites 1 and 2 are ready for occupation	Sites 1 and 2 ready	
15-Sep				T5	When the MIS system has been delivered	MIS delivered	
30-Sep		O7			When sites 1 and 2 are operational and procedures implemented	Procedures implemented	
30-Nov	P3				When a successful intermediate review has been conducted and roll-out plans revised and agreed	Intermediate review	
31-Mar			A4		When the last site is operational and procedures fully implemented	Roll-out implemented	
30-Sep	P4				When it has been shown, through a postimplementation audit that all benefit criteria have been met	Postcompletion audit	

TriMagi **Milestone List**	
Project: **Project Sponsor:** **Project Manager:**	CRMO Rationalization Project Steve Kenny Rodney Turner
Accommodation	A1: Estates plan A2: Sites 1 and 2 obtained A3: Sites 1 and 2 ready A4: Estates roll-out
Technology	T1: Technology design T2: MIS design T3: Technology plan T4: System in sites 1 and 2 T5: MIS delivered T6: Technology roll out
Organization	O1: Communications plan O2: Operational procedures O3: Job/management design O4: Staff allocation O5: Management changes O6: Redeployment and training O7: Procedures implemented
Project	P1: Project definition P2: Financial approval P3: Intermediate review P4: Postcompletion audit

Author:	JRT	Date:	2 March	Issue:	A

TriMagi **Milestone List**	
Project: **Project Sponsor:** **Project Manager:**	CRMO Rationalization Project Steve Kenny Rodney Turner
Milestone: **Milestone Manager:**	CRMO Rationalization Project John Wain
Scope:	The workpackage requires the preparation of high-level plans and estimates to be prepared, to enable resource budgets to be prepared and their availability agreed.
Possible work:	Identify key managers Hold launch workshop Finalize milestone plan and project responsibility chart Estimate resource requirements and durations Schedule resource requirements Discuss requirements with managers Plan and agree resource availability
Measure of completion:	Project plans approved by the steering committee Resource managers sign agreements to resource availability

Author:	JRT	Date:	2 March	Issue:	A

TriMagi Project Responsibility Chart

Project Schedule

Project:	Rationalization of the Customer Repair and Maintenance Organization
Project Sponsor:	Steve Kenny
Project Manager:	Rodney Turner

Period: Month **Target end:** 30-Jun-02

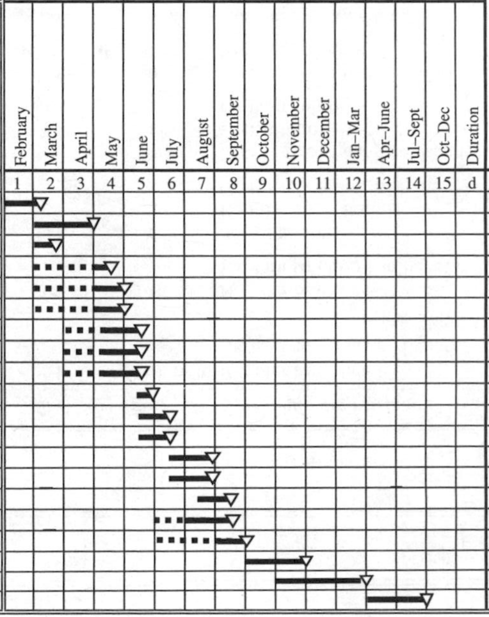

Code	Meaning
X	e**X**ecutes the work
D	takes **D**ecisions solely/ultimately
d	takes **d**ecisions jointly
P	manages **P**rogress
T	on-the-job **T**raining
I	must be **I**nformed
C	must be **C**onsulted
A	may **A**dvise

No	Milestone Name	Regional board	Operations director	CRMO managers	CRMO team leader	CRMO staff	Project manager	Project support office	Estates manager	Estates department	Network manager	Networks department	IS department	Operators	Personnel	Suppliers	End Date
P1	Project defintion	D	D	dX	dX	I	PX	X	X	I	X	I	C	C	C		5-Mar
T1	Technology design					C					PX	X	X	X		A	30-Apr
O1	Communicaton plan	I	D	d	PX												22-Mar
O2	Operational procedures	I	D	d	PX	X											15-May
O3	Job and management design	I	D	d	PX	C									TX		31-May
T2	MIS functional spec	I	D	d	dX						PX		X				31-May
O4	Staff allocation	I	D	d	PX	C									TX		15-Jun
T3	Technical roll-out plan	D	d	C			C	X	C	I	PX	X	X	I		C	15-Jun
A1	Estates roll-out plan	D	d	C	X	I	C	X	PX	X	C	I	ISD	I	I	C	15-Jun
P2	Financial approval	D	d	I			PX	X	C		C		C	A	A	C	30-Jun
A2	Sites 1 and 2 available		I	I	I				PX	X	I						15-Jul
O5	Management changes	I	DX	X	PX	I											15-Jul
T4	Systems in sites 1 and 2		I	I	I					X	PX	X	X	I		X	31-Aug
O6	Redeployment and training			D	PX										TX		31-Aug
A3	Sites 1 and 2 ready		I	I	X	X	P			X		X	X	I		X	15-Sep
T5	MIS delivered	I	D	I	X						PX		X			X	15-Sep
O7	Procedures implemented	D	D	PX	X							A	A	I	X		30-Sep
P3	Intermediate review	D	d	C	C		PX	X	A		A		A	A	A		30-Nov
A4	Roll-out implemented	I	D	dX	dX	X	PX	X	I	X	I	X	X	I	X	X	31-Mar
P4	Postcompletion audit	D	d	C	C		PX	X						C			30-Sep

Schedule periods: 1 February, 2 March, 3 April, 4 May, 5 June, 6 July, 7 August, 8 September, 9 October, 10 November, 11 December, 12 Jan–Mar, 13 Apr–June, 14 Jul–Sept, 15 Oct–Dec, d Duration

TriMagi
Stakeholder Register

Project: CRMO Rationalization Project
Project Sponsor: Steve Kenny
Project Manager: Rodney Turner

Stakeholder	*Objectives*	*For/Against*	*Influence*	*Informed*	*Communication strategy*
Board	Expand operations Improved customer service Improved profitability	For	Hi	Must be	Regular briefing Explain solution and benefits
Operations managers	Improved customer service Excellent support	For	Med	Must be	Regular briefing Explain solution and benefits
Maintenance managers	Operation that works Maintain position and influence	For	Hi	Yes	Seek opinions Regular consultation Confirm solution with them
Maintenance staff	Ease of operation Maintain jobs	For	Med	Not at start	Briefings/company newspaper Consultation Explain solution
Operations staff	Support their work Minimum disruption	Ambivalent	Low	Not at start	Briefings/company newspaper Explain solution
Customers	Good service	For	Low	Not at start	Customer newsletters
Local community	Minimum disruption to environment	Ambivalent	Low	Low	Local newspaper advertisements

Author: JRT Date: 2 March Issue: A

TriMagi Activity Plan

Project:	Rationalization of the Customer Repair and Maintenance Organization
Milestone:	D3: Personnel information pages designed
Manager:	Rodney Turner

X	e**X**ecutes the work
D	takes **D**ecisions solely/ultimately
d	takes **d**ecisions jointly
P	manages **P**rogress
T	on-the-job **T**raining
I	must be **I**nformed
C	must be **C**onsulted
A	may **A**dvise

Activity Schedule

Period: Week ending **Target end:** 5-Mar

Weeks: 1, 2 (5-Feb), 3, 4 (12-Feb), 5, 6 (19-Feb), 7, 8 (26-Feb), 9, 10 (25-March), 11, 12 (12-March), 13, 14, 15, d (Duration)

No	Activity Name	Regional board	Operations director	CRMO managers	CRMO team leader	Project manager	Project support office	Estates manager	Estates department	Network manager	Networks department	IS department	Operators	Personel	End Date
1	Produce project proposal	C	D	d	dX	PX	A	A		A		A	A	A	5-Feb
2	Hold definition workshop		DX		X	PX	X								9-Feb
3	Define required benefits	C	D	d	dX	PX									12-Feb
4	Draft definition report	C	D	d	dX	PX	X	I		I		I	I	I	17-Feb
5	Hold launch workshop	C		X	dX	PX	X	X		X		X			19-Feb
6	Finalize milestone plan		D	d	D	PX	X	C		C		C	A	C	24-Feb
7	Finalize responsibility chart		D	d	D	PX	X	C		C		C	A	A	24-Feb
8	Finalize risk assessment		D	d	dX	PX	X	C		C		C	C	C	26-Feb
9	Finalize time estimates				A	P	X	A		A		A		A	26-Feb
10	Finalize cost estimates				A	P	X	A		A		A		A	26-Feb
11	Finalize revenue estimates		A	A	A	PX									26-Feb
12	Assess project viability		D	d	d	PX									26-Feb
13	Finalize definition report	D	d	d	d	PX	X	C		C		C		C	5-Mar
14	Mobilize team		D	d	dX	PX	X	X	I	X	I	IX		I	5-Mar

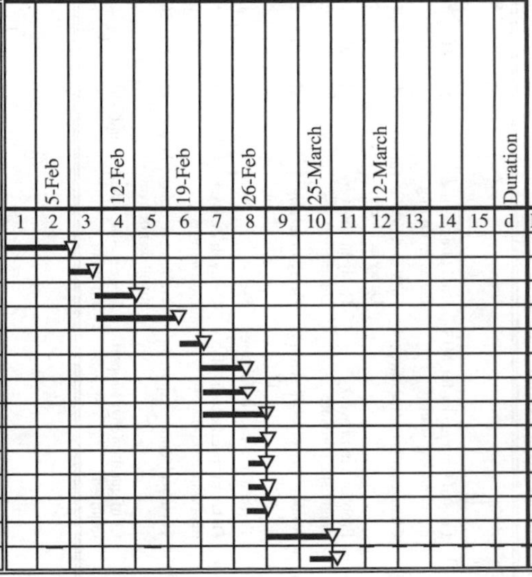

TriMagi
Risk Register

Project:	CRMO Rationalization Project
Project Sponsor:	Steve Kenny
Project Manager:	Rodney Turner

Rank	Mile-stone	Risk	Impact	L	C	Strategy
1	T5	Loss of team leader for MIS development team	Loss of expertise Delay in coding	M	H	Identify potential replacement
1	T5	Changes to user interface	H/W and S/W definition Delay delivery	H	M	Ensure user involvement in evaluation of prototype
2	T4	Problems in network diagnostic software	Delay in software completion	M	M	New version of software has fewer faults
2	T4	Availability of work stations for testing	Delay in testing	M	M	Expedite delivery with supplier
2	T3	Testbed interface definitions	Delay	M	M	Expedite definitions
2	T3	Delay in specification of data transmission	Delay in delivery of hardware	M	M	Meeting scheduled to consider alternatives
1	O2	Technical author required	Training & maintenance manuals not available	M	L	Contact agency
1	All	Configuration mgt support required	Poor quality systems and manuals	M	L	Contact agency

TriMagi Procedural Responsibility Chart

Project:	Procedure for project start
Project Sponsor:	Steve Kenny
Project Manager:	Rodney Turner

X	e**X**ecutes the work
D	takes **D**ecisions solely/ultimately
d	takes **d**ecisions jointly
P	manages **P**rogress
T	on-the-job **T**raining
I	must be **I**nformed
C	must be **C**onsulted
A	may **A**dvise

Period: Week Target end:

No	Milestone Name	Project management tasks: Background	Purpose	Objectives	Work breakdown	Responsibilities	Detailed plans	Resource allocation	PM system	Cooperation	Information		Responsibilities: Steering committee	Project manager	Project office	Participants	Consultants
	Project description	S	M	M	M	M							DX	C	C	I	
	Definition workshop	S	S	S	S	S	S	S	S	S	S			PX	X	X	T
	Draft issue of plan	M	S	S	S	S	M	M	M	M	M		I	PX	X	I	
	Meeting consultants			S		S				S				PX	X	X	A
	First issue of plan	M	M	M	M	M	S	S	S	S	S		I	PX	X	I	
	Invitation to start-up workshop														X		
	Start-up workshop																
	Part 1	M	M	M	M	M	S	S					DX	PX	X	X	T
	Part 2	S	S	S	S	S	M	M	M	M	M		DX	PX	X	X	T
	Final editing of plan	M	M	M	M	M	M	M	M	M	M		I	PX	X	I	
	Approval of plan	M	M	M	M	M	S	S	S	S	S		D	PX	I	I	
	Issue of definition report	M	M	M	M	M	M	M	M	M	M		D	PX	X	C	

M = Main issue
S = Subsidiary issue

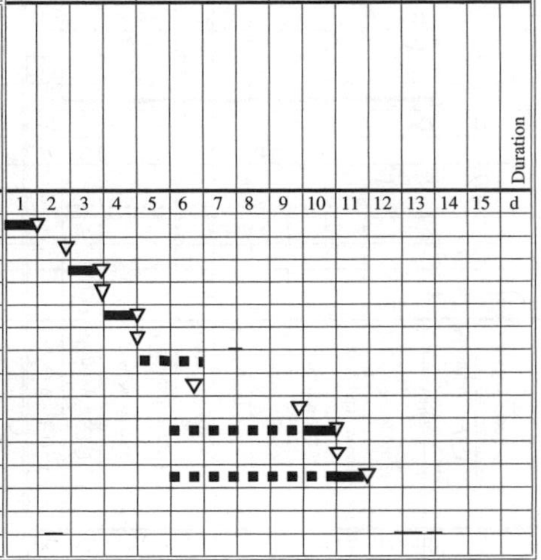

TriMagi Procedural Responsibility Chart

Project:	Procedure for monitoring and control
Project Sponsor:	Steve Kenny
Project Manager:	Rodney Turner

Period: Day **Target end:**

X	e**X**ecutes the work
D	takes **D**ecisions solely/ultimately
d	takes **d**ecisions jointly
P	manages **P**rogress
T	on-the-job **T**raining
I	must be **I**nformed
C	must be **C**onsulted
A	may **A**dvise

No	Milestone Name	Project manager	Team leaders	Project members	Project support office	Steering committee	Project sponsor
	Develop milestone plan	PX	X		I	d	D
	Create high-level network	PX			X		
	↓						
	Develop new activity schedules	DP	X	X	I		
	Update network	P	C	C	X		
	↓						
	Issue work-to lists			I	X		
	Do work	P	PX	X			
	Return turnaround documents		P	X	I		
	Activity review meeting	I	PX	X			
	Identify variances (activities)	I	PX	X			
	Plan recovery	DP	X	X	I		
	Issue activity progress reports	PI	X				
	↓						
	Review progress against milestones	PX			X	I	
	Milestone progress meeting	PX			X	X	
	Identify variances (milestones)	PX			X	DX	
	Plan recovery	PX	X		X	DX	
	Issue milestone progress report	PX			X	C	I
	↓						
	Approve progress						D

1	2	3	4	5	6	7	8	9	10	11	12	13	14	15	d
Friday week 0	Monday week 1	Tuesday week 1	Wednesday week 1	Thursday week 1	Friday week 1	Monday week 2	Tuesday week 2	Wednesday week 2	Thursday week 2	Friday week 2	Monday week 3	Tuesday week 3			Duration

TriMagi Project Responsibility Chart

Progress Report

Project:	Rationalization of the Customer Repair and Maintenance Organization
Project Sponsor:	Steve Kenny
Project Manager:	Rodney Turner

X	eXecutes the work
D	takes **D**ecisions solely/ultimately
d	takes **d**ecisions jointly
P	manages **P**rogress
T	on-the-job **T**raining
I	must be **I**nformed
C	must be **C**onsulted
A	may **A**dvise

No	Milestone Name	Operations director	CRMO managers	CRMO team leader	CRMO staff	Project manager	Project support office	Estates manager	Estates department	Network manager	Networks department	IS department	Operators	Personnel	Suppliers		Subcontract	Materials	Plant & equipment	Duration
		$K	$K	$K	$K	$K	$K	$K	$K	$K	$K	$K	$K	$K	$K		$K	$K	$K	$K
P1	Project defintion	3.0	3.0	4.0		14.0	12.0	2.0		2.0								25.0		
T1	Technology design				10.0					5.0	20.0	5.0	20.0							
O1	Communicaton plan			6.0														10.0		
O2	Operational procedures			10.0	40.0															
O3	Job and management design			10.0	20.0									50.0						
T2	MIS functional spec			10.0						10.0		40.0								
O4	Staff allocation			10.0	10.0									40.0						
T3	Technical roll-out plan			5.0		5.0	10.0	5.0		5.0	5.0	5.0					5.0			
A1	Estates roll-out plan			5.0		5.0	10.0	5.0	5.0	5.0							5.0			
P2	Financial approval			5.0		10.0	10.0	5.0		5.0		5.0					5.0	5.0		
A2	Sites 1 and 2 available							10.0	50.0									25.0		
O5	Management changes	10.0	10.0	20.0																
T4	Systems in sites 1 and 2								30.0	10.0	30.0	10.0					30.0			
O6	Redeployment and training			20.0										60.0				210.0		
A3	Sites 1 and 2 ready			10.0	10.0				10.0		10.0	10.0					20.0	180.0		
T5	MIS delivered			5.0						10.0		20.0					20.0	240.0		
O7	Procedures implemented		10.0	20.0										20.0				160.0		
P3	Intermediate review		10.0	10.0		10.0	40.0	10.0		10.0							10.0	45.0		
A4	Roll-out implemented		10.0	40.0	40.0	20.0	40.0	10.0	40.0	10.0	40.0	40.0		20.0			20.0			
P4	Postcompletion audit		20.0	40.0		40.0	40.0						20.0					900.0		
		13	73	230	130	104	162	47	135	72	105	135	40	190	0	0	115	1800	0	0

No	Milestone Name	Labour: Labour estimate	Labour: Actual	Labour: Estimated remaining	Labour: % Complete	Labour: Earned value	Labour: Calculated remaining	Cash: Cost total	Cash: Actual	Cash: Estimated remaining	Cash: % Complete	Cash: Earned value	Cash: Calculated remaining
		$K	$K	$K	%	$K	$K	$K	$K	$K	%	$K	$K
P1	Project defintion	40	40	0	1.0	40	0	25	25	0	1.0	25	0
T1	Technology design	60	75	0	1.0	60	0	0		0	1.0	0	0
O1	Communicaton plan	6	5	0	1.0	6	0	10	10	0	1.0	10	0
O2	Operational procedures	50	50	0	1.0	50	0	0	0	0	1.0	0	0
O3	Job and management design	80	80	0	1.0	80	0	0	0	0	1.0	0	0
T2	MIS functional spec	60	55	0	1.0	60	0	0	0	0	1.0	0	0
O4	Staff allocation	60	65	0	1.0	60	0	0	0	0	1.0	0	0
T3	Technical roll-out plan	40	40	0	1.0	40	0	5	10	0	1.0	5	0
A1	Estates roll-out plan	35	40	0	1.0	35	0	5	5	0	1.0	5	0
P2	Financial approval	40	20	20	0.5	20	20	10	0	5	0.5	5	5
A2	Sites 1 and 2 available	60	20	40	0.3	18	42	25	0	25	0.3	8	18
O5	Management changes	40	10	25	0.3	12	28	0		0	0.3	0	0
T4	Systems in sites 1 and 2	80		80	0.0	0	80	30		30	0.0	0	30
O6	Redeployment and training	80		80	0.0	0	80	210		210	0.0	0	210
A3	Sites 1 and 2 ready	50		50	0.0	0	50	200		200	0.0	0	200
T5	MIS delivered	35		35	0.0	0	35	260		260	0.0	0	260
O7	Procedures implemented	50		50	0.0	0	50	160		160	0.0	0	160
P3	Intermediate review	90		90	0.0	0	90	55		55	0.0	0	55
A4	Roll-out implemented	320		320	0.0	0	320	20		20	0.0	0	20
P4	Postcompletion audit	160		160	0.0	0	160	900		900	0.0	0	900
		1436	500	950		481	955	1915	50	1865		58	1858
	Forecast Cost at Completion			1450			1455			1915			1908

TriMagi Activity Plan

Activity Schedule

Project:	Rationalization of the Customer Repair and Maintenance Organization
Milestone:	D3: Personnel information pages designed
Manager:	Rodney Turner

Period: Week ending | **Target end:** 5-Mar

X	e**X**ecutes the work
D	takes **D**ecisions solely/ultimately
d	takes **d**ecisions jointly
P	manages **P**rogress
T	on-the-job **T**raining
I	must be **I**nformed
C	must be **C**onsulted
A	may **A**dvise

No	Activity Name	Regional board	Operations director	CRMO managers	CRMO team leader	Project manager	Project support office	Estates manager	Estates department	Network manager	Networks department	IS department	Operators	Personnel		Labour $K	Material $K	End Date
1	Produce project proposal				0.5	1.0	1.0									2.5		5-Feb
2	Hold definition workshop				0.5	1.0	0.5									2.0		9-Feb
3	Define required benefits		1.5		0.5	2.0										4.0		12-Feb
4	Draft definition report					2.0	3.0									5.0		17-Feb
5	Hold launch workshop		1.5	3.0	0.8	1.5	0.8	1.5		1.5						10.5	15.0	19-Feb
6	Finalize milestone plan					1.0	0.5									1.5		24-Feb
7	Finalize responsibility chart					1.0	0.5									1.5		24-Feb
8	Finalize risk assessment				0.5	1.0	0.5									2.0		26-Feb
9	Finalize time estimates						1.0									1.0		26-Feb
10	Finalize cost estimates						1.0									1.0		26-Feb
11	Finalize revenue estimates						0.5									0.5		26-Feb
12	Assess project viability					1.0										1.0		26-Feb
13	Finalize definition report				0.5	2.0	1.5									4.0		5-Mar
14	Mobilize team				0.3	0.5	0.3									1.0	10.0	5-Mar
	Contingency				0.5	1.0	1.0									2.5		
			3	3	4	15	12	2		2						40	25	

Schedule weeks: 1, 2 (5-Feb), 3, 4 (12-Feb), 5, 6 (19-Feb), 7, 8 (26-Feb), 9, 10 (5-Mar), 11, 12 (12-Mar), 13, 14, 15, d (Duration)

APPENDIX B

PROJECT CONTROL DOCUMENTS FOR THE CRMO RATIONALIZATION PROJECT

TriMagi **Control Forms**					
Project: **Project Sponsor:** **Project Manager:**	CRMO Rationalization Project Steve Kenny Rodney Turner				
CONTENTS				Page	
1. Milestone	Activity plan turnaround document Time sheet				
2. Project	Report against the milestone plan Milestone tracking form Tracked bar chart				
3. Project cost	Project cost report				
Author:	JRT	Date:	1 June	Issue:	A

TriMagi Activity Plan

Project:	Rationalization of the Customer Repair and Maintenance Organization
Milestone:	D3: Personnel information pages designed
Manager:	Rodney Turner

X	eXecutes the work
D	takes Decisions solely/ultimatel
d	takes decisions jointly
P	manages Progress
T	on-the-job Training
I	must be Informed
C	must be Consulted
A	may Advise

No	Activity Name	Regional board	Operations director	CRMO managers	CRMO team leader	Project manager	Project support office	Estates manager	Estates department	Network manager	Networks department	IS department	Operators	Personnel	Labour $K	Material $K	End Date
1	Produce project proposal				0.5	1.0	1.0								2.5		5-Feb
2	Hold definition workshop				0.5	1.0	0.5								2.0		9-Feb
3	Define required benefits		1.5		0.5	2.0									4.0		12-Feb
4	Draft definition report					2.0	3.0								5.0		17-Feb
5	Hold launch workshop		1.5	3.0	0.8	1.5	0.8	1.5		1.5					10.5	15.0	19-Feb
6	Finalize milestone plan					1.0	0.5								1.5		24-Feb
7	Finalize responsibility chart					1.0	0.5								1.5		24-Feb
8	Finalize risk assessment				0.5	1.0	0.5								2.0		26-Feb
9	Finalize time estimates						1.0								1.0		26-Feb
10	Finalize cost estimates						1.0								1.0		26-Feb
11	Finalize revenue estimates						0.5								0.5		26-Feb
12	Assess project viability					1.0									1.0		26-Feb
13	Finalize definition report				0.5	2.0	1.5								4.0		5-Mar
14	Mobilize team				0.3	0.5	0.3								1.0	10.0	5-Mar
	Contingency				0.5	1.0	0.3								2.5		
			3	3	4	15	12	2		2					40	25	

Activity Schedule

Period:	Week ending	Target end:	5-Mar-08

1	2	3	4	5	6	7	8	9	10	11	12	13	14	15	d
	5-Feb		12-Feb		19-Feb		26-Feb		5-Mar		12-Mar				Duration

Time Now

Report

Author:	Rodney Turner
Date:	13 February
Checked	Ian Simmons

No	Estimated Completion	Work done (d)	Work to do (d)	Quality acceptable	Responsibilities kept	Change required	External delay	Special problem
3	12-Feb	3	0					
4	14-Feb	4	1					
5	19-Feb							

TriMagi
Time-sheet

Act No	Description	Orig dur (d)	Rem dur (d)	Sched start	Sched finish	Actual start	Actual finish	M (h)	Tu (h)	W (h)	Th (h)	F (h)	WE (h)	Work rem (h)
P1A	Project proposal	5.0		01 Feb	05 Feb	01 Feb	05 Feb							
P1B	Definition workshop	2.0		08 Feb	09 Feb	08 Feb	09 Feb							
P1C	Define benefits	3.0	3.0	10 Feb	12 Feb	10 Feb	12 Feb	6	6	2				
P1D	Draft definition report	6.0	6.0	10 Feb	17 Feb	10 Feb	14 Feb			4	6	6		
P1E	Launch workshop	2.0	2.0	18 Feb	19 Feb									
P1F	Milestone plan	3.0	3.0	22 Feb	24 Feb									
P1G	Responsibility chart	3.0	3.0	22 Feb	24 Feb									
P1H	Assess risks	3.0	3.0	22 Feb	26 Feb									
P1I	Estimate time	2.0	2.0	25 Feb	26 Feb									
P1J	Estimate cost	2.0	2.0	25 Feb	26 Feb									
P1K	Estimate revenue	2.0	2.0	25 Feb	26 Feb									
P1L	Assess viability	2.0	2.0	25 Feb	26 Feb									
P1M	Finalize definition report	3.0	3.0	01 Mar	05 Mar									
P1N	Mobilize team	2.0	2.0	04 Mar	05 Mar									

TriMagi Milestone Plan — Milestone Report

Project:	Rationalization of the Customer Repair and Maintenance Organization
Project Sponsor:	Steve Kenny
Project Manager:	Rodney Turner

Date	P	O	A	T	Milestone Name	Report Date	Milestone Report
5-Mar	P1				When the project definition is complete, including benefits map, milestone plan, and responsibility chart	5-Mar	Project definition complete
30-Apr				T1	When the technical solution, including appropriate networking and switching technology has been designed and agreed	7-May	Technological design required additional work, completed a week late
22-Mar		O1			When a plan for communicating the changes to the CRM Orgaization has been agreed	22-Mar	Communication plan published
15-May		O2			When the operational procedures in the CRM Offices has been agreed	15-May	Completed. Not delayed by technical design
31-May		O3			When the job design and management design is complete and agreed	31-May	Completed. Not delayed by technical design
31-May				T2	When the functional specification for the supportiong management information system (MIS) has been agreed	7-Jun	Completed a week late. Delayed by the design
15-Jun		O4			When the allocation of staff to the new offices, and recruitment and redeployment requirements have been designed and agreed	22-Jun	Completed
15-Jun				T3	When the technical roll-out stratgey has been defined and agreed	22-Jun	Completed
15-Jun			A1		When the estates roll-out stratgey has been designed and agreed	22-Jun	Completed. Work on sites 1 & 2 will be delayed a week, but)7 can be completed on time.
30-Jun	P2				When the budget for implementation has been determined and provisional fianancial authority obtained	30-Jun	Delayed by 1 week by delay of technical design.
15-Jul			A2		When sites 1 and 2 are available	30-Jun	Work started 22 Jun
15-Jul		O5			When the management changes for sites 1 and 2 are in place (first call receipt and first diagnostic offices)	30-Jun	Work started 22 Jun
31-Aug				T4	When the system is ready for service in sites 1 and 2	30-Jun	Work expected to start 22 Jul
31-Aug		O6			When a minumum number of staff have been recruited and redeployed and their training is complete	30-Jun	Work expected to start 22 Jul
15-Sep			A3		When sites 1 and 2 are ready for occupation		
15-Sep				T5	When the MIS system has been delivered		
30-Sep		O7			When sites 1 and 2 are operational and procedures implemented		
30-Nov	P3				When a successful intermediate review has been conducted and roll-out plans revised and agreed		
31-Mar			A4		When the last site is operational and procedures fully implemented		
30-Sep	P4				When it has been shown, through a postimplementation audit that all benefit criteria have been met		

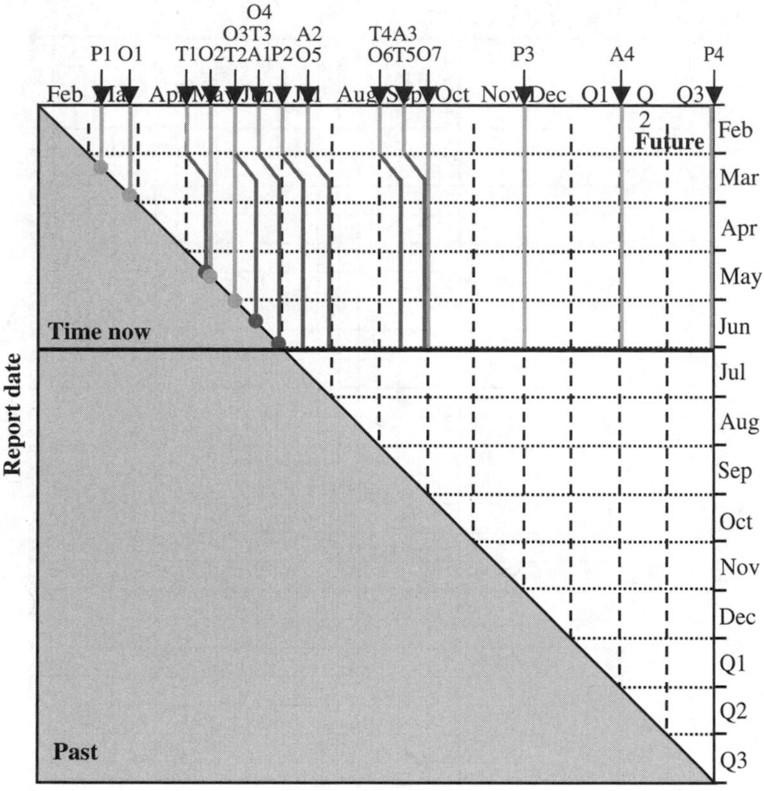
O4
O3T3
A2
T4A3
P1 O1
T1O2T2A1P2 O5
O6T5O7
P3
A4
P4
Feb
Mar
Apr
May
Jun
Jul
Aug
Sep
Oct
Nov
Dec
Q1
Q2
Q3
Future
Time now
Past
Report date
Planned date

TriMagi Project Responsibility Chart

Project:	Rationalization of the Customer Repair and Maintenance Organization
Project Sponsor:	Steve Kenny
Project Manager:	Rodney Turner

X	e**X**ecutes the work
D	takes **D**ecisions solely/ultimately
d	takes **d**ecisions jointly
P	manages **P**rogress
T	on-the-job **T**raining
I	must be **I**nformed
C	must be **C**onsulted
A	may **A**dvise

No	Milestone name	Regional board	Operations director	CRMO managers	CRMO team leader	CRMO staff	Project manager	Project support office	Estates manager	Estates department	Network manager	Networks department	IS department	Operators	Personnel	Suppliers	End Date
P1	Project defintion	D	D	dX	dX	I	PX	X	X	I	X	I	C	C	C		5-Mar
T1	Technology design					C					PX	X	X	X		A	30-Apr
O1	Communicaton plan	I	D	d	PX												22-Mar
O2	Operational procedures	I	D	d	PX	X											15-May
O3	Job and management design	I	D	d	PX	C									TX		31-May
T2	MIS functional spec	I	D	d	dX						PX		X				31-May
O4	Staff allocation	I	D	d	PX	C									TX		15-Jun
T3	Technical roll-outplan	D	d				C	X	C	I	PX	X	X	I		C	15-Jun
A1	Estates roll-outplan	D	d	CRI	X	I	C	X	PX	X	C	I	IS I	I	I	C	15-Jun
P2	Financial approval	D	d	I	CRMO		PX	X	C		C		C	A	A	C	30-Jun
A2	Sites 1 and 2 available		I	I	I				PX	X	I						15-Jul
O5	Management changes	I	DX	X	PX	I											15-Jul
T4	Systems in sites 1 and 2		I	I	I					X	PX	X	X	I		X	31-Aug
O6	Redeployment and training			D	PX										TX		31-Aug
A3	Sites 1 and 2 ready		I	I	X	X	P			X		X	X	I		X	15-Sep
T5	MIS delivered	I	D	I	X						PX		X			X	15-Sep
O7	Procedures implemented	D	D	PX	X							A	A	I	X		30-Sep
P3	Intermediate review	D	d	CRI	CRMO		PX	X	A		A		A	A	A		30-Nov
A4	Roll-out implemented	I	D	dX	dX	X	PX	X	I	X	I	X	X	I	X	X	31-Mar
P4	Postcompletion audit	D	d	CRI	CRMO		PX	X						C			30-Sep

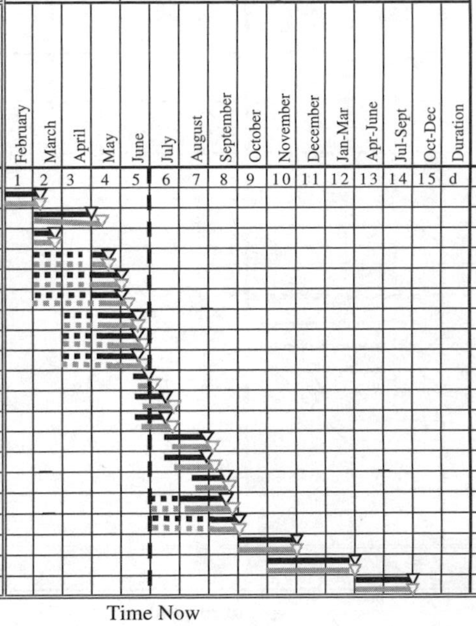

Project Schedule

Period:	Month	**Target end:**	30-Jun-02

February	March	April	May	June	July	August	September	October	November	December	Jan-Mar	Apr-June	Jul-Sept	Oct-Dec	Duration
1	2	3	4	5	6	7	8	9	10	11	12	13	14	15	d

Time Now

TriMagi Project Responsibility Chart

Progress Report

Project:	Rationalization of the Customer Repair and Maintenance Organization
Project Sponsor:	Steve Kenny
Project Manager:	Rodney Turner

X	e**X**ecutes the work
D	takes **D**ecisions solely/ultimately
d	takes **d**ecisions jointly
P	manages **P**rogress
T	on-the-job **T**raining
I	must be **I**nformed
C	must be **C**onsulted
A	may **A**dvise

No	**Milestone Name**	Operations director	CRMO managers	CRMO team leader	CRMO staff	Project manager	Project support office	Estates manager	Estates department	Network manager	Networks department	IS department	Operators	Personnel	Suppliers		Subcontract	Materials	Plant & equipment	Duration
		$K	$K	$K	$K	$K	$K	$K	$K	$K	$K	$K	$K	$K	$K		$K	$K	$K	$K
P1	Project defintion	3.0	3.0	4.0		14.0	12.0	2.0		2.0								25.0		
T1	Technology design				10.0					5.0	20.0	5.0	20.0							
O1	Communicaton plan			6.0														10.0		
O2	Operational procedures			10.0	40.0															
O3	Job and management design			10.0	20.0									50.0						
T2	MIS functional spec			10.0						10.0		40.0								
O4	Staff allocation			10.0	10.0									40.0						
T3	Technical roll-out plan			5.0		5.0	10.0	5.0		5.0	5.0	5.0					5.0			
A1	Estates roll-out plan			5.0		5.0	10.0	5.0	5.0	5.0							5.0			
P2	Financial approval			5.0		10.0	10.0	5.0		5.0		5.0					5.0	5.0		
A2	Sites 1 and 2 available							10.0	50.0									25.0		
O5	Management changes	10.0	10.0	20.0																
T4	Systems in sites 1 and 2								30.0	10.0	30.0	10.0					30.0			
O6	Redeployment and training			20.0										60.0				210.0		
A3	Sites 1 and 2 ready			10.0	10.0				10.0		10.0	10.0					20.0	180.0		
T5	MIS delivered			5.0						10.0		20.0					20.0	240.0		
O7	Procedures implemented		10.0	20.0										20.0				160.0		
P3	Intermediate review		10.0	10.0		10.0	40.0	10.0		10.0							10.0	45.0		
A4	Roll-out implemented		10.0	40.0	40.0	20.0	40.0	10.0	40.0	10.0	40.0	40.0		20.0			20.0			
P4	Postcompletion audit		20.0	40.0		40.0	40.0						20.0					900.0		
		13	73	230	130	104	162	47	135	72	105	135	40	190	0	0	115	1800	0	0

		Labour						Cash					
No	**Milestone Name**	Labour estimate	Actual	Estimated remaining	% Complete	Earned value	Calculated remaining	Cost total	Actual	Estimated remaining	% Complete	Earned value	Calculated remaining
		$K	$K	$K	%	$K	$K	$K	$K	$K	%	$K	$K
P1	Project defintion	40	40	0	1.0	40	0	25	25	0	1.0	25	0
T1	Technology design	60	75	0	1.0	60	0	0		0	1.0	0	0
O1	Communicaton plan	6	5	0	1.0	6	0	10	10	0	1.0	10	0
O2	Operational procedures	50	50	0	1.0	50	0	0	0	0	1.0	0	0
O3	Job and management design	80	80	0	1.0	80	0	0	0	0	1.0	0	0
T2	MIS functional spec	60	55	0	1.0	60	0	0	0	0	1.0	0	0
O4	Staff allocation	60	65	0	1.0	60	0	0	0	0	1.0	0	0
T3	Technical roll-out plan	40	40	0	1.0	40	0	5	10	0	1.0	5	0
A1	Estates roll-out plan	35	40	0	1.0	35	0	5	5	0	1.0	5	0
P2	Financial approval	40	20	20	0.5	20	20	10	0	5	0.5	5	5
A2	Sites 1 and 2 available	60	20	40	0.3	18	42	25	0	25	0.3	8	18
O5	Management changes	40	10	25	0.3	12	28	0		0	0.3	0	0
T4	Systems in sites 1 and 2	80		80	0.0	0	80	30		30	0.0	0	30
O6	Redeployment and training	80		80	0.0	0	80	210		210	0.0	0	210
A3	Sites 1 and 2 ready	50		50	0.0	0	50	200		200	0.0	0	200
T5	MIS delivered	35		35	0.0	0	35	260		260	0.0	0	260
O7	Procedures implemented	50		50	0.0	0	50	160		160	0.0	0	160
P3	Intermediate review	90		90	0.0	0	90	55		55	0.0	0	55
A4	Roll-out implemented	320		320	0.0	0	320	20		20	0.0	0	20
P4	Postcompletion audit	160		160	0.0	0	160	900		900	0.0	0	900
		1436	500	950		481	955	1915	50	1865		58	1858
	Forecast Cost at Completion			1450			1455			1915			1908

INDEXES

SUBJECT INDEX

AUTHOR AND SOURCE INDEX

PROJECT INDEX

Examples and Companies

Types and Industries

Locations